T0136854

Springer Climate

Series Editor

Springer Climate is an interdisciplinary book series dedicated to climate research. This includes climatology, climate change impacts, climate change management, climate change policy, regional climate studies, climate monitoring and modeling, palaeoclimatology etc. The series publishes high quality research for scientists, researchers, students and policy makers. An author/editor questionnaire, instructions for authors and a book proposal form can be obtained from the Publishing Editor.

Now indexed in Scopus® !

More information about this series at http://www.springer.com/series/11741

Md. Nazrul Islam • André van Amstel
Editors

India: Climate Change Impacts, Mitigation and Adaptation in Developing Countries

Editors
Md. Nazrul Islam
Department of Geography and Environment
Jahangirnagar University
Dhaka, Bangladesh

André van Amstel
Environmental Systems Analysis SENSE Research School
Wageningen University
Wageningen, The Netherlands

ISSN 2352-0698 ISSN 2352-0701 (electronic)
Springer Climate
ISBN 978-3-030-67863-0 ISBN 978-3-030-67865-4 (eBook)
https://doi.org/10.1007/978-3-030-67865-4

This Springer imprint is published by the registered company Springer Nature Switzerland AG
The registered company address is: Gewerbestrasse 11, 6330 Cham, Switzerland

Dedicated
To my Daughter
SABABA MOBASHIRA ISLAM
(Only Daughter of Professor Dr. Md. Nazrul Islam)

Preface

India has started various remedial measures to minimize the emission of greenhouse gases. It has launched various schemes to promote the production of renewable sources of energy (solar energy, wind energy, geothermal energy, tidal energy, hydroelectric energy, and biomass energy) on a large scale as well as on a small scale in rural area. The government of India has promoted numerous steps to resolve global warming and climate change. They have facilitated the production of renewable sources of energy on a major scale so that the consumption of fossil fuel is minimized, which helps in reducing the carbon load in the atmosphere. Climate change will significantly impact agriculture production, and water will be a major constraint in the future for food production. Developing countries such as India are more vulnerable, and the ramification of climate change on rain-fed agriculture areas has been affecting the rural population owing to the large dependency on agriculture for livelihood. Understanding how the cardinal and sensitive climate variables such as temperature and precipitation changes will affect natural resources, and developing management policies for mitigation and adaption strategies with regional-scale assessments, are crucial steps.

Most of the major cities and towns in India are located in river floodplains, which causes additional stress to the river basins. During the past decades, more developmental activities have taken place in urban and agricultural lands, which has led to additional stress in these floodplains, resulting in the loss of flood spread areas and increase in flood risk in the downstream portions of the basin. For example, most major European cities in England and Germany are located in floodplains that have induced the burden in these rivers, and preventive measures have been undertaken to recreate floodplain storage. In Indian cities such as Delhi, Vijayawada, Hoshangabad, and Surat thers located along the banks of rivers often experience floods for this reason, aggravated by several other factors as discussed subsequently. Climate change may have direct effects on increase in temperature, which may have a negative impact on fruit crops, temperate fruit crops as well as tropical and subtropical fruits. Elevated CO_2 and changed precipitation are also considered important, with direct effects on production technology, delay or early harvest, reduced available irrigation water, increased irrigation cost, increased insect pest attacks, increased physiological disorders, inferior fruit quality, and the lack of suitable cultivars, and also an impact on the soil because of the excess rainfall and changes in temperature that negatively impact fruit crops in India.

Changes in two important economic activities, namely, agriculture and fish/shrimp farming practices, were analyzed among the sampled households. Adaptation in agricultural activities was addressed through varied responses, such as no adaptation strategy, crop diversification, change in cropping calendar, adapting of new farming techniques, and transformation of land for fishery adaptation strategies among the sampled households in India. A recent study revealed that climate change leads to alterations in the precipitation and evapotranspiration rate of the country, so it may be considered as an additional major threat to this resource. Hence, understanding the behavior and dimensions of the groundwater regime under future climatic and other changes is significant for adopting an accurate water management strategy. This review presents an outline of the groundwater resource base of India and its susceptibility to ongoing and future trends of changes in climatic variables. The Indian climate is controlled by the monsoon wind; 75% to 90% of the rainfall in the Indian subcontinent is determined by the summer monsoon, which contributes about 60% of the gross water demand in agriculture. Indian summer monsoon rainfall greatly influences the Indian economy because about 55% of the rural workforce is directly linked with agriculture, and 60 percentage of the net area shown belongs to the rain-fed zone.

Forests are an integral part of the terrestrial ecosystem, maintaining biodiversity, carbon flux, and ecosystem services and even supporting livelihoods. Rampant exploitation of forest resources has resulted in deforestation and the loss of associated benefits. It is estimated that deforestation accounts for 11% of global carbon emissions. Their function in carbon sequestration has helped international bodies to create programmes such as reducing emissions from deforestation and forest degradation (REDD+). The phenomena of climate change adversely affect the functioning of forest species, changing the timing of their flowering and fruiting habits, and causing changes in the ecophysiology of the species. However, such changes depend on the species and their climatic conditions to a large extent. Phenology, the temporal order of the annual cycle of plant functions, is quite sensitive to changes in climate. In this chapter we have explored the relationships between climate change and its impact on tree phenology and mitigation and adaptation strategies. Some species may tend to exhibit a certain amount of resilience against climate change.

Three micro-watersheds, undisturbed, semidisturbed, and disturbed catchments covering a distance of 12 km at Geyzing, located in West Sikkim, were selected for this research work. Hydro-meteorological instruments were installed in three micro-watersheds and recorded data for 4 years to monitor the impact of forest cover on stream discharge and water quality. A composite cum rectangular weir was installed at field level to accommodate measuring stream discharge during the monsoon period as well as the pre-monsoon period. El Nino and La Nina events, which are based on sea surface temperature (SST), take place at the Pacific Ocean along the

Equator. The El Nino Southern Oscillation is like a pendulum, as El Nino takes place in one side (west) and La Nina takes place along the other side (east), and vice versa. Its major cause is the trade winds that move from east to west, especially in Ecuador, Peru, and South America. By correlating the rainfall data and El Nino, we can identify the effects along coastal Karnataka. As El Nino and La Nina have major impacts on agriculture, food supply, and the economy of the country, it is necessary to study the related patterns and influence of rainfall.

The energy sector in the world needs major changes for reduction in greenhouse gas emissions and climate change mitigation. It has been believed the current rate in increase of greenhouse gas emissions will cause an increase in sea levels and changes in global climate patterns. These effects will hinder adaptation for climate change mitigation; hence, it is important to incorporate and understand the models that will be implemented in the energy sector for decarbonization. According to this perspective, the geothermal resources are considered as one of the most productive energy sources that can be adopted for the control of greenhouse gas emissions. The purpose of this study is to understand the global status of geothermal energy and its important role in the mitigation of climate change.

Dhaka, Bangladesh
Wageningen, The Netherlands
Md. Nazrul Islam

Dhaka, Bangladesh
Wageningen, The Netherlands
André van Amstel

Contents

About the Editors

Md. Nazrul Islam is a permanent Professor at the Department of Geography and Environment in Jahangirnagar University, Savar, Dhaka, Bangladesh. Prof. Nazrul has completed his Ph.D. from the University of Tokyo, Japan. Also, he has completed a Two-Year Standard JSPS Postdoctoral Research Fellowship from the University of Tokyo, Japan. Prof. Nazrul's fields of interest include environmental and ecological modeling, climate change impact on aquatic ecosystems, modeling of phytoplankton transition, harmful algae, and marine ecosystems regarding hydrodynamic ecosystems-coupled models on coastal seas, bays, and estuaries, and the application of computer-based programming for numerical simulation modeling. Prof. Nazrul is an expert on scientific research techniques and methods to develop models for environmental systems analysis research. Prof. Nazrul has also visited as an invited speaker in several foreign universities in Japan, the USA, Australia, UK, Canada, China, South Korea, Germany, France, the Netherlands, Taiwan, Malaysia, Singapore, and Vietnam. Prof. Nazrul has been awarded the "Best Young Researcher Award" by the International Society of Ecological Modeling (ISEM) for outstanding contribution to the Ecological Modeling Field, 2013, Toulouse, France, and he has also been awarded "Best Paper Presenter Award 2010" by SautaiN in Kyoto, Japan. He has also been awarded a "Best Poster Presenter Award" at the Techno Ocean Conference, Kobe City Exhibition Hall, at Kobe in Japan.

Prof. Nazrul has made more than 40 scholarly presentations in more than 20 countries around the world and authored more than 100 peer-reviewed articles and 15 books and research volumes. Recently, Prof. Nazrul has published an excellent textbook entitled *Environmental Management of Marine Ecosystems* jointly with Prof. Sven Erik Jorgensen by the CRC Press (Taylor & Francis). He has also currently published an excellent book entitled *Bangladesh I: Climate Change Impacts, Mitigation and Adaptation in Developing Countries*, a Springer publication (Netherlands and Germany). Prof. Nazrul is currently serving as an "Executive Editor-in-Chief" for the journal *Modeling Earth Systems and Environment*, Springer International Publications (journal no. 40808). E-mail: nazrul_geo@juniv.edu

André van Amstel is an Assistant Professor at Wageningen University and Research, the Netherlands. Andre van Amstel studied physical geography and planning in Amsterdam and has long-standing expertise in integrated environmental assessment. He contributed to the development of an Integrated Model to Assess the Global Environment at the Institute of Public Health and the Environment in the Netherlands. He is particularly interested in the global environment and the risk of a runaway greenhouse effect by methane from melting of the permafrost in the Arctic. Van Amstel contributed to the IPCC 2006 Guidelines on Agriculture, Forestry and Land use and as such contributed to the Nobel Peace Prize 2007 for Al Gore and the IPCC. Since 2014 he has been a member of the Nordforsk Scientific Advisory Board for the Arctic Council of Ministers on Arctic Integrated Environmental Research. E-mail: andre.vanamstel@wur.nl

Chapter 1
New Challenges on Natural Resources and their Impact on Climate Change in the Indian Context

Sunil Kumar Srivastava

Abstract Global warming resulting from industrialization, excess consumption of fossil fuel, and agricultural activity is a new challenge for human civilization in the twenty-first century. The impact of global warming in the melting of glaciers and the rise in sea level is being observed worldwide (Ayala A, Farias-Barahona D, Huss M, Pellicciotti F, McPhee J, Farinotti D. 2020. Cryosphere 14:2005. doi:10.5194/tc-14-2005-2020). India is a developing country suffering from excess human population and a lack of awareness among the people regarding the impact of global warming on human civilization. India has started various remedial measures to minimize the emission of greenhouse gases. It has launched various schemes to promote the production of renewable sources of energy (solar energy, wind energy, geothermal energy, tidal energy, hydroelectric energy, and biomass energy) on a large scale as well as on a small scale in rural areas. The government of India has promoted numerous steps to resolve global warming and climate change. They have facilitated the production of renewable sources of energy on a major scale so that consumption of fossil fuel is minimized, which helps reduce carbon load in the atmosphere. India has also taken steps to fix carbon by increasing plantation. It has been shown that forest and tree cover in the country has increased from 14% in 1950–1951 to 24.01% in 2011–2012. India is a country with a rich diversity of socioeconomic conditions, geography, and climate: hence, the implementation of a uniform policy throughout the country is always challenging. A policy on a small scale (zone-wise) is needed so that the diversity of socioeconomic conditions, geography, and climate is considered seriously in polices. Hence, it is essential that every section of society also participates to resolve these issues and to cooperate with policies implemented by the government of India.

Keywords Natural resource · Climate change · Agriculture · Energy consumption · India

S. K. Srivastava (✉)
Department of Chemistry and Environmental Sciences, Jaypee University of Engineering and Technology, Guna, MP, India
e-mail: sunil.srivastava@juet.ac.in; sunil16sster@gmail.com

M. N. Islam, A. van Amstel (eds.), *India: Climate Change Impacts, Mitigation and Adaptation in Developing Countries*, Springer Climate,
https://doi.org/10.1007/978-3-030-67865-4_1

Introduction

India is a densely populated developing Asian country located in the Northern Hemisphere at latitude 8°4′N to 37°6′N and longitude 68°7′E to 97°25′E. Geographically, India is surrounded on three sides by oceans (the Bay of Bengal, the Arabian Sea, and the Indian Ocean) and on the north sides by the Himalayas, which significantly influence its climatic conditions and local weather. The total length of the coastline of India is approximately 7517 km. The Indian peninsula tapers southward, dividing the Indian Ocean into two water bodies: the Bay of Bengal and the Arabian Sea. India is divided geographically almost into two halves through the Tropic of Cancer (latitude 23°30′N). Land use patterns indicate approximately 46% of area is occupied by arable land, approximately 24% is under forest and tree cover, approximately 23% is not cultivable, and approximately 7% is fallow land (MOEFCC 2015). India's land use pattern has been influenced by demographic needs, industrial growth, urbanization, livestock grazing, diversion of forest lands for development purposes, the creation of irrigation facilities, and natural calamities such as flood and drought (Ranjan et al. 2017; Srivastava 2019). India possesses much variety in geography, climate, and local weather, which has influenced its biodiversity significantly.

India is the second most densely populated developing country in the world after China. India supports approximately 17% of the world population and possesses only about 2.16% in land resources of the total area available to the world. Its population density is approximately 382 persons/km^2 higher than most of the developing and developed countries. This large population load has created a burden on nature. The high population density has forced overexploitation of natural resources such as water, soil, forests, and food resources. India is quite rich in natural resources, but overexploitation and mismanagement have caused concern in Indian society, which also influences the world communities. The excessive consumption of fossil fuel for electricity, transportation, industries, and other domestic and commercial activities has produced an increased quantity of greenhouse gases: this may be one of the reasons for global warming and further climate change (Srivastava 2020).

Global warming caused by industrialization, excess fossil fuel consumption, and agricultural activity is a new challenge for human civilization in the twenty-first century. The impact of global warming in the melting of glaciers and the rise in sea level is being observed worldwide (Ayala et al. 2020). India is a developing country suffering excess human population and a lack of awareness among the people regarding the impact of global warming on human civilization. India has started various remedial measures to minimize the emission of greenhouse gases. It has launched various schemes to promote the production of renewable sources of energy (solar energy, wind energy, geothermal energy, tidal energy, hydroelectric energy, and biomass energy) on a large scale as well as on a small scale in rural areas. Statistics of the recent past in the development and consumption of renewable energy in India show positive growth. People have begun using solar, biomass, and wind energy significantly in rural areas for domestic, agricultural, and other small-scale industrial purposes. India has also promoted afforestation on a large scale since 1984, which helps fix atmospheric carbon to reduce the proportion of carbon in the atmosphere.

The carrying capacity of India is quite low in comparison to other developed world countries. Thus, obtaining food and other natural resource requirements of the Indian people becomes a challenging task considering the environmental impact. The availability of natural land resource per person in India is quite poor based on carrying capacity, only about 0.5 hectares (ha). Municipal solid waste (MSW) has contributed approximately 46% (1.7–1.9 billion tonnes) of total solid waste, produced mainly by urban settlement (Srivastava and Ramanathan 2012).

One report suggested solid waste and wastewater worldwide accounted for approximately 1.5 Gt CO_2 equivalent in the year 2010 (MOEFCC 2014). Greenhouse gas (GHG) emissions from the building sector have more than doubled since 1970, reaching 9.18 Gt CO_2 equivalents in 2010. In 2010, the building sector accounted for approximately 32% of global final energy consumption and approximately 19% of energy-related GHG emissions (including electricity related), which occurred because many modern buildings are highly energy intensive. A rough assessment suggests that to produce 1 tonne of Portland cement roughly 1 tonne of CO_2 is released to the atmosphere. In 2013, cement production accounted for 9.5% of global CO_2 emissions. A study reported approximately 67% greenhouse gas emitted by the energy sector including the cement industry, approximately 23% greenhouse gas by agriculture activities, approximately 4% greenhouse gas by landfill area, and about 6% greenhouse gas by other industries, in India in the year 2000 (MOEFCC 2015). The emitted GHG is classified as approximately 67.25% CO_2, approximately 26.73% CH_4, approximately 5.34% N_2O, approximately 0.34% CHF_3, approximately 0.02% SF_6, and approximately 0.42% PFC.

As per the Intergovernmental Panel on Climate Change (IPCC), human-generated GHG emission occurs as the result of population size, economic activities, lifestyle, energy consumption, agriculture and land use, technology, and climate policy (MOEFCC 2015). Further, vulnerability to climate change, GHGs, and the capacity for adaptation and mitigation have been strongly influenced by livelihood, lifestyle, behavior, and culture. These emissions can be substantially lowered through changes in consumption patterns, adoption of energy-saving measures, dietary changes, reduction in food wastes, and the reuse, reducing, and recycling of natural resources. India has a history of low-carbon footprint because of the nature-loving attitudes of the people. These national characteristics and the inclination of the people should be encouraged through mass education and awareness, rather than being replaced by more modern but unsustainable practices and technology (MOEFCC 2015).

Natural Resources in India

India is very rich in natural resources. The natural resources available in India are classified into five major categories: water resources, mineral resources, forest resource (includes agriculture activity), land resources (includes soil), and energy. India has been rich in natural resources from ancient times, as supported by published historical data. The current scenario has been changed by the increased population and further overexploitation of natural resources to fulfil the requirements of these growing populations. In the recent past, changes in the living patterns and lifestyles of the people have also influenced the rate of consumption of natural resources. Such practices stress natural resources, causing side effects such as drops in the groundwater table, increased GHGs, shortages in food resources, higher salinity of agricultural land, and soil, air, and water pollution (Prasad et al. 2006; Srivastava and Ramanathan 2018a). Global warming and further climate change are caused by excessive emission of GHGs in the atmosphere through anthropogenic activities. Global data indicate that transport accounts for approximately 23% of total energy related to CO_2 emissions, with road transport itself accounting for 17–18%. In urban areas and cities, transportation becomes challenging because of the increased number of vehicles used for transportations of goods and passengers (MOEFCC 2015).

Water Resources

India possesses approximately 1123 billion cubic meters (BCM) of water resources, of which about 61.4% is surface water and the remaining approximately 38.6% is subsurface water. India has approximately 4.63% of its total geographic area available as wetland area (MOEFCC 2015). Statistically, India has a surface water area of approximately 360,400 km^2: this area is under stress from excessive water consumption for fulfilling various industrial, domestic, and commercial requirements and for irrigation during agriculture activity (https://en.wikipedia.org/wiki/Natural_resources_of_India). The methods of irrigation adopted in India during agriculture activities have caused approximately 55% freshwater wastage. Further contamination of freshwater through various anthropogenic activities has also reduced available freshwater and created stress over water resources (Srivastava and Ramanathan 2008).

India annually receives about 4000 BCM water, of which approximately 1123 BCM is available for utilization. Of this, about 690 BCM is available as a surface water resource and the rest, approximately 433 BCM, is the groundwater resource. India receives water through 12 major rivers occupying a total catchment area of about 252.8 million hectares (ha) that covers more than 75% of the total area of the country. Indian rivers are classified as the Himalayan rivers, peninsular rivers, coastal rivers, and inland drainage basin.

India receives maximum rainfall mostly through the southwest (SW) monsoon, which accounts for about 75% of the total rainfall received by the country. About 11% of rainfall takes place pre monsoon and about 10% post monsoon. Interannual variability in monsoon onset, seasonal rainfall distribution, and rainfall intensity impact the agriculture and water resources of the country. The phase of the El Niño Southern Oscillation (ENSO) often impacts rainfall amount and distribution: the warm phase (El Niño) tends to result in lower rainfall, both during and beyond the monsoon period. Variations in the onset, withdrawal, and amount of precipitation during the monsoon season affect the water resources, agriculture, power generation, and ecosystems of the country.

Daily rainfall observations during the period 1901–2004 indicate that the frequency of extreme rainfall events (precipitation rate >100 mm/day) has a significant positive trend of 6% per decade. Frequency and duration of rainstorms (synoptic weather systems that have the potential for causing the floods) have also increased during the past 60 years.

Mineral Resources

India is rich in mineral resources. Of these resources, petrochemicals are mainly responsible for global warming and further influence the rate of climate change. India fulfills some of its basic requirements for petrochemicals, but a large proportion is imported from the Arabian countries. By combustion, petrochemicals emit a significant amount of greenhouse gas to the atmosphere. These GHGs are mainly responsible for global warming and further climate change. Petrochemical products are used in industrial activities, the energy sector, transportation, and domestic requirements. The cement industry also emits greenhouse gas, because limestone, the major raw material of the cement industry, releases CO_2 in calcination during clinker formation in the rotary kiln.

$$CaCO_3 \longrightarrow CaO + CO_2 \uparrow$$

Other metallic, nonmetallic, economic, and strategic minerals produce less GHC. Hence, those minerals are not discussed in this chapter.

Forest, Food, and Agriculture

As per the State of Forest Report, forest and tree cover in the country has increased from approximately 14% in 1950–1951 to around 24.01% in the year 2011–2012, indicating an appreciable approach by India to reduce atmospheric carbon by fixing CO_2 through afforestation. The State of Forest Report in 2013 showed the forest cover of the country as approximately 697,898 km^2, compared to 692,027 km^2 in

2011, an increase of 5871 km^2 recorded within 2 years. A study also reported approximately 23% of GHG is emitted through agriculture activities in India (MOEFCC 2015).

Food production, processing, marketing, consumption, and disposal have an impact on the environment because these activities consume energy and natural resources, emitting GHGs (Parashar et al. 2018). The contribution of wetlands, landfill, and other agriculture waste to GHGs during microbial decomposition is also noticeable (Parashar et al. 2019a, b). The global volume of food wastage is estimated at approximately 1.6 gigatonnes (Gt) of 'primary product equivalents,' whereas the total wastage for the edible part of food is about 1.3 Gt (MOEFCC 2015). The global carbon footprint of food wastage, excluding land use change, has been estimated as about 3.3 Gt CO_2 equivalent (FAO 2013). Upstream wastage volumes, including production, post-harvest handling, and storage, represent approximately 54% of total wastage, and downstream wastage volumes, including processing, distribution, and consumption, are approximately 46% (MOEFCC 2015).

The forest in India occupies approximately 24.01% of the total land area resource. Forest resources contribute significantly to fulfill the basic needs of rural and urban people throughout the country in the form of food, fuel, wood, and fodder. The forests help in regulating the water cycle, nutrient biogeochemical cycle, and oxygen supply, as well as conserving the soil by reducing soil erosion, and also provide natural habitat for wild animals, insects, birds, reptiles, and other species. The medicinal value for forest resources has been recognized in India from ancient times. A list of medical products worldwide utilizes forest resources. Conservation of biodiversity also helped to reduce atmospheric carbon. Forest resources neutralized approximately 12% of total GHG emitted in India through various anthropogenic activities (MOEFCC 2015).

In India, agricultural activity is mostly dependent on rainfall during the monsoon period. Overdependence of farmers on the monsoon poses uncertainties in crop production and the income of farmers, and sometimes excess floods, as well as drought, affect agriculture activity (Srivastava 2019; Srivastava and Ramanathan 2018b). A published report indicated declines (~18.9% in 2011–2012 to ~17.6% in 2014–2015) in the contribution of agriculture to the national gross domestic product (GDP) may be caused by changes in job prospects or the mobilization of formers to the city from the villages. Climatic uncertainties affect specific crops in different years. India ranks first in the world in total livestock population. The waste released by livestock emits a significant amount of methane to the atmosphere. Hence, a large quantity of waste can be used as raw material for biomass energy. First, animal waste can be used as clean energy, the sludge can be used as biofertilizer, and further release of methane to the atmosphere can be minimized. Second, biomass energy is a renewable source of energy; it reduces the stress on fossil fuels and further reduces the emission of GHG to the atmosphere (Srivastava 2020).

Land and Soil

India possesses approximately 1,945,355 km^2 arable land area, which is approximately 56.78% of the total available land resource in India (https://en.wikipedia.org/wiki/Natural_resources_of_India). This area has begun to shrink because of the population explosion and rapid urbanization in India. The Indian government promoted a better infrastructure to reduce the traffic load, which helps in reducing the consumption of fossil fuels and hence reduces the carbon load in the atmosphere. The government also promoted that villagers utilize a renewable source of energy to fulfill their basic energy requirement, which also reduced the emission of GHGs to the atmosphere. Solar and wind energy are promoted on a large scale in the whole country through various social groups, committees, the government, and nongovernment organizations.

Glaciers constitute the land resource having maximum effects on global warming and climate change. Glacier melting is a better indicator of global warming worldwide; Indian glaciers are also affected by global warming. The Indian Space Research Organisation (ISRO) has monitored the advance and retreat of 2018 glaciers across the Himalayan region using satellite data of 2000–2001 and 2010–2011. This study indicated that 1752 glaciers showed no change, 248 glaciers were retreating, and 18 glaciers were advancing. This scenario has raised concern in scientific communities in India, and the government has planned for further inspection of the existing policy on global warming and climate change.

Energy

The consumption of energy in India has been increasing rapidly during the past few decades. Considering the needs of people, industry, and other activities, a large proportion of energy requirements in India is fulfilled by fossil fuels in such forms as coal, petrol, kerosene, diesel, or naphtha. Coal has been used extensively in the production of electricity as an energy source in thermal powerplants. After fractional distillation, petroleum has been used for transportation and other industrial requirements. A study reported approximately 67% GHG is emitted by the energy sector, including the cement industry (MOEFCC 2015). The cement industry uses limestone, which releases CO_2 to the atmosphere during calcination, as the raw material; calcination occurs during clinker formation in the rotary kiln. India is the third largest producer of cement in the world, and cement is important for infrastructure, building materials, and development of settlements. Once the infrastructure is properly developed the consumption of cement is also reduced, which also reduces GHG in the atmosphere.

The contribution of renewable energy sources is still less than expectations or requirements. Renewable energy sources used in India include solar, wind, tidal, biomass, and geothermal. Solar energy and wind energy are more successfully used in India for reasons of widespread and easy availability; these sources are more efficient in comparison to other renewable sources of energy. Better technology and widespread awareness regarding renewable energy will help to reduce GHG in the atmosphere, which will also result in a better and healthy environment for the present populations.

Energy Status of India

The utilization of energy for various activities in India has been reported from ancient times. People have learned to use biomass energy for cooking food, light, and heat. A recently published report stated the total energy production capacity in India is approximately 3254.15 billion kWh. Of this total, ~2287.57 billion kWh (~70.3%) of energy is produced through consumption of fossil fuel, ~64.44 kWh (~1.98%) by nuclear power plants, ~386.63 kWh (~11.88%) by hydroelectric power plants, and ~515.51 kWh (~15.84%) through renewable energy (Table 1.1, Fig. 1.1).

In comparison to European countries, India consumed more fossil fuel to fulfill the energy requirements of the people (Table 1.1), possibly because India has a large population. Details of energy consumption and production as compared between India and European countries are given in Table 1.1 and Fig. 1.2, indicating India needs a massive effort to achieve sustainable growth in energy. The consumption of energy in India, approximately 2,405.82 kWh per capita, is much lower than the per capita energy consumption in European countries, approximately 16,542.38 kWh (Table 1.1). This difference in energy consumption observed between European and Indian society includes both urban and rural populations. Gujarat contributed approximately 25.04% (36,956 MW), Karnataka approximately 13.08% (19,315 MW), and Tamil Nadu approximately 11.17% (16,483 MW) of total renewable energy. The report indicated that access to electricity has increased from about 67% (2010) to about 80% (2020) of the population.

The high consumption of fossil fuel can be minimized through heavy taxation on crude oil, vehicles, and road tax to reduce unnecessary movements of people and

Table 1.1 Summary of energy consumption and production in India

Energy source	Total in India (billion kWh)	Percentage in India (%)	Percentage in Europe (%)	Per capita in India (kWh)	Per capita in Europe (kWh)
Fossil fuels	2287.57	70.3	49.2	1691.22	8125.79
Nuclear power	64.44	1.98	7	47.64	1155.77
Hydroelectric power	386.63	11.88	24.1	285.84	3982.3
Renewable energy	515.51	15.84	19.7	381.12	3278.52
Total production capacity	3254.15	100	100	2405.82	16542.38

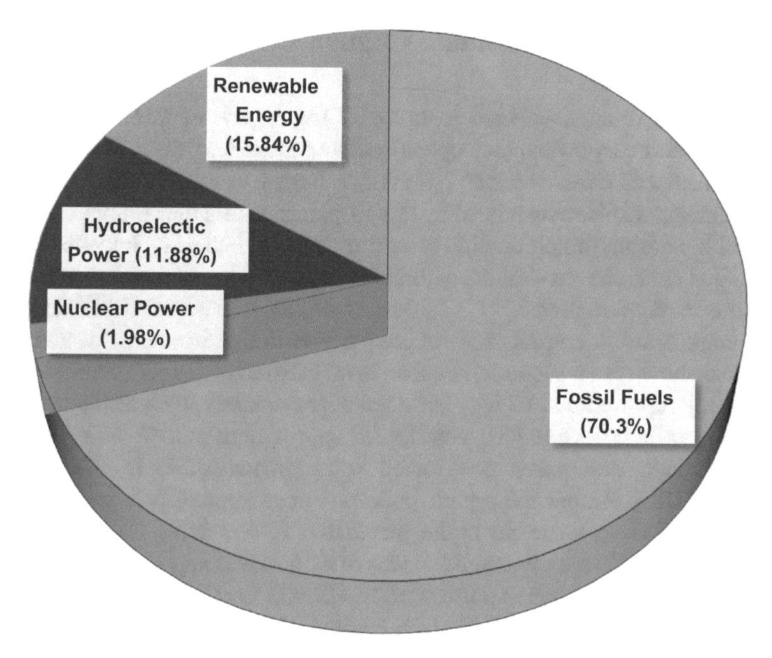

Fig. 1.1 Total energy production in India

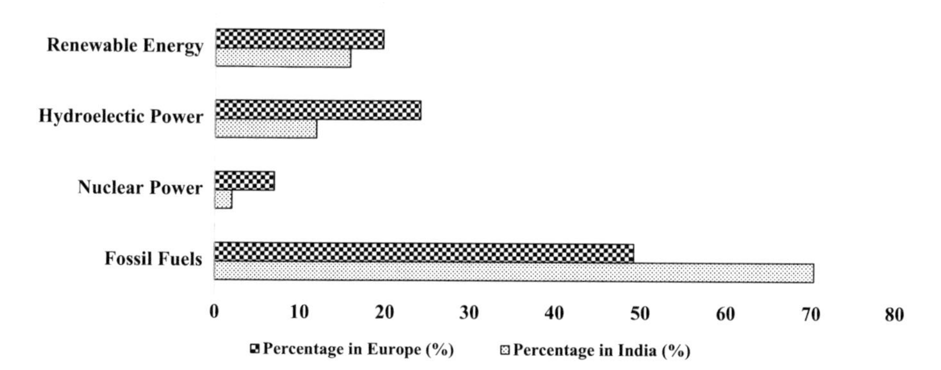

Fig. 1.2 Comparison chart for energy sources

vehicle use. Effective reduction of fossil fuel consumption thus reduces GHG emission and the impact of global warming on climate change. The government has initiated such steps as promotion of afforestation, an environmental awareness program, conservation of biodiversity, and heavy taxation on petroleum products, vehicles, and roads. Anton (2020) also suggested taxing crude oil can help to mitigate climate change effectively because GHG emission increases global risk by the adverse impact on climate.

Global Warming and Climate Change

Climate change is most affected in India by the release of GHG during cement production, thermal power plant operation, combustion of fossil fuels, and other pyro-treatment activities. Recently, India has begun various initiatives to reduce carbon in the atmosphere by educating the people, checking the combustion of fossil fuels, and increasing the production of nonconventional energy. Statistics of Indian metropolitan areas show a significant increase in the number of days when maximum temperature was more than 35 °C (Table 1.2). Notably, these data indicated a sharp rise in temperature in coastal cities in comparison to that in towns located in hilly areas. Also shown is an increase in ocean level caused by glacier melting.

The study reported that CO_2 contributed approximately 74% of the total greenhouse gases in India, while CH_4 contributed approximately 19%, NO_2 contributed 5%, and halogenated gases contributed approximately 2%, in the year 2010 (MOEFCC 2015). As per the report, India produced approximately 2,136,841 Gg CO_2 equivalent greenhouse gas in the year 2010 (Table 1.3).

The emitted GHG contributed by oxides of carbon was approximately 1,574,362 Gg CO_2 equivalent, methane approximately 412,086 Gg CO_2 equivalent, oxides of nitrogen approximately 114,365 Gg CO_2 equivalent, and halogenated gases approximately 36,028 Gg CO_2 equivalent (Table 1.3). Further, other industrial activities released 1.43 Gg CH_3F, 2.13 Gg CF_4, 0.58 Gg C_2F_6, and 0.0042 Gg SF_6. All these together account for 36,027.53 Gg CO_2 equivalent emissions.

The energy sector contributed approximately 70.7% (1,510,120.76 Gg CO_2 equivalent) of total greenhouse gas emitted and the agriculture sector only 18.3% (390,165.14 Gg CO_2 equivalent). The industrial sector contributed approximately 8.0% (171,502.87 Gg CO_2 equivalent) of total greenhouse gases whereas solid waste

Table 1.2 Comparison between metro-cities of India

	Number of days (with maximum temperature >35 °C)	
Cities	1959–1968	2009–2018
Delhi	1350	1465
Mumbai	113	313
Kolkota	453	449
Chennai	1009	1613
Hyderabad	1078	1137
Bengaluru	249	286

Table 1.3 Composition of emitted greenhouse gas in India (2010)

Gas	Composition greenhouse gas (Gg CO_2 equivalent)
Carbon oxide	1,574,362
Nitrogen oxide	114,365
Halogenated gases	36,028
Methane	412,086

landfill areas emitted approximately 3% (65,052.47 Gg CO_2 equivalent) in the year 2010. LULUCF (Land Use, Land Use Change, and Forestry) sectors acted as a net sink that offset approximately 12% of the total greenhouse gas emissions.

To minimize the impact of greenhouse emitted during the combustion of fossil fuels, the Indian government set a target for the production of renewable energy. Available information suggests India will be able to produce approximately 175,000 MW electricity through renewable sources by the year 2022: solar energy will contribute approximately 100,000 MW, wind energy approximately 60,000 MW, biomass energy approximately 10,000 MW, and other small hydropower loci approximately 5,000 MW. Afforestation and water conservation programs are promoted in the country through schemes such as the Mahatma Gandhi National Rural Employment Guarantee Scheme (MGNREGA) and the National Afforestation Programme. The schemes launched such as the Green India Mission (GIM), forest restoration, afforestation, agroforestry, and urban forestry have promoted reducing carbon load in the atmosphere. The production of paddy crops also minimizes the production of methane. These agricultural lands have been utilized for another crop with higher economic value.

The National Action Plan on Climate Change (NAPCC) was launched by the government in the year 2008 to address climate change concerns and to promote sustainable development. The NAPCC promotes development activities while addressing climate change. Eight National Missions form the core of the NAPCC, representing multi-pronged, long-term, and integrated strategies for achieving the goals in the context of climate change. Several of the programs described under NAPCC have undertaken various schemes of the government of India, but in the present context require a change in direction, enhancement of scope, and accelerated implementation. Positive response toward the effort has been observed in a few localities on a regional basis.

Future of India in the Twenty-First Century

The government of India has been promoting the production of energy through renewable sources such as solar energy, wind energy, tidal energy, biomass energy, and geothermal energy since the last century. An appreciable improvement in renewable energy in India has been observed in the past few decades, but a major improvement is still needed, considering other European countries (Fig. 1.2). The Energy Conservation Act and National Electricity Policy came in 2001 and 2005, respectively, for resolving the energy problems in India. The Energy Conservation Act was enacted in 2001 with the objective of energy security through conservation and efficient use of energy. Some of the major provisions of the act include Standard and Labelling of Appliances, Energy Conservation Building Codes, setting up of the Bureau of Energy Efficiency (BEE), and Establishment of Energy Conservation Fund. The National Electricity Policy of 2005 stipulates that progressively the share of electricity from nonconventional sources needs to increase.

Some researchers have reported an increased frequency of heavy rainfall, particularly in central India, in the past 60 years (Rajeevan et al. 2008; Krishnamurthy et al. 2009; Sen Roy 2009; Pattanaik and Rajeevan 2010). Goswami et al. (2006) have reported a decrease in the trend of light rainfall. Further, Krishnan et al. (2013) have reported reduction in moderate to heavy precipitation in the area surrounding the Western Ghats in India.

The data published in worlddata.info suggested renewable sources of energy produced approximately 36% of the total energy in the year 2015 (www.worlddata.info), which indicated a genuine effort of the Indian government in the energy sector. GHG emissions increased by 1301.2 million tonnes, to 1884.3 million tonnes between 2000 and 2010. These results suggested an appreciable effort of the government to resolve the issue of global warming and climate change. Still, serious effort is needed by the people and the government to implement a sustainable approach to minimize GHG in the atmosphere to reduce the impact of climate change. The government has developed the policies, but the common man needs to understand the importance of these policies and to try to reduce the use of fossil fuel: this will be effectively implemented with the proper awareness programs in society through various print media and mass media regarding the impact of GHG. A few localities across the country have shown positive growth that needs to be appreciated as a role model for other regions or localities. The government needs an appropriate policy for every zone considering its socioeconomic conditions along with the geographic and climatic situation of the region. India is a country with a rich diversity of socioeconomic conditions, geography, and climate; hence, implementation of a uniform policy throughout the country is always challenging. Policies on a small scale are needed so that the diversity of socioeconomic conditions, geography, and climate can be considered seriously.

The National Action Plan on Climate Change (NAPCC) was launched in 2008 to address climate change concerns and promote sustainable development. The Ministry of Agriculture of India has taken the major initiative to resolve the issue under the National Initiative on Climate Resilient Agriculture (NICRA) in 2011: it has four main sections of resource management, improving soil health, improving crop production, and livestock to make the farmers self-reliant for adaptation under a changing climate. All these programs help to minimize the emission of GHG in various ways including fixing CO_2 from the atmosphere. In summary: the photosynthesis potential of C_3 plants is generally lower (~40%) in comparison to C_4 plants (Monteith 1978).

Analysis of the annual temperature of India indicated a rise of approximately 0.54 °C in the average temperature of the country in the past century (Table 1.4). A similar pattern of increased temperature will create an explosive situation in India from population growth, overexploitation of natural resources, and further global warming that will impact and seriously affect the whole of human society in the form of climate change and changes in the lifestyle of the people.

The trend of temperature increases as shown in Fig. 1.3 has caused concern for Indian society considering future impact. Climate change is a continuous process that will have effect through changes in nutrient biogeochemical cycles, the

Table 1.4 Recorded changes in average temperature in India

Decade	Average temperature (°C)
1900–1910	25.26
1911–1920	25.27
1921–1930	25.4
1931–1940	25.38
1941–1950	25.39
1951–1960	25.53
1961–1970	25.44
1971–1980	25.46
1981–1990	25.57
1991–2000	25.73
2001–2010	25.74
2011–2020	25.84

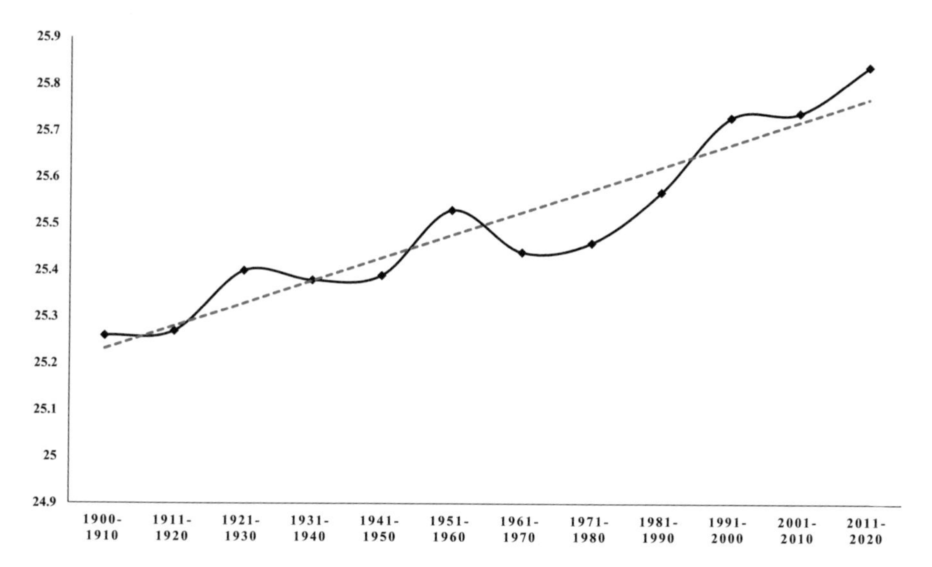

Fig. 1.3 The change in average temperature in India in the past century

hydrological cycle, the natural habitat of all living beings, and further geographic situations. Hence, every section of society needs a serious approach to resolve these issues and cooperate with the policy implemented by the government of India.

Conclusion

India is a member of the United Nations Framework Convention on Climate Change (UNFCCC), and it supports approximately 17% of total human populations on the earth. This instrument is highly responsive to control of greenhouse gases and takes remedial measures to minimize greenhouse gas emission. The average annual temperature of India has increased approximately 0.54 °C in the past century. Indian metro-cities have significantly more days when the maximum temperature is more than 35 °C. The most important point about these data is the sharp rise in temperature in coastal cities in comparison to towns in the hilly areas, indicating an increased level of ocean water from climate change.

The government of India has promoted numerous steps to resolve global warming and climate change. It has facilitated the production of renewable sources of energy on a major scale so that the consumption of fossil fuel is minimized, reducing the carbon load in the atmosphere. India has taken steps to fix carbon by increasing plantation: the forest and tree cover in the country has increased from 14% in 1950–1951 to 24.01% in 2011–2012.

India has a rich diversity of socioeconomic conditions, geography, and climate; hence, the implementation of the same policy throughout the country is always challenging. India needs policies on a small scale (zone-wise) so that the diversity of socioeconomic conditions, geography, and climate can be considered seriously. Hence, it is essential that every section of society also participates to resolve these issues and cooperates with the policy implemented by the government of India.

References

Anton A (2020) Taxing crude oil: a financing alternative to mitigate climate change. Energy Policy 136:111031

Ayala A, Farias-Barahona D, Huss M, Pellicciotti F, McPhee J, Farinotti D (2020) Glacier runoff variations since 1955 in the Maipo River basin, in the semi-arid Andes of Central Chile. Cryosphere 14:2005. https://doi.org/10.5194/tc-14-2005-2020

FAO (2013) Food and agricultural commodities production. https://www.faostat.fao.org

Goswami BN, Venugopal V, Sengupta D, Madhusoodanan MS, Xavier Prince K (2006) Increasing trend of extreme rain events over India in a warming environment. Science 314 (5804):1442–1445. https://doi.org/10.1126/science.1132027

https://www.livemint.com/. Accessed on 26 June 2020

https://www.worlddata.info/asia/india/energy-consumption.php. Accessed on 26 June 2020

Krishnamurthy CKB, Upmanu L, Hyun-Han K (2009) Changing frequency and intensity of rainfall extremes over India from 1951 to 2003. J Clim 22:4737–4746. https://doi.org/10.1175/2009JCLI2896

Krishnan R, Sabin TP, Ayantika DC, Kitoh A, Sugi M, Murakami H, Turner AG, Slingo JM, Rajendran K (2013) Will the South Asian monsoon overturning circulation stabilize any further? Clim Dynam 40:187–211. https://doi.org/10.1007/s00382-012-1317-0

MOEFCC (2014) Meeting on GHG inventory for BUR chaired by secretary, MoEFCC on 11th November 2014

MOEFCC (2015) India first biennial update report to the United Nations Framework Convention on Climate Change. https://unfccc.int/resource/docs/natc/indbur1.pdf

Monteith JL (1978) Reassessment of maximum growth rates for C3 and C4 crops. Exp Agric 14:1–5

Nation Action Plan on Climate Change (NAPCC) (2008) National Action Plan on climate change. PMs council on climate change. Government of India, New Delhi

Natural Resource. https://en.wikipedia.org/wiki/Natural_resources_of_India. Accessed on 22 May 2020

Parashar SK, Srivastava SK, Dutta NN, Garlapati VK (2018) Engineering aspects of immobilized lipases on esterification: a special emphasis of crowding, confinement, and diffusion effects. Eng Life Sci 18:308–316

Parashar SK, Srivastava SK, Garlapati VK, Dutta NN (2019a) Production of microbial enzyme triacylglycerol acylhydrolases by *Aspergillus sydowii* JPG01 in submerged fermentation using agro residues. Asian J Microbiol Biotechnol Environ Sci 21(4):1076–1079

Parashar SK, Srivastava SK, Garlapati VK (2019b) The mathematical modeling for the optimization of triacylglycerol acylhydrolases production through artificial neural network and genetic algorithm. Int J Pharm Bio Sci 10(3):135–143

Pattanaik DR, Rajeevan M (2010) Variability of extreme rainfall events over India during southwest monsoon season. Met Apps 17:88–104. https://doi.org/10.1002/met.164

Prasad MBK, Ramanathan AL, Srivastava SK, Anshumali SR (2006) Metal fractionation studies in surfacial and core sediments in the Achankovil River basin, India. Environ Monit Assess 121 (1–3):77–102

Rajeevan M, Bhate J, Jaswal A (2008) Analysis of variability and trends of extreme rainfall events over India using 104 years of gridded daily rainfall data. Geophys Res Lett 35:L18707. https://doi.org/10.1029/2008GL035143

Ranjan R, Srivastava SK, Ramanathan AL (2017) Assessment of hydro-geochemistry of two wetlands located in Bihar state in the subtropical climatic zone of India. Environ Earth Sci 76 (16):1–17. https://doi.org/10.1007/s12665-016-6330-x

Sen Roy S (2009) A spatial analysis of extreme hourly precipitation patterns in India. Int J Climatol 29:345–355. https://doi.org/10.1002/joc.1763

Srivastava SK (2019) Assessment of groundwater quality for the suitability of irrigation and its impacts on crop yields in the Guna district, India. Agric Water Manag 216:224–241. https://doi.org/10.1016/j.agwat.2019.02.005

Srivastava SK (2020) Advancement in biogas production from the solid waste by optimizing the anaerobic digestion. Waste Dispos Sustain Energy 2:85–103. https://doi.org/10.1007/s42768-020-00036-x

Srivastava SK, Ramanathan AL (2008) Geochemical assessment of groundwater quality in the vicinity of Bhalswa landfill, Delhi, India, using graphical and multivariate statistical methods. Environ Geol 53:1509–1528

Srivastava SK, Ramanathan AL (2012) Groundwater in the vicinity of the landfill. Application of the graphical and multivariate statistical method for hydrogeochemical characterization of groundwater. Lambert Publication, Germany 1–380. ISBN-13:978-3847328858, ISBN-10:3847328859

Srivastava SK, Ramanathan AL (2018a) Assessment of landfills vulnerability on the groundwater quality located near floodplain of the perennial river and simulation of contaminant transport. Model Earth Syst Environ 4(2):729–752. https://doi.org/10.1007/s40808-018-0464-7

Srivastava SK, Ramanathan AL (2018b) Geochemical assessment of fluoride enrichment and nitrate contamination in groundwater in hard rock aquifer by using graphical and statistical methods. J Earth Syst Sci 127:104(1–23). https://doi.org/10.1007/s12040-018-1006-4

Chapter 2
Regional Assessment of Impacts of Climate Change: A Statistical Downscaling Approach

Nagraj S. Patil and Rajashekhar S. Laddimath

Abstract Water resource sectors are impacted by industrialization, urbanization, and climate change, making these vital resources more vulnerable and posing substantial challenges to water resources planners. Climate change significantly impacts agricultural production, and water will be a major constraint in the future for food production. Developing countries such as India are more vulnerable, and the ramification of climate change on rain-fed agriculture areas has been affecting the rural population because of their great dependency on agriculture for their livelihood. For understanding how the changes in cardinal and sensitive climate variables such as temperature and precipitation will affect natural resources, and for developing management policies for mitigation and adaption strategies, regional-scale assessments are crucial. This chapter showcases regional assessment of the impacts of climate change by the statistical downscaling technique, the Change Factor method, using the CMIP-5 General Circulation Model data in India with a case study in Bhima sub-basin. Spatial cross-correlation of maximum temperature and precipitation across the Indian Meteorological Department (IMD) gauge grids was captured adequately, and the downscaling satisfactorily captured the cross-correlation between rainfall grids of Bhima sub-basin. Future projections reports, average daily maximum temperature and rainfall are likely to increase in response to climate change under different representative concentration pathway (RCP) scenarios, with maximum changes in RCP 8.5.

Keywords Climate change · General Circulation Model · Regional assessment · Change Factor Method · Bhima sub-basin

---

N. S. Patil (✉)
Department of Civil Engineering, Center for P.G. Studies, Visvesvarayya Technological University, Belagavi, Karnataka, India
e-mail: nspatil@vtu.ac.in

R. S. Laddimath
School of Civil Engineering, REVA University, Bengaluru, Karnataka, India

Regional Research Center, Visvesvarayya Technological University, Belagavi, India

M. N. Islam, A. van Amstel (eds.), *India: Climate Change Impacts, Mitigation and Adaptation in Developing Countries*, Springer Climate,
https://doi.org/10.1007/978-3-030-67865-4_2

Introduction

Water, being the most vital resource, has been under growing demand for ages. As an important natural resource, water is naturally influenced by the environment, weather, geology, and topography. These factors can cause difficulties in evaluating future water resources under a changing climate. Nineteenth-century industrialization has extensively utilized natural resources for construction activities, raising the level of CO_2 in the atmosphere, and thus global warming has triggered climate change. The manifest drafted by the experts of the Intergovernmental Panel on Climate Change (IPCC)[1] reports that increase in greenhouse gas (GHG) emission can cause sea level rise, increasing the frequency of storms, heavy rainfall events, seasonal shifts, heat waves, floods, and droughts.

The major consumption of water in India is by the agriculture sector, using 83% of the total freshwater annually. Climate change significantly impacts agricultural production, and water will be a major constraint for food production in the future (Aggarwal 2008). Assessment of impacts of climate change on the agriculture sector has been a thrust area in research around the world. Developing countries such as India are more vulnerable, and the ramification of climate change on rain-fed agriculture areas has been affecting rural populations owing to their great dependency on agriculture for livelihood. Agriculture primarily depends on monsoon and changing precipitation patterns; intensity combined with duration will probably influence the balance in water supply and demand, thus affecting the hydrological cycle and leading to frequent droughts and floods (Gleick et al. 2010). India is the second largest country by population count, with 17% of the world's population but with substantially as little as 4% of world freshwater resources. This available water under increasing demand by various sectors is at high risk of pollution, and its quality is being diminished day by day due to anthropogenic activities. The facts and figures reported by Dhawan (2017) show India is a water-stressed country.

Climate change adaption and mitigation actions require an integrated approach to create a win–win situation among policymakers and stakeholders. Such an integrated approach by Shrestha and Dhakal (2019) for Nepal that proposed an integrated approach at the national level carry an important message. Pasimeni et al. (2019) justify the contribution of urban factors for adaption and mitigation strategies toward changing climate; well recognized by the policymakers, this has led to several policy initiatives. In the context of Indian agriculture, these policy incentives have a vital role in the adoption of climate change strategies, especially in the rain-fed agriculture sectors (Venkateswarlu and Singh 2015). One of the most important impacts of climate change on society will be changes in water availability, particularly at the regional level (Buytaert et al. 2009). Governments in the developing nations have

---

[1]The Intergovernmental Panel on Climate Change (IPCC) is an intergovernmental body of the United Nations that is dedicated to providing the world with objective, scientific information relevant to understanding the scientific basis of the risk of human-induced climate change, its natural, political, and economic impacts and risks, and possible response options.

not been able to effectively address challenges posed by climate change. Recent studies have used General Circulation Models (GCMs)[2] to assess the impacts of climate change at the regional level using statistical downscaling[3] techniques (Zhang et al. 2019; Woldemeskel et al. 2016; Rehana and Mujumdar 2012; Ghosh and Mujumdar 2006).

Regional-Scale Assessment

Understanding how the changes in cardinal and sensitive climate variables such as temperature and precipitation will affect natural resources, and also for developing management policies for mitigation and adaption strategies, regional-scale assessments are crucial. Buotte et al. (2016) captured the regional variability of changing climate, vulnerable across subregions to regional scale, to elicit and synthesize our understanding of its impacts. The regional climate model is engaged to understand the climate change scenario of the twenty-first century through assessment at the watershed scale to evaluate the water budget and flood conditions of Peninsular Malaysia (Amin et al. 2017). Coastal risks and vulnerabilities related to the physical and socioeconomic impacts of climate change at the regional scale in Mediterranean coastal zones were analyzed using the Coastal Risk Index model (Satta et al. 2017). The impact of changing climate on socioeconomic scenarios on water quality of the Mahanadi River was assessed to reflect the implicit implications of the people's livelihood (Jin et al. 2018). Spatiotemporal trend detection of hydrometeorological parameters, namely temperature and precipitation, for assessing climate change at river basin scale in India was useful for watershed management and policy decisions for climate change adaptation (Chandole et al. 2019). Thus, it is important to carry out investigations of regional climate change to assess the impacts at the regional scale, which in the due course of time has received considerable attention.

Climate Change: Indian Scenario

India is an agricultural country, with two-thirds of its population depending on rain for agriculture and allied activities for their livelihood and sustenance. Further, agriculture forms the backbone of the country's economy, accounts for nearly

---

[2]General Circulation Models (GCMs), mathematical models representing physical processes in the atmosphere, ocean, cryosphere, and land surface, are the most advanced tools currently available for simulating the response of the global climate system to increasing greenhouse gas concentrations.

[3]Statistical downscaling involves the establishment of empirical relationships between historical and/or current large-scale atmospheric and local climate variables. Once a relationship has been determined and validated, future atmospheric variables that GCMs project are used to predict future events.

17% of the country's gross domestic product (GDP), and feeds 1.3 billion people. India is situated roughly between 20°N and 20°S as a tropical area; the monsoon mechanism is greatly influenced by the differential heating and cooling of land and water. The periodic change in pressure conditions over the Indian Ocean is connected with El Nino[4], causing a negative pressure difference known as the Southern Oscillation (SO) that affects the monsoon pattern; this phenomenon is referred to as ENSO (El Nino Southern Oscillations). Any small deviation in the monsoon rains causes huge damage to the country's economy.

India receives much of its rainfall from June to September every year, known as the Indian summer monsoon rainfall (ISMR). Approximately 80% of annual rainfall during June–September is the annual cycle of more than 44 regions (Singh et al. 2019), with a key role in the country's economy. ISMR exhibits large spatiotemporal variation (Revadekar et al. 2018), making it more accountable for climate change studies. The impacts of climate change have decreased the intensity and occurrence of the ISMR, and its distribution has become more extreme. The declining ISMR tendency is dependent on the weakening of the easterly jet stream[5] and the southern oscillation, besides warming of the equatorial Indian Ocean, more than northern latitudes of the tropics, thus affecting the sea surface temperature gradient around the equatorial region (Naidu et al. 2009). An increase in surface temperature has already adversely affected Indian crop yields (Sedova and Kalkuhl 2020), as well as imposing restrictions on water supply, industry, and associated activities to impede the economy. This problem needs to be accounted for and addressed through regional projections to implement critical adaptation and policy framing (Shashikanth et al. 2013).

Because of global warming, the country has become calamity prone. Of 35 states, 27 states are prone to the aftereffects of global warming. Recent studies by the scientific community accept the major role of anthropogenic activity as the primary cause of global warming and climate change for the twentieth century (Madhusoodhanan et al. 2017). An investigation on 12 Indian river basins by Gosain et al. (2006) using HadRM2 climate data under current and probable future emission of GHG shows severe flood and drought will result during the twenty-first century. Unsustainable use of land and water, rapid urbanization, economic growth, and poor infrastructure pose serious threats to India's natural resources. Economic development influences the consumption pattern and leads to increasing demands for land, water, and food. As a result, water resource sectors in India since the 1970s are greatly influenced by competition from allied sectors. India exhibits diverse weather conditions, and the impacts of climate change exhibit complex diversity with many

---

[4]El Nino is a Spanish word meaning 'the child,' referring to the baby Christ because it starts flowing during Christmas. This name is assigned to the periodic development of a warm ocean current along the coast of Peru to temporarily replace the cold Peruvian current. The presence of El Nino causes an increase in sea surface temperature and weakening of the trade wind in the region.

[5]Jet streams are the narrow belt of high altitude (above 12,000 m) westerly winds in the troposphere. Mid-latitude and subtropical jet streams are the most constant jet streams with speed approximately 110 km/h during summer to about 184 km/h in winter.

parts of the country, namely, Rajasthan, Madhya Pradesh, Maharastra, Karnataka, Andhra Pradesh, and Bihar, being more likely vulnerable in terms of extreme events (Mall et al. 2006). Therefore, understanding the regional scenario is a prerequisite for assessing the impacts of changing climate.

Climate Change: Regional Scenario

The Krishna, an eastward draining interstate river basin, is one of the fastest developing regions and ranked as the second-largest river in peninsular India. The basin covers parts of four South Indian states: Karnataka, Andhra Pradesh, Telangana, and Maharashtra. The Krishna basin is divided into 12 subbasins as per the National Water Development Agency of India (NWDA) report (2017–2018). The Bhima river basin is 1 of the 12 subbasins of the Krishna basin, the eastward draining river of India (Biggs et al. 2007). The watershed designated as Upper Bhima (K5) and Lower Bhima (K6) is situated between 15 °N and 20 °N latitude and 73 °E and 78 °E longitude, with a catchment area of approximately 70,263 km^2 (Fig. 2.1). The pattern of rainfall, shifts in the seasons, and nonsustainable development resulting from urbanization have disturbed the demand and supply mechanism of water resources in the Bhima River basin (Garg et al. 2012). This imbalance has also hampered the availability of surface sources and caused stakeholders to depend

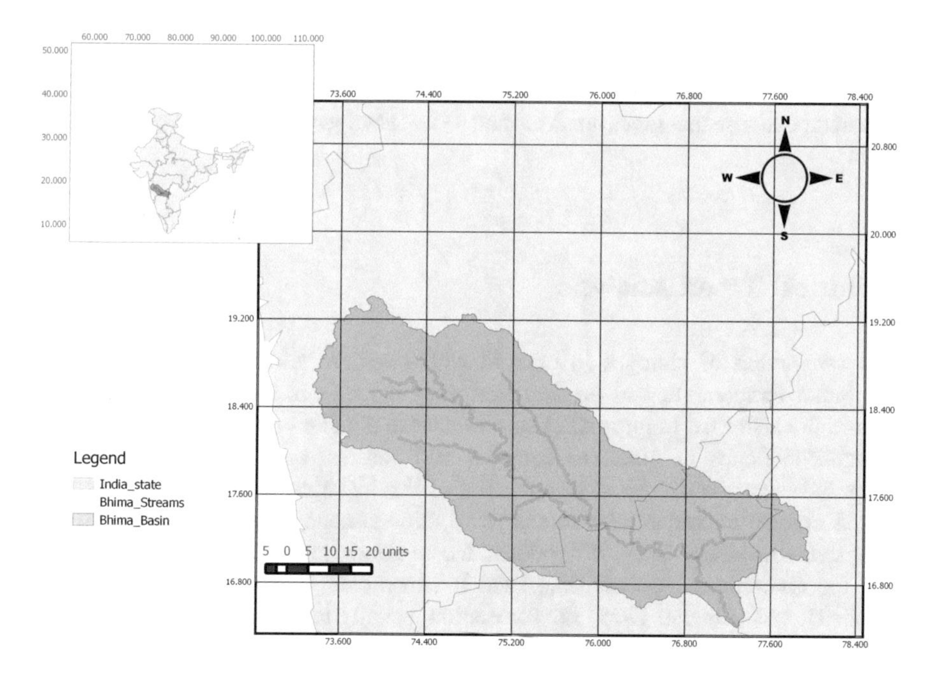

Fig. 2.1 Study area: Bhima sub-basin (sub-basin of Krishna)

on groundwater. A report released by the Ministry of Environment and Forests (2010) has also cautioned that the river basin can suffer from a chronic shortage of water in the future. Most of the upper region of the Bhima sub-basin (K5), owing to its proximity to the Western Ghats, receives abundant rainfall, but the stringent policy and interstate tribunal obstruct the complete usage and claims to share among the neighboring states to cater to the need caused by low rainfall situations. In addition, the Bhima sub-basin has lingering issues of water quality and overdraft of groundwater that hamper the basaltic aquifer (Surinaidu et al. 2013; Udmale et al. 2014) and has lacked effective monitoring of watershed management programs for decades (Biswas 1987).

It is estimated that by 2035 the people in this region may have to depend on external food sources because of a huge difference in food production and demand: also, climate change may burden the water resources by influencing water demand, exclusively in the agricultural sector (Chanapathi and Thatikonda 2020). Robust and resilient strategies for managing the natural resources and climate change mitigation measures for the Bhima sub-basin need to be sensitized to the local administration through policy documents. Watershed management policies set the guidelines and directives for the optimal utilization of resources. For instance, water policy guides the allocation priorities for agriculture, industry, and domestic uses, as well as awards for incentives and penalties toward conservation and detonation of water resources. Thus, impact assessment of climate change is vital in the framework of watershed management policies. One of the most important sources of information for impact assessment studies is the output from general circulation models of the climate system (Robock et al. 1993). It has been observed that the lack of research on these events of natural or human-induced disasters, drought, and flood should be examined to understand the past historical events. Historical trend analysis is an important exercise that gives an idea of historical changes over a period of time in the basin.

Historical Trend Analysis

The assessment of changes in climate with sensitive climate variables such as maximum temperature and precipitation over the past historical period not only picture the historical trends and changes of these climate variables but will also be valuable evidence in future to compare and contrast the climate changes. The magnitude and the statistical significance of the historical trend in the aforementioned climate variables with reference to climate change are ascertained by the nonparametric Mann–Kendall trend test. Mann (1945) presented this nonparametric test for randomness against time, which constitutes a particular application of Kendall's test (Kendall 1975) for correlation, usually identified as the Mann–Kendall (MK) test.

Trends in Maximum Temperature

Temperature is an important climate variable in assessing the impact of increasing global warming, and human-induced global warming is a key component in the management of water resources over a river basin (Jhajharia et al. 2014). Therefore, examining temperature trends is an important manifestation of global climate change. Annual and daily temperature events were examined to detect the variability and trends of temperature for 1969–2005. Spatial plots (Fig. 2.2a, b) illustrate the average daily maximum temperature and maximum temperature, respectively, in degrees Celsius (°C) over the Bhima sub-basin for the period 1969–2005.

It can be inferred from these figures that both upward and downward trends were experienced for maximum temperature at upper and lower regions of the Bhima sub-basin. The maximum temperature recorded for the basin was 45 °C with most of the region on Lower Bhima (K6) in this category. The average daily temperature for most of the lower region is 33 °C, whereas the Upper Bhima (K5) experiences relatively less heat as the maximum temperature recorded during this period is 42 °C with an average daily maximum temperature of 32 °C.

The trend of the mean daily maximum temperature and the average annual maximum temperature over the basin is increasing consistently. The M-K trend test for mean daily maximum temperature (Fig. 2.2c) in the upper region of the basin has shown the maximum positive tau (τ) value, indicating a trend of increasing temperature. The daily maximum temperature trend line has been increasing

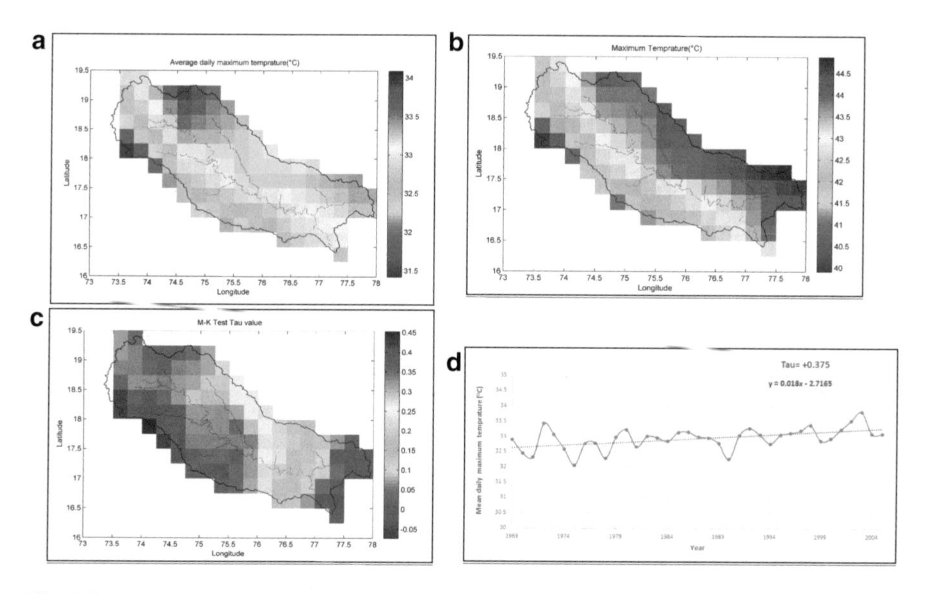

Fig. 2.2 (**a**) Average daily maximum temperature (in °C); (**b**) maximum temperature (in °C); (**c**) M-K trend test tau values; (**d**) mean daily maximum temperature series. Linear best fit indicated by *dotted line*

consistently since 1969. The slope of the trend line considering the linear equation is 0.018 (Fig. 2.2d), which means the daily maximum temperature is increasing at a rate of 0.005% per decade. For the period 1969–2005, the annual temperature has risen by 1.3 °C and has been increasing constantly at 0.001% per decade as seen from the slope of the best fit line of 0.0358. These facts are evidence for the changes in climate and certainly explain the current scenarios of global warming and the occurrence of natural calamities (floods and droughts).

Rainfall Trends

Indian Meteorological Department (IMD) $0.25^0 \times 0.25^0$ gridded data for a period of 1969 to 2016 is an input for generating annual mean rainfall over the entire Bhima sub-basin. Figure 2.3a, b illustrates annual mean rainfall and maximum daily rainfall observed during 1969–2016, respectively. The M-K trend test tau value (Fig. 2.3c) explains the trend for the annual rainfall series of the IMD gridded data over the basin. The tau value of the M-K test is quite similar and as good as the correlation coefficient, with its value ranging from −1 to +1. A positive (+) tau value shows the increasing trend of the series of data (rainfall in this case) and its counterpart, a negative (−) tau value, represents a decreasing trend. Figure 2.3d illustrates the average annual rainfall over the basin (with tau value).

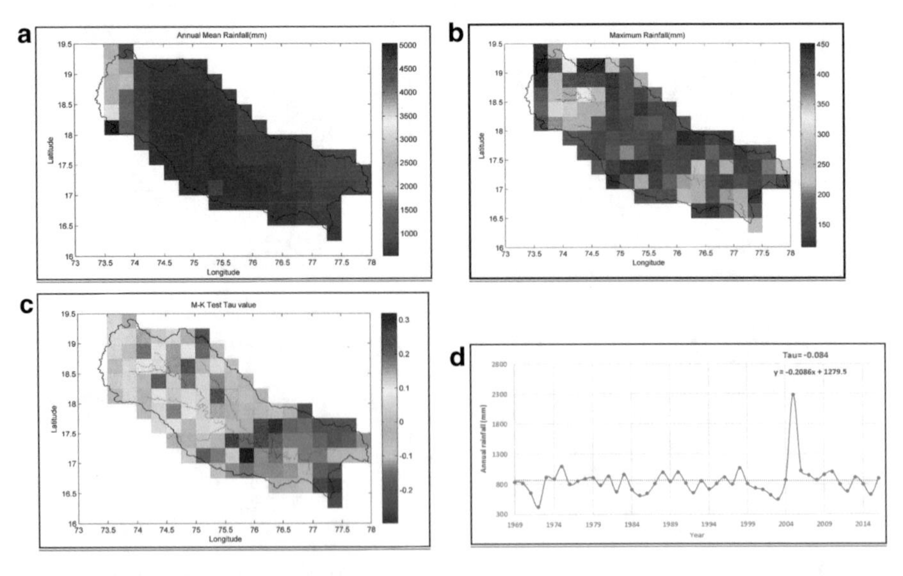

Fig. 2.3 (**a**) Annual mean rainfall (mm); (**b**) maximum daily rainfall (mm); (**c**) M-K trend test tau values; (**d**) annual rainfall plot. Linear best fit indicated by *dotted line*

Upon examining the spatial plots and analyzing the trends in annual rainfall, it is clear that there is neither a fixed increase nor a decrease of annual rainfall over the basin. However, the grids toward the Western Ghats region (Maharashtra state) lying in the Upper Bhima (K5) show the increasing trends (positive tau values) in rainfall. Annual mean rainfall varies between 2000 and 2500 mm, and the maximum daily rainfall of 450 mm is also evidence of the occurrence of extreme events; the Lower Bhima (K6) has altogether a decreasing trend of annual rainfall. The spatial grids on the downstream side (Karnataka state) have annual mean rainfall values between 500 and 1000 mm with maximum daily rainfall of 100–150 mm. Further, as we examine the annual rainfall, the best fit line for the range of data has a mean value of 850 mm with a decreasing rainfall trend up to 2004. Interestingly, in the year 2005, the annual maximum rainfall reached 2300 mm and declined subsequently. The average slope of yearly time series data as per the linear equation is −0.2086, resulting in 0.044% of decline of annual rainfall for every decade. The annual maximum rainfall trends, however, show a positive trend and with an occurrence of an extreme event during 2005. Flooding was a norm in Upper Bhima as the tau values are positive and even more in the magnitude than Lower Bhima (negative values of tau); this could have resulted in drought conditions in the Lower Bhima regions.

Materials and Methods

This study uses the Coupled Model Intercomparison Project Phase-5 (CMIP5), a GCM model used extensively in studies of climate changes associated with scenarios at global and regional scales (Zhang et al. 2019). IPCC developed four new pathways under representative concentration pathway (RCP) scenarios for long-term and near-term climate modeling experiments based on future climate projections until 2100, with radiative forcing values in the range 2.6–8.5 W/m^2 and named sequentially as RCP 2.6–8.5. These RCPs include one alleviation scenario priming a low driving level (RCP 2.6), two stabilization scenarios (RCP 4.5 and RCP 6), and one scenario with high GHG emissions (RCP 8.5): scenarios are based on GHG and socioeconomic development (Nilawar and Waikar 2019). The RCP development process and main characteristics have been summarized by van Vuuren et al. (2011). GCM models used for this study for climate projections are at coarse resolution (150–250 km) and poor in the ability to resolve important subgrid scale features required at the regional scale (20–25 km), so that the projections may not be reliable for regional impact assessment studies. Downscaling methods are robust techniques to overcome this problem to obtain regional-scale weather and climate information, particularly at the surface level. Downscaling techniques are broadly classified as dynamic downscaling and statistical downscaling (Wilby et al. 1998).

Statistical Downscaling

Statistical downscaling techniques are reliable to correlate the cardinal and sensitive climate variable maximum temperature and precipitation obtained from GCMs at coarse resolution and fine-scale spatial resolution for impact assessment at the regional scale (Ahmed et al. 2013). Statistical downscaling, despite having limitations (Wangsoh et al. 2017), is a practical technique requiring relatively less computational effort than dynamic downscaling (Tripathi et al. 2006); it finds its major applications for the assessment of climate change on the themes of water resources in response to hydrological modeling. Weather classification, weather generators, and transfer functions are among the prominent subcategories (Anandhi et al. 2008). Wilby et al. (1998) compared and contrasted various statistical downscaling methods.

Among the sophisticated methods under statistical downscaling techniques, the change factor (CF) method is a simple and linear spatial downscaling method for impact analysis studies widely accepted for rapid assessment of climate change. The CF method simulated daily times series data of temperature and precipitation and further compared the performance of simulation in three mountainous basins of the United States (Hay et al. 2000). The CF method was used to compare the bias corrections-based methods for downscaling projections of extreme flow indices to quantify the impacts from floods (Hundecha et al. 2016). Owing to its ease of implementation to assess the impact of climate change at hydrological scale (Boé et al. 2007) and enabling the representation of multiple GCMs and emission scenarios with minimal computing effort, CF estimates values of climate variables at future time scales and at spatial scales that are appropriate for regional climate change impact assessment (Anandhi et al. 2011). The difference in these projections from future and historical simulations are considered the means of simple addition/multiplication or scaling the mean climatic CF to each baseline observation datum (Fowler et al. 2007). This method is also termed constant scaling as the changes in the different percentile of climate variables (such as precipitation) in magnitude are assumed equal. Scaling approaches are simple and consider multiple GCMs for the downscaling (Wang et al. 2017).

Change Factor

The change factor (CF) method is a perturbation method based on the assumptions of relative changes in future and historical climate projections by GCMs even though the GCM is biased. Additive change factor (ACF) and multiplicative change factor (MCF) are the classifications for assessing temperature and precipitation (Akhtar et al. 2008). ACF calculates the arithmetical difference between a GCM variable derived from a current climate simulation and from a future climate scenario considered at the same GCM grid location, as shown in Eq. 2.1:

$$CF_{add} = \left(\overline{GCM_f} - \overline{GCM_b}\right) \tag{2.1}$$

The MCF method is similar to an additive CF except that the ratio between the future and current GCM simulations is calculated, instead of arithmetical difference (Eq. 2.2):

$$CF_{mul} = \left(\overline{GCM_f}/\overline{GCM_b}\right) \tag{2.2}$$

where CF_{add} and CF_{mul} are additive and multiplicative change factors, respectively; $\overline{GCM_f}$ is a GCM future climate scenario, and $\overline{GCM_b}$ is the value from the GCM baseline considered for a temporal domain.

The next step is to obtain the local scaled future values ($LSf_{add,i}$ and $LSf_{mul,i}$) by relating the change factors obtained from Eqs. 2.1 and 2.2:

$$LSf_{add,i} = LOb_i + CF_{add} \tag{2.3}$$

$$LSf_{mul,i} = LOb_i \times CF_{mul} \tag{2.4}$$

where LOB_i are the observed values of climate variables considered at the ith time step on an individual meteorological station, or the averaged meteorological time series for a catchment for the designated temporal domain. $LSf_{add,i}$ and $LSf_{mul,i}$ are the values of future scenarios of climate variables obtained using additive and multiplicative formulation of the CF method.

To evaluate and validate the performance of CF, the daily average rainfall during these southwest (SW) monsoon months (June, July, August, September) is evaluated considering the observed daily rainfall data obtained from the Indian Meteorological Department (IMD) through spatial map plots (Fig. 2.4). The average daily rainfall calculated from the CF downscaled data is closely similar to that of observed daily rainfall values simulated from the IMD. Further, the CF downscaling technique is good at simulating daily average rainfall during monsoon months.

Results and Discussion

Temperature and Precipitation Analysis

Global warming has recently become is an important concern to the world because of the enhanced emission of greenhouse gas, triggering temperature rise that unbalances the global water cycle and water resource sector. Climate change leads to severe impacts on agricultural ecology by inducing imbalance in the ecosystem and causing frequent and extreme events of droughts and floods. A number of recent studies assessing the impact of climate change on hydrology (Mujumdar and Kumar 2012; Knutti 2008; Di Baldassarre and Uhlenbrook 2011), long-term water availability (Yu and Wang 2009), and water quality (Rehana and Mujumdar 2011) in

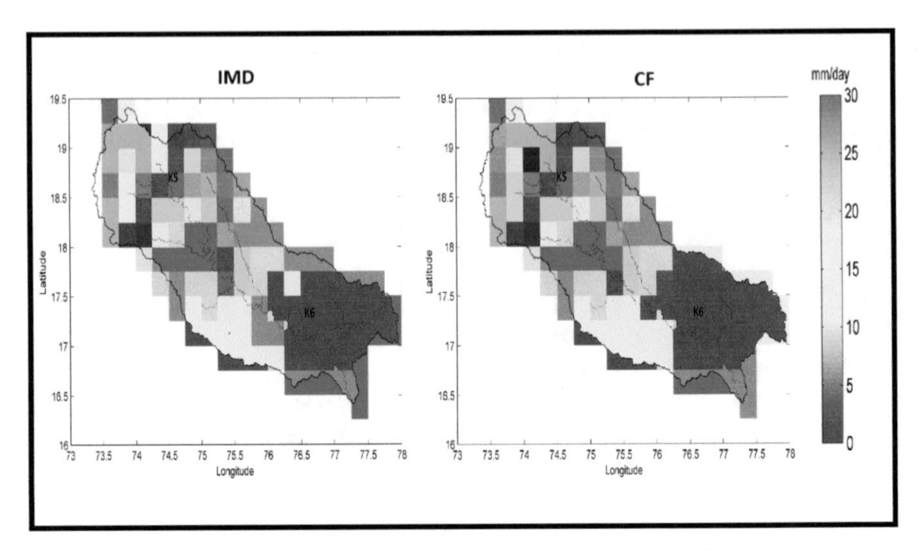

Fig. 2.4 Performance of the CF method over the daily observed rainfall data of IMD during the SW monsoon season (June–September)

river basins are evidence to why one has to consider the impacts of climate change. These assessments will leverage policymakers to identify adaption and mitigation strategies to strengthen agriculture and other allied sectors to safeguard the livelihood of deprived communities.

Maximum Temperature Projections over the Bhima Sub-Basin

Maximum temperature over the Bhima sub-basin is downscaled using the ACF method. The climate data obtained from five climate models of GCMs and IMD observed data regridded (because temperature data have a spatial resolution of $1^0 \times 1^0$) to size $0.25^0 \times 0.25^0$ gridded locations using a two-dimensional linear interpolation technique. The change factors of all the IMD gauge stations in the Bhima basin are calculated by applying the additive CF method for four different RCP scenarios with respect to different time series data for 2021–2040, 2041–2060, 2061–2080, and 2081–3100. The projected daily maximum temperature over the basin for five GCMs in four RCP scenarios (2.6, 4.5, 6.0, 8.5) is expressed in terms of the multimodel ensemble (averaged), the percentage change over average daily maximum temperature for the different RCP scenarios with respect to the historical (1986–2005) observation period. The spatial map of the historical period (1986–2005) refers to the average daily maximum temperature data for the entire Bhima basin (Fig. 2.5). It can be inferred that the Upper Bhima (K5) region during 1986–2005 experienced a relatively lower record of daily maximum temperature as

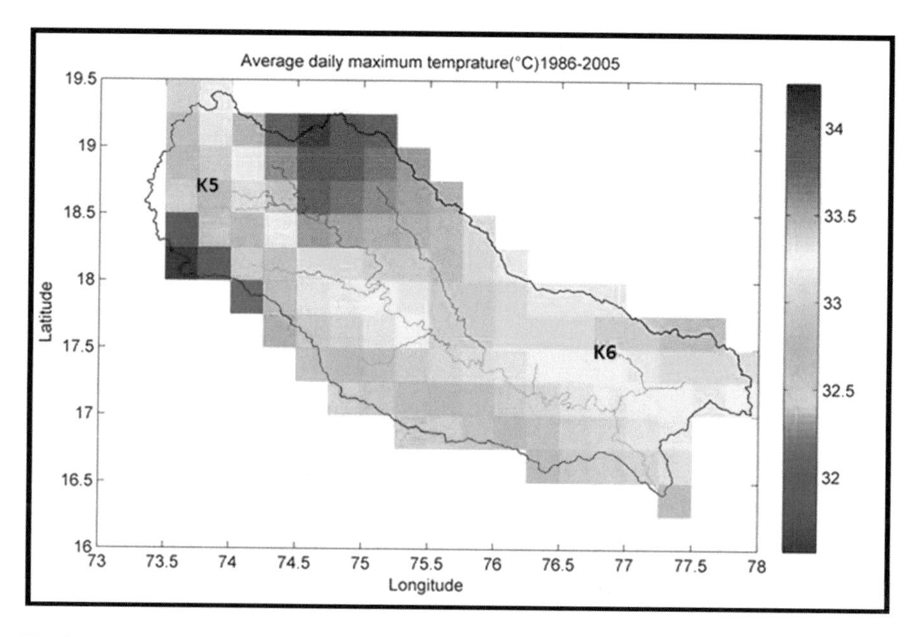

Fig. 2.5 Average daily maximum temperature (in °C) over the Bhima sub-basin during 1986–2005

compared to the Lower Bhima (K6) region, where the temperature was in the range 33 –33.5 °C.

Daily maximum temperature of the Bhima sub-basin is increasing in comparison with historical observed data. This increasing trend is reflected by the percentage (%) changes in average daily maximum temperature over the subbasins and is represented through the spatial maps and histograms plots for the different RCP scenarios in Figs. 2.6 and 2.7. A histogram below each of the spatial maps is used to summarize climate data. The histogram plot illustrates a visual interpretation of numerical data by showing the number of data points that fall within a specified range of values. A redline in the histogram indicates a mean percentage change. The ranges of percentage (%) change in average daily maximum temperature are ascending in order from RCP 2.6 to RCP 8.5 scenarios. However, there is a marginal change of 1% for the RCP 2.6 scenario as the variation is of the order of 3.3–4.3%. Further, an identical percentage (%) change of average daily maximum temperature for RCP 4.5 and RCP 6.0 occurs in the range of 6.4–8.0%. As expected, the highest percentage (%) change in the range 12.6–14.6% changes in the average daily maximum temperature is seen for the RCP 8.5 scenario.

Monthly average changes in percentage (%) over the basin for daily maximum temperature based on the ensemble average of projections from five GCMs are presented in Table 2.1.

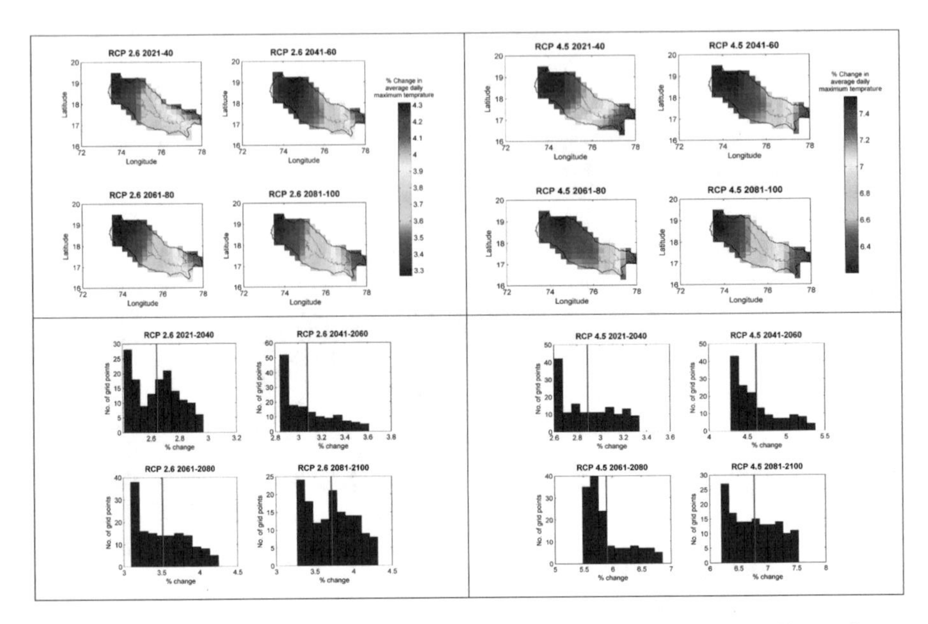

Fig. 2.6 Spatial maps and histograms of the percentage change in average daily maximum temperature projected from the ensemble average in RCP 2.6 and 4.5 scenarios. *Red line* in the histogram indicates a mean percentage change

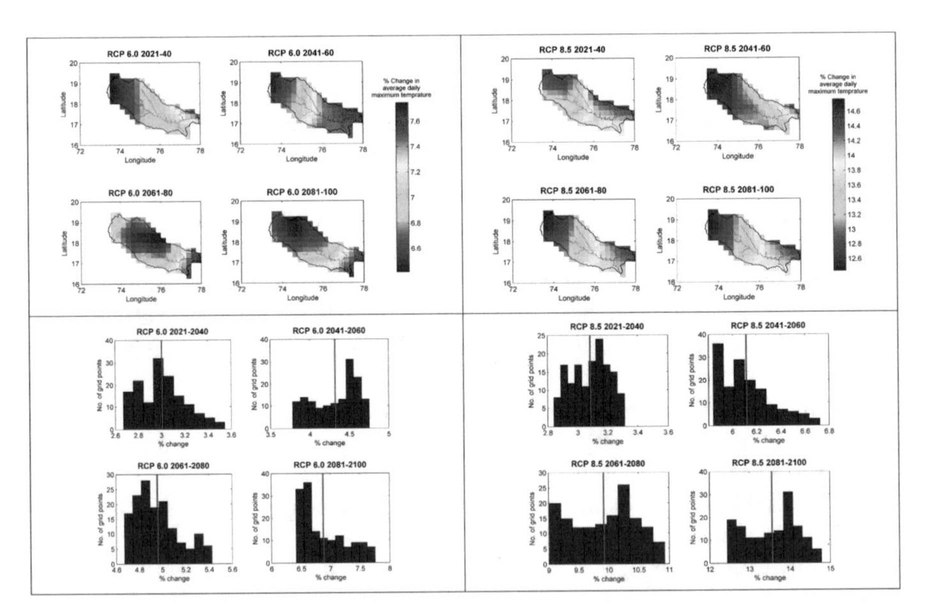

Fig. 2.7 Spatial maps and histograms of the percentage change in average daily maximum temperature projected from the ensemble average in RCP 6.0 and 8.5 scenarios

Table 2.1 Average change (%) over the basin in daily maximum temperature based on the ensemble average of projections from five scenarios

Scenario (s)	Month→Time Series↓	Jan	Feb	Mar	Apr	May	Jun	Jul	Aug	Sep	Oct	Nov	Dec
RCP 2.6	2021–2040	3.8653	3.122	3.207	2.8841	2.4481	0.672	3.3794	2.1499	2.2653	2.1987	2.6765	2.8974
	2041–2060	4.611	5.2731	4.7511	3.5865	2.0243	0.677	1.991	2.352	2.7585	2.619	2.6411	3.7665
	2061–2080	6.259	7.7535	6.2767	4.4325	0.8008	−0.1989	1.9586	2.8087	3.5074	2.6583	2.4565	3.8273
	2081–2100	7.2631	9.1063	7.663	4.8106	−0.2171	−1.5932	1.9632	4.2137	3.7971	2.2019	1.9904	4.0255
RCP 4.5	2021–2040	3.7908	3.8483	3.7662	3.2455	2.5677	1.6365	2.6836	1.6563	2.4796	2.8904	2.9185	3.1705
	2041–2060	7.0113	6.9841	6.1747	5.0177	2.7023	2.8409	3.3624	3.59	4.0857	3.7971	4.5351	5.6755
	2061–2080	8.9732	10.339	8.5405	6.9228	3.542	1.355	3.5868	5.7548	5.2864	4.6322	5.1532	7.0548
	2081–2100	11.702	12.93	11.154	7.6784	3.0351	1.3234	3.6775	6.0747	6.5861	4.9935	5.2803	7.435
RCP 6.0	2021–2040	3.2226	3.968	4.1302	3.5163	2.6373	1.1588	3.1737	2.4411	2.6523	3.0643	2.8905	3.0926
	2041–2060	5.4993	6.478	6.4698	5.5385	3.3607	1.6507	3.8773	4.0828	4.385	3.4667	3.0242	3.806
	2061–2080	8.3187	8.7689	7.3834	4.8936	2.4757	1.5212	3.4806	4.7741	4.7538	3.9843	4.0104	5.7352
	2081–2100	11.098	11.959	10.452	7.0826	3.4672	2.3935	4.9069	6.6681	6.6514	5.6591	4.919	7.8966
RCP 8.5	2021–2040	3.8027	3.674	3.6641	3.4197	2.4978	2.2226	3.5359	2.2106	2.9169	2.9125	2.9158	3.2638
	2041–2060	8.1648	9.1388	8.0185	6.7569	4.9066	3.1424	4.8662	4.5688	5.6772	5.4279	5.5097	7.3654
	2061–2080	13.411	14.028	13.374	10.915	7.0952	6.4982	6.9504	8.3353	8.9864	8.7911	9.3043	11.599
	2081–2100	18.81	19.889	17.941	14.348	8.6964	8.1496	11.5	12.48	12.61	12.082	11.639	15.36

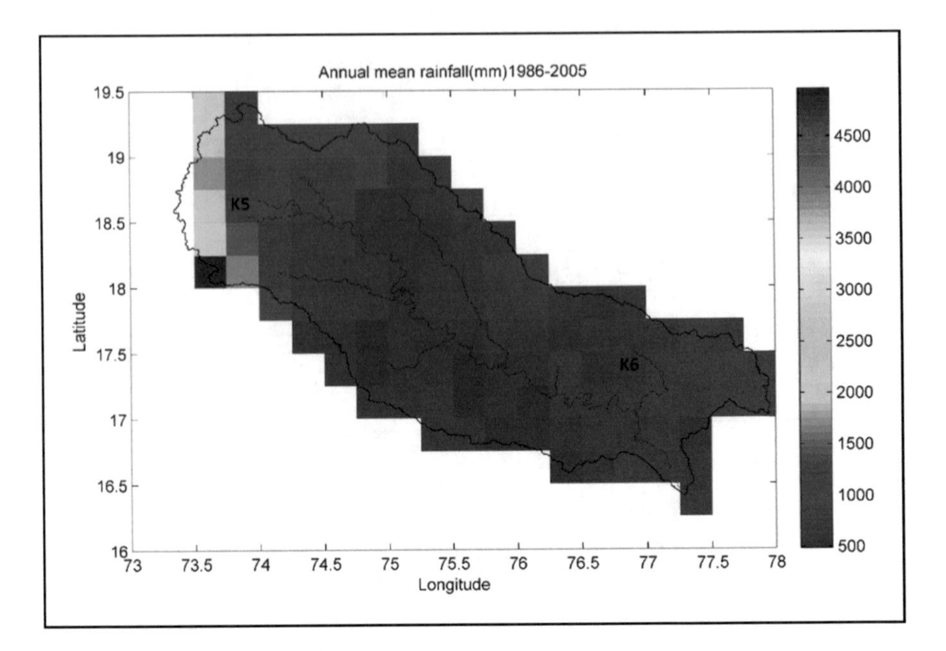

Fig. 2.8 Annual mean rainfall (in mm) over the basin during 1986–2005

Rainfall Projections

The mean annual rainfall over the basin for the four RCP scenarios is shown in Figs. 2.8, 2.9 and 2.10. It is inferred from these spatial maps and histograms that the percentage (%) change in mean annual rainfall is positive and is increasing for most of the grids lying in both the Upper Bhima and Lower Bhima regions for all four RCP scenarios. The percentage change in mean annual rainfall varies from 2% to 30% for the grids in the basin under scenarios RCP 2.6, 4.5, and 6.0. Further, for RCP 8.5 scenarios this variation is slightly higher, with the range extending to 35%, and during 2061–2100 almost all the grids in the basin under RCP8.5 scenarios depict extreme changes (30–35% variation) in the record of mean annual rainfall. However, this may be caused by rising radiative forcing pathway RCP 8.5 scenarios leading to very high greenhouse gas emissions (Stocker 2013).

The average changes in percentage (%) over the basin according to months for precipitation based on the ensemble average of projections from five GCMs are outlined in Table 2.2. It is observed in almost all the RCP scenarios for the different periods that the percentage changes in precipitation value based on multimodel projections are slightly higher for the month of March.

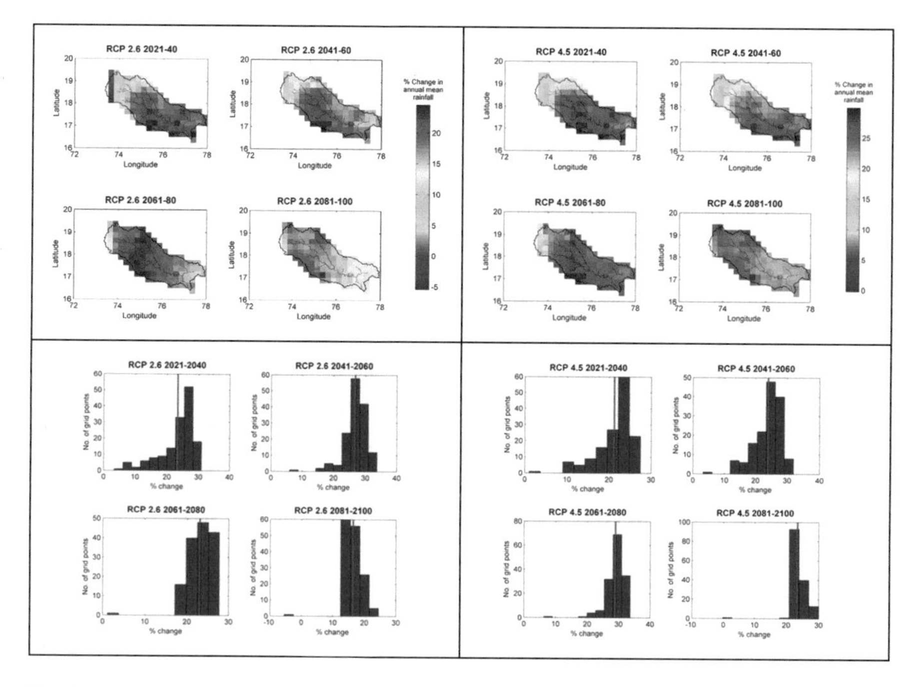

Fig. 2.9 Spatial maps and histograms of the percentage change in mean annual rainfall projected from the ensemble average in RCP 2.6 and 4.5 scenarios

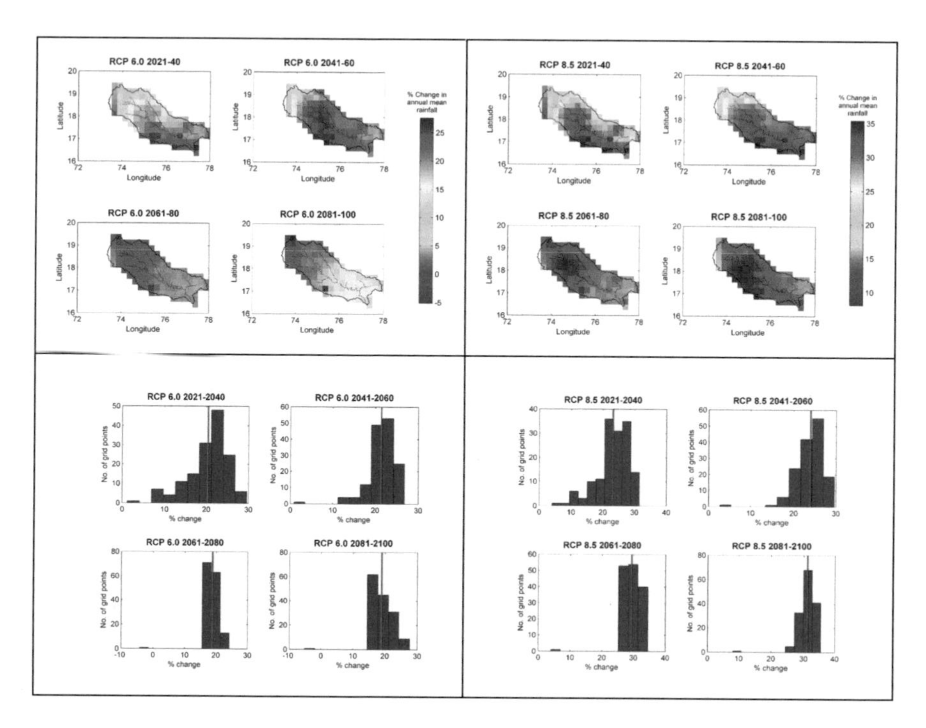

Fig. 2.10 Spatial maps and histograms of the percentage change in mean annual rainfall projected from the ensemble average in RCP 6.0 and 8.5 scenarios

Table 2.2 Average change (%) over the basin in monthly precipitation based on the ensemble average of projections from five GCMs

Scenario (s)	Month→Time Series↓	Jan	Feb	Mar	Apr	May	Jun	Jul	Aug	Sep	Oct	Nov	Dec
RCP 2.6	2021–2040	5.3177	−8.0965	38.438	14.881	1.9272	−8.1729	−3.4697	14.89	28.034	105.22	81.867	−10.7
	2041–2060	−11.849	−12.077	51.59	25.368	16.912	13.43	15.213	2.1904	26.354	102.14	39.415	−15.045
	2061–2080	−9.0212	−13.148	112.77	37.856	57.703	39.105	9.507	6.6049	14.656	49.818	40.381	−34.59
	2081–2100	16.368	−2.2717	113.2	55.793	80.393	43.846	8.707	−2.698	5.7018	15.546	32.29	−24.092
RCP 4.5	2021–2040	−24.175	−21.687	−7.818	2.2046	−4.4468	−11.752	1.5142	16.279	32.009	96.54	26.64	−32.543
	2041–2060	0.389	12.635	−2.7851	7.0784	15.952	−3.3284	6.7154	15.149	28.909	92.412	54.735	−30.975
	2061–2080	14.474	−26.518	125.55	51.775	32.552	32.162	14.552	9.6737	28.586	72.137	34.042	−27.755
	2081–2100	0.9783	−24.606	177.4	67.636	47.372	38.241	20.583	5.7741	15.626	38.6	21.718	−24.27
RCP 6.0	2021–2040	−11.912	−5.7138	−3.5592	−1.4602	2.3942	0.4273	−2.7658	12.619	21.524	86.958	98.327	−24.747
	2041–2060	13.918	−6.1541	27.628	8.8437	−1.3724	7.4322	6.8257	9.4104	24.903	83.828	17.795	−20.839
	2061–2080	−25.511	−51.439	119.75	6.8468	16.398	24.867	16.096	8.5443	15.631	42.699	19.795	−24.161
	2081–2100	−19.001	−23.618	186.32	24.27	29.671	30.118	23.34	6.6665	6.6081	38.057	22.097	−27.606
RCP 8.5	2021–2040	11.536	−12.427	30.248	12.284	−2.0738	−14.416	−1.6894	12.167	26.838	119.8	71.529	−0.9812
	2041–2060	−8.9766	−29.492	12.529	3.8044	3.6023	9.754	10.139	17.192	21.492	85.183	46.437	−34.123
	2061–2080	−39.962	−33.739	107.87	25.361	−0.2554	12.435	33.343	11.363	28.801	89.178	35.065	−29.709
	2081–2100	−41.773	−33.493	91.442	16.265	39.192	26.44	22.773	19.198	34.09	60.725	88.098	−33.042

Conclusion

The impacts of climate change on agriculture and the water resources sector must be urgently emphasized for remedial policies to be drafted by local planners and managers. The expected changes in temperature and precipitation will hamper water availability, especially for the rain-fed agriculture regions. In this study, regional impact assessment has been carried out to predict future climate scenarios through projected changes in temperature and precipitation. The projections of these climate variables are obtained by using the change factor downscaling technique through the climate data obtained from the CMIP5 global model for RCP scenarios. The evaluation of CF downscaling techniques over the Bhima sub-basin indicates satisfactory results in comparison to baseline historical data. Spatial cross-correlations of maximum temperature and precipitation across the IMD gauge grids were captured adequately. Further, it is observed that this downscaling satisfactorily captured cross-correlations between rainfall grids of the Upper Bhima (K5) and Lower Bhima (K6) regions.

The comparison of historical and future downscaled maximum temperature indicates that the average daily minimum temperature (for the historical period 1986–2005) of the Bhima sub-basin is likely to increase in response to climate change scenarios. A marginal change of 1% is indicated for RCP 2.6; an identical change of average daily maximum temperature for RCP 4.5 and RCP 6.0, in the range of 6.4% to 8.0%, is also evident from the study, and the highest percentage change, in the range of 12.6% to 14.6% in the average daily maximum temperature, is seen for the RCP 8.5 scenario. In a similar context, downscaled projections for precipitation are also increased under changing climate scenarios. RCP 2.6, 4.5, and 6.0 depicted changes in the range 2% to 30%, whereas the RCP 8.5 scenario accounted for more than 35% change to the historical baseline climate (1986–2005). However, this may be the result of rising radiative forcing pathway RCP 8.5 scenarios, thus leading to very high greenhouse gas emissions.

References

Aggarwal PK (2008) Global climate change and Indian agriculture: impacts, adaptation, and mitigation. Indian J Agri Sci 78:911–919

Ahmed KF, Wang G, Silander J, Wilson AM, Allen JM, Horton R, Anyah R (2013) Statistical downscaling and bias correction of climate model outputs for climate change impact assessment in the U.S. northeast. Global Planet Change 100:320–332. https://doi.org/10.1016/j.gloplacha.2012.11.003

Akhtar M, Ahmad N, Booij M (2008) The impact of climate change on the water resources of Hindukush–Karakorum–Himalaya Region under different glacier coverage scenarios. J Hydrol 355:148–163. https://doi.org/10.1016/j.jhydrol.2008.03.015

Amin MZM, Shaaban AJ, Ercan A, Ishida K, Kavvas ML, Chen ZQ, Jang S (2017) Future climate change impact assessment of watershed-scale hydrologic processes in Peninsular Malaysia by a regional climate model coupled with a physically-based hydrology models. Sci Total Environ 575:12–22. https://doi.org/10.1016/j.scitotenv.2016.10.009

Anandhi A, Srinivas VV, Nanjundiah S, Kumar DN (2008) Downscaling precipitation to river basin in India for IPCC SRES scenarios using support vector machine. Int J Climatol 420 (2007):401–420

Anandhi A et al (2011) Examination of change factor methodologies for climate change impact assessment. Water Resour Res 47:1–10

Biggs T, Gaur A, Scott C, Thenkabail P, Gangadhara Rao P, Gumma MK, Turral H (2007) Closing of the Krishna basin: irrigation, streamflow depletion, and macroscale hydrology, vol 111. IWMI, Colombo

Biswas SK (1987) Regional tectonic framework, structure and evolution of the western marginal basins of India. Tectonophysics 135(4):307–327. https://doi.org/10.1016/0040-1951(87)90115-6

Boé J, Terray L, Habets F, Martin E (2007) Statistical and dynamical downscaling of the Seine basin climate for hydro-meteorological studies. Int J Climatol 27(12):1643–1655

Buotte PC, Peterson DL, McKelvey KS, Hicke JA (2016) Capturing subregional variability in regional-scale climate change vulnerability assessments of natural resources. J Environ Manag 169:313–318. https://doi.org/10.1016/j.jenvman.2015.12.017

Buytaert W, Célleri R, Timbe L (2009) Predicting climate change impacts on water resources in the tropical Andes: effects of GCM uncertainty. Geophys Res Lett 36(7). https://doi.org/10.1029/2008gl037048

Chanapathi T, Thatikonda S (2020) Investigating the impact of climate and land-use land cover changes on hydrological predictions over the Krishna river basin under present and future scenarios. Sci Total Environ 721:137736. https://doi.org/10.1016/j.scitotenv.2020.137736

Chandole V, Joshi GS, Rana SC (2019) Spatiotemporal trend detection of hydrometeorological parameters for climate change assessment in Lower Tapi river basin of Gujarat state, India. J Atmos Sol Terr Phys 195:105130. https://doi.org/10.1016/j.jastp.2019.105130

Dhawan V (2017) Water and agriculture in India. [online] Available at: https://www.oav.de/fileadmin/user_upload/5_Publikationen/5_Studien/170118_Study_Water_Agriculture_India.pdf. Accessed 12 May 2020

Di Baldassarre, G., & Uhlenbrook, S. (2011). Is the current flood of data enough? A treatise on research needs for the improvement of flood modelling. Hydrological Processes, 26(1), 153–158. https://doi.org/10.1002/hyp.8226

Fowler HJ, Blenkinsop S, Tebaldi C (2007) Linking climate change modelling to impacts studies: recent advances in downscaling techniques for hydrological modelling. Int J Climatol 27 (12):1547–1578

Garg KK, Bharati L, Gaur A, George B, Acharya S, Jella K et al (2012) Spatial mapping of agricultural water productivity using the SWAT model in Upper Bhima catchment, India. Irrig Drain 61(1):60–79. https://doi.org/10.1002/ird.618

Ghosh S, Mujumdar P (2006) Future rainfall scenario over Orissa with GCM projections by statistical downscaling. Curr Sci 90(3):396–404

Gleick PH, Adams RM, Amasino RM, Anders E, Anderson DJ, Anderson WW, Anselin LE, Arroyo MK, Asfaw B, Ayala FJ, Bax A, Bebbington AJ, Bell G, Bennett MVL, Bennetzen JL, Berenbaum MR, Berlin OB, Bjorkman PJ, Blackburn E, Zoback ML (2010) Climate change and the integrity of science. Science 328(5979):689–690. https://doi.org/10.1126/science.328.5979.689

Gosain AK, Rao S, Basuray D (2006) Climate change impact assessment on the hydrology of Indian river basins. Current Sci 90:346–353

Hay LE, Wilby RL, Leavesley GH (2000) A comparison of delta change and downscaled GCM scenarios for three mountainous basins in the United States. 1. J Am Water Resour Assoc 36 (2):387–397

Hundecha Y, Arheimer B, Donnelly C, Pechlivanidis I (2016) A regional parameter estimation scheme for a pan-European multi-basin model. J Hydrol Reg Stud 6:90–111. https://doi.org/10.1016/j.ejrh.2016.04.002

Jhajharia D, Kumar R, Dabral PP, Singh VP, Choudhary RR, Dinpashoh Y (2014) Reference evapotranspiration under changing climate over the Thar Desert in India. Meteorol Appl 22(3):425–435. https://doi.org/10.1002/met.1471

Jin L, Whitehead PG, Rodda H, Macadam I, Sarkar S (2018) Simulating climate change and socio-economic change impacts on flows and water quality in the Mahanadi River system, India. Sci Total Environ 637–638:907–917. https://doi.org/10.1016/j.scitotenv.2018.04.349

Kendall MG (1975) Rank correlation methods. Charles Griffin, London

Knutti R (2008) Why are climate models reproducing the observed global surface warming so well? Geophys Res Lett 35(18) https://doi.org/10.1029/2008gl034932

Madhusoodhanan CG, Sreeja KG, Eldho TI (2017) Assessment of uncertainties in global land cover products for hydro-climate modeling in India. Water Resour Res 53(2):1713–1734

Mall R, Singh R, Gupta A, Srinivasan G, Rathore L (2006) Impact of climate change on Indian agriculture: a review. Clim Change 78:445–478

Mann HB (1945) Nonparametric tests against trend. Econometrica 13:245–259

Ministry of Environment and Forests (MoEF) Report, Government of India (2010). State of environment (SoE) report: Maharashtra, final draft. http://moef.nic.in/. Accessed on 27 Mar 2015

Mujumdar P, Kumar DN (2012) Floods in a changing climate: hydrologic modeling. Cambridge University Press, Cambridge. https://doi.org/10.1017/CBO9781139088428

Naidu CV, Durgalakshmi K, Muni Krishna K, Ramalingeswara Rao S, Satyanarayana GC, Lakshminarayana P, Malleswara Rao L (2009) Is summer monsoon rainfall decreasing over India in the global warming era? J Geophys Res 114(D24). https://doi.org/10.1029/2008jd011288

National Water Development Agency (Ministry of Water Resources, River Development and Ganga Rejuvenation, Government of India) http://www.nwda.gov.in/upload/uploadfiles/files/Annual.pdf

Nilawar AP, Waikar ML (2019) Impacts of climate change on streamflow and sediment concentration under RCP 4.5 and 8.5: a case study in Purna river basin, India. Sci Total Environ 650:2685–2696

Pasimeni MR, Valente D, Zurlini G, Petrosillo I (2019) The interplay between urban mitigation and adaptation strategies to face climate change in two European countries. Environ Sci Pol 95:20–27. https://doi.org/10.1016/j.envsci.2019.02.002

Rehana S, Mujumdar PP (2011) River water quality response under hypothetical climate change scenarios in Tunga-Bhadra river, India. Hydrol Process 25(22):3373–3386

Rehana S, Mujumdar PP (2012) Climate change induced risk in water quality control problems. J Hydrol 444–445:63–77. https://doi.org/10.1016/j.jhydrol.2012.03.042

Revadekar JV, Varikoden H, Murumkar PK, Ahmed SA (2018) Latitudinal variation in summer monsoon rainfall over Western Ghat of India and its association with global sea surface temperatures. Sci Total Environ 613–614:88–97. https://doi.org/10.1016/j.scitotenv.2017.08.285

Robock A, Turco RP, Harwell MA et al (1993) Use of general circulation model output in the creation of climate change scenarios for impact analysis. Clim Change 23:293–335. https://doi.org/10.1007/BF01091621

Satta A, Puddu M, Venturini S, Giupponi C (2017) Assessment of coastal risks to climate change related impacts at the regional scale: the case of the Mediterranean region. Int J Disaster Risk Reduct 24:284–296. https://doi.org/10.1016/j.ijdrr.2017.06.018

Sedova B, Kalkuhl M (2020) Who are the climate migrants and where do they go? Evidence from rural India. World Dev 129:104848. https://doi.org/10.1016/j.worlddev.2019.104848

Shashikanth K, Salvi K, Ghosh S, Rajendran K (2013) Do CMIP5 simulations of Indian summer monsoon rainfall differ from those of CMIP3? Atmos Sci Lett 15(2):79–85. https://doi.org/10.1002/asl2.466

Shrestha S, Dhakal S (2019) An assessment of potential synergies and trade-offs between climate mitigation and adaptation policies of Nepal. J Environ Manag 235:535–545. https://doi.org/10.1016/j.jenvman.2019.01.035

Singh AK, Tripathi JN, Kotlia BS, Singh KK, Kumar A (2019) Monitoring groundwater fluctuations over India during Indian Summer Monsoon (ISM) and Northeast Monsoon using GRACE satellite: impact on agriculture. Quat Int 507:342–351. https://doi.org/10.1016/j.quaint.2018.10.036

Stocker T (2013) Climate change 2013: the physical science basis: summary for policymakers. A report of Working Group I of the IPCC, technical summary. Report accepted by Working Group I of the IPCC but not approved in detail and frequently asked questions: part of the Working Group I contribution to the fifth assessment report of the Intergovernmental Panel on Climate Change. Intergovernmental Panel on Climate Change, New York

Surinaidu L, Bacon CGD, Pavelic P (2013) Agricultural groundwater management in the upper Bhima Basin, India: current status and future scenarios. Hydrol Earth Syst Sci 17(2):507–517. https://doi.org/10.5194/hess-17-507-2013

Tripathi S, Srinivas VV, Nanjundiah RS (2006) Downscaling of precipitation for climate change scenarios: a support vector machine approach. J Hydrol 330(3–4):621–640

Udmale P, Ichikawa Y, Kiem A, Panda S (2014) Drought impacts and adaptation strategies for agriculture and rural livelihood in the Maharashtra state of India. Open Agric J 8:41–47. https://doi.org/10.2174/1874331501408010041

van Vuuren DP, Edmonds J, Kainuma M et al (2011) The representative concentration pathways: an overview. Clim Change 109:5. https://doi.org/10.1007/s10584-011-0148-z

Venkateswarlu B, Singh AK (2015) Climate change adaptation and mitigation strategies in rainfed agriculture. In: Climate change modelling, planning and policy for agriculture. Springer, New Delhi, pp 1–11. https://doi.org/10.1007/978-81-322-2157-9_1

Wangsoh N, Watthayu W, Sukawat D (2017) A hybrid climate model for rainfall forecasting based on combination of self-organizing map and analog method. Sains Malaysiana 46 (12):2541–2547. https://doi.org/10.17576/jsm-2017-4612-32

Wang J, Nathan R, Horne A, Peel MC, Wei Y, Langford J (2017) Evaluating four downscaling methods for assessment of climate change impact on ecological indicators. Environ Model Softw 96:68–82. https://doi.org/10.1016/j.envsoft.2017.06.016

Wilby RL, Wigley TML, Conway D, Jones PD, Hewitson BC, Main J, Wilks DS (1998) Statistical downscaling of general circulation model output: a comparison of methods. Water Resour Res 34(11):2995–3008. https://doi.org/10.1029/98wr02577

Woldemeskel FM, Sharma A, Sivakumar B, Mehrotra R (2016) Quantification of precipitation and temperature uncertainties simulated by CMIP3 and CMIP5 models. J Geophys Res Atmos 121 (1):3–17

Yu PS, Wang YC (2009) Impact of climate change on hydrological processes over a basin scale in northern Taiwan. Hydrol Process Int J 23(25):3556–3568

Zhang Q, Shen Z, Xu C-Y, Sun P, Hu P, He C (2019) A new statistical downscaling approach for global evaluation of the CMIP5 precipitation outputs: model development and application. Sci Total Environ 690:1048. https://doi.org/10.1016/j.scitotenv.2019.06.310

Chapter 3
Flood Risk Assessment for a Medium Size City Using Geospatial Techniques with Integrated Flood Models

Surendar Natarajan and Nisha Radhakrishnan

Abstract Flood is defined as the high stage of a river that flows across the river banks. Floods are classified based upon the nature of the occurrence, usually as riverine floods, urban floods, lake fluctuations, and rise of sea level. Factors responsible for floods are topography, rainfall intensity, climate change, and improper design of drainage facilities. In this study, we discuss urban flooding, because urban flooding usually differs from other types of flooding where the increase in flood peaks of 1.8 to 8 fold results in flood volume up to 6 fold. An urban peak flood occurs quickly as compared to other types of floods. In urban region the factors responsible for high peak discharge are land use/land cover (LULC) changes, climate change, and increase in rainfall intensity within a given duration. The recent floods in major cities in India have raised the awareness for flood modeling studies. In India, most flood modeling studies are done for major urban river basins passing through cities, but for this study the Koraiyar River basin that passes through the medium-sized city of Tiruchirappalli in South India was chosen. The parameter adopted for analyzing flood volume in this basin is LULC and its impact on urban surface runoff. LULC changes and their impact on surface runoff are studied by integrating remote sensing and geographic information systems (GIS) with hydrologic HEC-HMS (Hydrologic Engineering Centre-Hydrologic Modeling System) and hydraulic HEC-RAS (Hydrologic Engineering Centre-River Analysis System). Runoff is generated by using LULC, slope, hydrologic soil group, curve number (CN) maps, and rainfall intensity. The overall performance of the hydrologic model during calibration was satisfactory, based on the Nash–Sutcliffe Efficiency criteria of with values of 0.5–0.6. The generated peak discharge is used for developing floodplain and hazard maps from 1986 to 2036. The floodplain and hazard maps estimate flood depth and risk in the basin area for changing LULC conditions of different return periods. Flood plain and hazard maps of 2, 5, 10, 50, and 100 years return periods were generated to identify flood extent and hazard level in the basin.

S. Natarajan (✉) · N. Radhakrishnan
Department of Civil Engineering, National Institute of Technology Tiruchirappalli, Tiruchirappalli, India
e-mail: nisha@nitt.edu

M. N. Islam, A. van Amstel (eds.), *India: Climate Change Impacts, Mitigation and Adaptation in Developing Countries*, Springer Climate,
https://doi.org/10.1007/978-3-030-67865-4_3

The developed maps are used as a tool for effective flood forecasting and warning of flood hazards in the basin.

Keywords Land use/land cover · HEC-HMS · RAS · NSE · Flood risk map

Introduction

The migration of people from rural to urban areas causes changes in land use and land cover (LULC) in the urban region (Dewan and Yamaguchi 2009). The changes in LULC in the urban region impact surface runoff, which may reduce the travel time of peak flow of the urban region so that flood frequency is increased. For flood risk assessment, it is necessary to understand the dynamics of LULC change and its influence on the urban hydrologic cycle for predicting the floodplain and hazard maps (Suriya and Mudgal 2012). The LULC change and its implications for the urban flooding is one of the upcoming research topics in the field of urban hydrology (Fox et al. 2012).

For efficient flood risk assessment in the urban area, the floodplain delineation should be done properly. In the traditional method of flood forecasting, statistical testing was used to determine the frequency of flood depth. In recent times, geospatial techniques are integrated with hydrologic and hydraulic modeling to identifying the impact of LULC changes (Zope et al. 2015).

The major causes of floods include an increase in precipitation in a short span of time, changes in LULC, and other related hydrologic impacts. For flood assessment, various approaches such as rational formulae and empirical methods are used, but these methods cannot be applied over large extents and are not suitable for all regions. The other advanced method is modeling techniques used for simulation, planning, and the management process. Flood modeling is designed for quantitative and qualitative estimations in the hydrologic and hydraulic modeling process.

The factors that influence floods and the modeling steps are included in this chapter in detail. Numerous factors are responsible for flood in the urban region: prominent among these are meteorological, hydrologic, and human factors. The meteorological factors include rainfall, cyclonic storms, and small-scale storms; the hydrologic factors include soil moisture, groundwater level fluctuations, impervious soil cover, channel cross sections, and high tides in the case of coastal cities; and manmade factors are LULC change, occupation of floodplains, climate change effects, urban microclimate change, and sudden release of water from dams. All these are prominent factors that influence flooding in urban regions.

Encroachment on Floodplains

Most major cities and towns are located in river floodplains, which causes additional stress to river basins. During the past decades more developmental activities have taken place in urban and agricultural lands, leading to additional stress in these floodplains and resulting in loss of flood spread areas and increase in flood risk in downstream portions of the basin. For example, most major European cities in England and Germany are located in floodplains that have induced burdens in the rivers, and preventive measures have been made to recreate floodplain storage (Wheater 2006). Indian cities such as Delhi, Vijaywada, Hoshangabad, Surat, and others located along the banks of rivers often experience floods for this reason, aggravated by several other factors.

Land Use Change

In natural catchments, peak flow is usually prolonged by the pervious nature of the basin. The direct and indirect LULC changes that replace green cover in urban regions have effects on surface runoff. The LULC changes, such as occur in the construction of built-up areas increases imperviousness in the basin, causing rapid overland flow and reducing the time to peak reduction, which increases flood peaks. However, urban catchments are not affected to the same extent by antecedent conditions and respond more rapidly to rainfall. Hence, intense rainfall in any season may become a major cause of flooding, leading to a change in flood seasonality (Aliani et al. 2019).

Urban Flooding

Increased urbanization alters the hydrologic cycle in the urban basin, affecting the rainfall–runoff relationship. During urbanization, small streams joining the main streams are affected and subsequently affect the natural drainage of rainwater that flows in the basin. Thus, proper drainage structures should be designed to overcome these overflow processes. The development of storm water drainage systems involves two steps: estimation of runoff hydrography and routing the runoff hydrograph through the drainage system to obtain the outflow hydrograph. The calculation of surface water runoff is quite a challenging task. So, initially the rainfall intensity has to be designed, being known as storm design, an appropriate method adopted for obtaining the hydrograph storm over the catchment and recommending suitable remedial methods, as discussed next. However, the methods adopted for estimating storm runoff in the catchment and the superiority of flood modeling over these methods is explained in this chapter in which an overall view of flood modeling parameters and the techniques used for generating flood modeling is presented. For further reading, the reader may refer to Chow et al. (1988), Mays and Tung (1992), and Linsley et al. (1992).

Rational Method

The rational method is the most prominently used method for estimating storm runoff in the basin. Although this method has some merits and demerits, it is still widely used because of its simplicity. The following assumptions are made in a rational formula. Maximum flood flow is produced by a certain rainfall intensity that lasts for a time equal to or greater than the period of concentration time. When a storm continues beyond concentration time, every part of the catchment would be contributing to the runoff at the outlet and therefore it represents the condition of peak runoff. The concentration time is the time required for the surface runoff from the remotest part of the catchment area to reach the basin outlet. The runoff rate to this condition is given by Eq. (3.1):

$$Q_p = CIA \tag{3.1}$$

where A is the area of the catchment in km^2, and I is the intensity of rainfall in cm/h.

The maximum rainfall intensity depends on duration and frequency. The intensity of rainfall should therefore be corresponding to a duration equal to concentration time and desired return period. The rational method is found to give good results for small areas up to about 50 km^2. It is used to estimate the peak flood in the design of urban drainage systems, storm sewers, airport drainage, and in the design of small culverts and bridges.

Structural and Nonstructural Measures

Structural measures are involved in the construction of hydraulic structures such as levees, dykes, flood divide walls, and detention and retention ponds to reduce the peak discharge in the basin. The nonstructural measures used to reduce flood risks include adopting master drainage plans, identifying flood-prone areas, early flood forecasting and warning, and proper land use plans in urban regions. For accurate flood forecasting, flood modeling studies are required to estimate the flood spread and depth by using integrated flood modeling tools. In such cases, remote sensing data combined with GIS have been found to provide valuable tools for flood risk modeling and defining possible areas of flood extent in the basin. The relationship between flood conditions and the spatial distribution of urban development with an optimization-based approach is used for investigations by flood modeling studies. Catchment treatments such as terracing and afforesting to reduce water velocities and rainwater harvesting to reduce peak runoffs are also commonly proposed, but so far have not attracted the importance they deserve in India.

Flood Inundation Modeling

For flood modeling studies in a river basin, Saint-Venant equations of gradually varied unsteady flow are generally used. The flood modeling approach is done in two ways, by one-dimensional (1-D) and two-dimensional (2-D) methods. The 1-D flood modeling approach is simple and the 2-D approach is appropriate. In urban flood modeling the modeling approach requires high-resolution data to distinguish complicated urban topography. This type of modeling requires high-resolution grids of the order of 1–5 m to acquire appropriate topographic features (Mark et al. 2004). The high-resolution remote sensing data available are LIDAR (light detection and ranging) and Ordnance Survey (OS) Master Map digital map data (Mason et al. 2007), recently used for providing accurate LULC classes, and these types of data can be used in hydrologic and hydraulic models in urban areas.

A number of studies have considered the application of 2-D hydraulic models to urban problems, including numerical solutions of the full 2-D shallow-water equations (Mignot et al. 2006), 2-D diffusion wave models (Hsu et al. 2000; Yu and Lane 2006a; Yu and Lane 2006b), and analytical approximations to the 2-D diffusion wave equations using uniform flow formulae (Schmitt et al. 2004). The most common approach to urban flood modeling is to employ a 2-D approach at high resolution and calibrate the friction parameters to observed data. The dynamic wave models are found to produce more efficient results in urban flood modeling simulations. These models can be coupled to 1-D flow models for accurate simulations.

Modeling Urban Drainage Systems

As per European Standards EN 752, 'flooding' is defined as a 'condition' where wastewater or surface water escapes from or cannot enter a drain or sewer system and either remains on the surface or enters buildings. Flooding in an urban drainage system is classified as separate or combined sewers. For design of flood flow in urban drainage systems, urban storm water drainage modeling is required. The modeling approach may be hydrologic or hydraulic or a combined modeling approach. The hydrologic model simulates surface runoff from rainfall and the hydraulic model describes the flood extent and depth in the basin. For computations of surface runoff, a standard rainfall runoff simulation model is required. The one-dimensional urban storm drainage models such as MOUSE, info works, Storm Water Management Model (SWMM), EPA SWMM, MIKESHE, HEC-HMS, and HEC-RAS are used for flood modeling purposes (Hunter et al. 2007).

Sustainable Drainage Systems

Previously, to reduce peak discharges in urban regions, hydraulic structures such as detention storage in the form of storage reservoirs have been constructed. In recent years, a new approach such as a sustainable urban drainage system is adopted to manage urban floods. These structures are more effective and convenient in reducing the peak discharges. Conventionally, these systems are used to reduce runoff volumes from small storms. In larger storms, a sustainable drainage system can be improved by including flow routing systems. Some of the structures commonly used in sustainable systems are pervious concrete, green roof, and storm water detention.

Integrated Flood Management System

An integrated management system considers the drainage system as a whole, covering all structural, hydraulic, and environmental aspects to make best use of both existing and future facilities. In conventional flood management methods, the different components in the hydrologic cycle are modeled individually by either hydrologic or hydraulic components. The integrated flood management practices provide some solutions to real-time flood control. In a conventional drainage design, the gravity method is the only method used to route water through the system. However, real-time control is still far from being widely practiced. To be effective, real-time control needs reliable hardware (sensors, regulators, controllers, data transmission systems) as well as objective and sound decision-making tools. Models have to be used to simulate the rainfall–runoff processes to supplement the measured data and to forecast future rainfall, as well as runoff developments, to provide the data necessary for a decision. Radar rainfall data are important in short-term rainfall forecasting to detect the areal distribution of rainfall patterns. The latest developments focus on the use of deterministic models in conjunction with stochastic approaches to investigate the potentials of real-time control for given systems and objectives. In densely populated cities, a cost-effective alternative to additional structural control measures for regulating flooding is operating existing flood gates using suitable controllers and control strategy (Haghizadeh et al. 2012).

Distributed Flow Routing

In distributed flow routing models, the rainfall is transformed into runoff over the basin to produce a unit hydrograph at the basin outlet. This hydrograph is used as a tool to produce a flow hydrograph at the outlet, and then this hydrograph is taken as an input at the upstream end of a drainage system for storm water disposal. The unit hydrograph is applied to storm design to yield the design flood hydrograph, and the storm with a specific return period for flood estimation is called the design storm. In the absence of a unit hydrograph, a synthetic unit hydrograph may be developed.

In absence of sufficient data, the design storm may be used by the transposition method. In case of large and important projects the probable maximum precipitation is used as the design storm. If probable maximum precipitation is used in design flood estimation, the resulting flood hydrograph gives the maximum probable flood. The unit hydrograph method cannot be applied for large areas, more than 5000 km^2. In such cases the basin is subdivided into a number of subbasins corresponding to each tributary. These flood hydrographs of subbasins are then routed through the channel reaches and superimposed with a common time scale to obtain the flood hydrograph at the point of interest. In case of large areas, it is assumed that the design storm produces rainfall over the entire basin that is simultaneous. The unit hydrograph not only produces flood peak but also the complete flood hydrograph, which is essential in fixing the spillway capacity after incorporating the effort of storage of the reservoir on flood peak through flood routing. If the area of basin is large, it is better to use frequency analysis methods (Ojha et al. 2008).

Trend Analysis of Rainfall

The rainfall data are collected, and trend analysis of rainfall determines the existing trends in the data (Ghosh et al. 2009) by parametric and nonparametric methods. In nonparametric methods, the Mann–Kendall test is used to determine the rainfall trend (Ahamad et al. 2015). In the nonparametric test, the data outliers can be evaded easily. The next test is Sen's slope estimator, commonly used to sense a linear trend. As the M-K test, the time series are evaluated by their ranks, known as Spearman's rho test (Siegel and Castellan 1998). To check the statistical changes in rainfall trend, the CUSUM test is used (Chowdhury and Beecham 2010).

Frequency Analysis of Rainfall

Rainfall frequency can be analyzed by parametric methods such as log-normal, log-Pearson type III, generalized extreme value (GEV), and Gumbel (GUM) and Weibull probability distributions. The probability distributions are used to evaluate the best fit probability distribution from the rainfall distribution analysis. The goodness of fit is evaluated by the chi-square test, the Kolmogorov–Smirnov test, the Cramer von Mises test, and the probability plot correlation coefficient (PPCC) test. This method is applied for testing the null hypothesis whether the maximum daily rainfall data follow the specified distribution. Tests such as Anderson–Darling and Kolmogorov–Smirnov were used with the chi-square test at the $\alpha = 0.01$ level of significance for the selection of the best fit probability distribution. The Gumbel distribution is known as the extreme value type I distribution used for extreme analyses such as flood and drought. In 1967, the United States Water Resources Council suggested that log-Pearson type III distribution should be adopted as the standard flood frequency distribution by all U.S. federal government agencies (Ojha et al. 2008).

Goodness-of-Fit Test for Different Probability Distribution Methods

The particular variables follow a specific distribution, and the most commonly used technique for testing frequency distribution is the chi-square test (Haan 1977). Kolmogorov–Smirnov is an alternative to the chi-square test in which the data are arranged in descending order of magnitude (Majumdar 2012).

Estimation of Short-Duration Rainfall

In the absence of hourly rainfall data, the IMD method can be adopted. The daily maximum rainfall data can be converted into hourly rainfall data by the "Indian Meteorological Department (IMD)" empirical formula in Eq. (3.2)

$$\mathrm{Pt} = \mathrm{P24}\sqrt[3]{\frac{t}{24}} \tag{3.2}$$

where Pt is the available rainfall depth in mm, t is time in hour intervals, P24 is the day-to-day rainfall in mm, and t is the period of rainfall in hours. This method is used to assess short-interval rainfall from daily rainfall data, and it is a standard that gives the best results in the absence of hourly data.

Derivation of Intensity Duration Frequency Equations

Initially, the proposed intensity duration frequency (IDF) curves are derived using an empirical equation developed by Kothyari and Garde (1992) by using a probability distribution for annual maximum rainfall. In the case of Mumbai city, the IDF curves are derived from the extreme rainfall event that occurred in the city (Zope et al. 2016). The empirical, as well as IDF, curves, are derived for Madinah city from the long records of rainfall data (Subyani and Al-Amri 2015).

LULC Analysis

LULC changes are continuous processes that occur following natural and human-made activities (Sarma et al. 2008). According to the National Remote Sensing Agency (NRSA) for Indian conditions (Reddy 2002), LULC classes are wasteland, vegetation, water bodies, grass, built-up land, and agricultural land. Several studies have been conducted regarding satellite imagery, including Landsat MSS, TM, and ETM+, to do change analysis. LULC change is analyzed by using many multidate images to evaluate the changes from human interventions and environmental changes at the time of image acquisition (Yang and Lo 2002).

The modeling approach includes two types: regression-based and spatial transition-based models (Lambin 1997). The spatial transition models are also called cellular automation (CA) simulation models as these predict future land development based on probabilistic estimations such as the Monte Carlo technique (Clarke and Gaydos 1998). The Markov model is an appropriate tool for simulating future LULC changes (Logsdon et al. 1996). From the various studies on the Markov model, the accuracy and the prediction accuracy of the model was 98.5% (Jianping et al. 2005; Zhang et al. 2011).

SCS-CN Approach

The challenging tasks in the field of hydrology are prediction and quantification of surface runoff from the catchment area. The curve number (CN) is a critical factor in defining runoff from the Soil Conservation Service (SCS)-based hydrologic modeling method. The curve number is a function of soil type and land use (Table 3.1).

Rainfall–Runoff Model: HEC-HMS

The HEC-HMS (Hydrologic Engineering Centre-Hydrologic Modeling System) is an open-source software developed by the U.S. Army Corps of Engineering Hydrologic Engineering Center. It is designed to simulate the rainfall–runoff processes of a dendritic watershed. The meteorological model in HEC-HMS is the most significant component because it defines the meteorology boundary conditions for the subbasins. The model includes precipitation, evapotranspiration, and snowmelt methods to be used in simulations. The control specifications in the model control the time span of a simulation, including starting and end date and time and a time interval. The simulation run is created by combining a basin model, a meteorological model, and control specifications. The simulation results include information on peak flow and total volume. Model calibration is required to obtain optimal parameters of different methods used in subbasin and junction elements (USACE 2000).

Table 3.1 SCS value based on soil cover type

	Land use		
Soil type	Cultivated	Pasture	Forest
With above average infiltration rates; usually sandy or gravelly	0.20	0.15	0.10
With average infiltration rates; no clay pans; loams and similar soils	0.40	0.35	0.30
With below-average infiltration rates; heavy clay soils or soils with a clay pan near the surface; shallow soils above impervious rock	0.50	0.45	0.40

Selection of Hydrologic Model

The model selection depends on available data, site conditions, gauged or ungauged catchment, and size of the watershed (HEC 2010).

General Applicability and Limitations

The significant advantages of this software are the wide selection of different hydrologic models suitable for different environmental conditions. Further, it includes options in the calibration of the chosen models against measured precipitation and runoff data. The limitation of the model is that continuous simulations cannot be performed, so the modeling is limited to single events.

Hydraulic Model-HEC RAS

HEC-RAS is a hydraulic model developed by the U.S. Army Corps of Engineers Hydrologic Engineering Center. The software is capable of performing one-dimensional (1-D) steady- and unsteady-flow simulations and also has graphical user interface data storage and separate hydraulic analysis components (HEC 2010).

Data Requirements

The data required for flow simulation in HEC-RAs are geometric details and the relative surface roughness of the analyzed flow channel, over banks, flow data, and data about hydraulic structures such as levees or weirs, and bridges. The base of any geometric model in HEC-RAS is the river system schematic.

Applications and Limitations of HEC-RAS

In this study, HEC-RAS is used for the determination of floodplains based on unsteady-flow simulations of flood hydrographs with statistical return periods of 2, 5, 10, 50, and 100 years. The software is selected primarily for its wide application in the stable and practical flow model. The model is easy to use and provides complete practical documentation (Hu and Walton 2008). Its suitability in different condition and its application have been widely examined (HEC 2010).

In major cities, flooding is one of the most dangerous phenomena, one that cause loss of human life and their properties to a large extent. The major cities are centers of economic activities with vital infrastructure that needs to be protected 24×7. In most of the cities, damage to vital infrastructure not only has a bearing for the state and the country but could even have global implications. Major cities in India have witnessed loss of life and property, disruption in transport and power, and the incidence of epidemics. In this literature study, flood modeling studies adopted for various major cities nationally and internationally are discussed, and a suitable methodology for developing flood modeling studies in a medium size city is recommended.

Case Studies in India

Mumbai

Zope et al. (2016) analyzed LULC changes and their effects on urban floods in the expanding urban catchment of the Oshiwara River in Mumbai, India. In this study, LULC change was determined for the years 1996, 2001, and 2009 by acquired topographic and satellite images. The analysis proved that LULC change between the years 1966 and 2001 was much slower that that between the years 2001 and 2009, when it was observed to be rather rapid. The results obtained between 1966 and 2009 proved that there was a 74.84% increase in the built-up area and a 42.8% decrease of open space in response to urbanization. The effects of LULC on the flood hydrograph for various return periods were evaluated using the HEC-HMS model. It was noted that over the past 43 years, the increase in peak runoff and volume was 3.0% and 4.45%, respectively, for the 100-year return period. The results also reported that increase in intensity of rainfall with short span and the lower return period had a maximum peak discharge and volume of runoff when compared to higher return periods for changing land use conditions.

Guwahati

A flood hazard map for a 100-year return period was developed for Guwahati city, and hazard ranks identified the areas susceptible to flood inundation in the city This study quantifies the reason for flood risk and identifies the parameters responsible for peak runoff. From the analysis it is identified that LULC and microclimate change are among the factors responsible for runoff changes in the city. The LULC and their influence on surface runoff were analyzed for the years 2006 and 2011. Almost 46.5% of the city was inundated. The flood risk was found to increase from 2006 to 2011 because of increased imperviousness. For reducing peak discharge, it is recommended to adopt a detention pond that reduces almost 43% of inundated area in the city. Thus, detention ponds were found to be effective in reducing the effect of urban flood, and hence can be proposed as a suitable measure for flood management (Sahoo and Sreeja 2017).

Bengaluru

Large-scale land conversion, loss of lakes, and the occupation of floodplains have resulted in many problems such as urban heat islands, depletion of groundwater, and more importantly, flash floods. The impervious surfaces of urbanization led to accelerated runoff and reduced water recharge in Bengaluru city. The LULC changes and conversion of land surface to impervious surface and the loss of water bodies led to urban flood. Recently, 280 locations in the city were identified as areas vulnerable to flash floods. Apart from LULC changes, the other factors responsible for frequent floods are dumping of solid waste and the allotment of floodplains for housing purposes. Increased urbanization with intensified vertical growth of the city has led to liquid waste disposal along the storm water drains, making the runoff models unpredictable. Because of LULC changes, even a short rainfall of 30 mm/h causes greater flood depth. For the proper assessment of vulnerable areas affected by floods, a multicriteria evaluation of spatial layers in GIS was adopted for flood assessment in the city. For proper flood risk assessment, self-sufficient eco-friendly buildings with proper conservation of rainwater by harvesting structures are recommended (Rama Prasad and Narayan 2016).

Hyderabad

In Hyderabad city, LULC changes and their impacts on surface runoff were studied by a 2-D hydraulic model integrated with GIS. In this study a simple approach was adopted to analyze the flood events that occurred in the city by analyzing extreme rainfall events with changing LULC conditions. Extreme rainfall can occur in a single day in the city. The synthetic hyetographs were developed from IDF curves and HEC-RAS freely available hydraulic model is used with GIS to generate flood depth over underlying terrain and hazard maps for different return periods. From the analysis it was identified that 17% of total area is liable to floods, of which 9% of the area indicates high risk, 52% of the area shows medium risk, and the remaining 35% has low risk of flooding (Rangari et al. 2018).

Chennai

In Chennai city, flood studies were done in the Tirusoolam area, a part of Chennai city. This method examines the different available methods to trace the flood damage and illustrates the methodology to explore the economic loss through social investigation in the Tirusoolam area. In this study the CN was generated for Indian conditions, and flood peak discharge was generated through HEC-HMS software. From the generated peak discharge, flood hazard maps were generated for the study area. The flood risk maps generated were used to identify the areas subject to greater risk of flood. In 2015 a high intensity of rainfall of 40 cm occurred within an hour, resulting in vast damage to properties in the city that was mainly caused by LULC and climate change patterns in the city (Suriya and Mudgal 2012).

International Level

Hafr Al-Batin City

Hafr Al-Batin city, located in the northeastern part of Saudi Arabia, was put in danger from floods that caused serious damage to public and private properties. The city is located in the dry valley of Wadi Al-Batin, part of Wadi Al-Rummah, where three valleys meet, which causes more city flooding during intense rain with short duration of rainfall. The estimated average rainfall in the basin was about 125 mm, and recently maximum rainfall events generated flood water flow from the valleys toward the city. In this study, HEC-HMS and HEC-RAS models were used to simulate flood occurrence in the city. The average flow depth occurred during this study period were 3.02 m, 3.26 m, 3.45 m, 3.76 m, 4.04 m, and 4.34 m for the simulated 2-, 5-, 10-, 25-, 50-, and 100-year design floods, respectively. Flood hazard maps are also generated to identify the areas within the city with high risk of flooding (Aliani et al. 2019).

Dhaka City

The greatest flood that had occurred in Greater Dhaka in Bangladesh in the year 1998 was studied by using synthetic aperture data (SAR) with GIS. Flood frequency and floodwater depth were calculated by SAR data for generation of flood hazard maps. The flood hazard maps were created by a ranking matrix with land cover, geomorphic units, and elevation data as GIS components. From the results it was observed that the major portion of Greater Dhaka is subject to very high hazard zones whereas only a small portion is not exposed to the potential flood hazard. The proposed flood hazard maps are useful for mitigation of the lives of people and property of Dhaka city (Dewan and Makoto 2006).

Ho Chi Minh City, Vietnam

Rapid and unplanned urbanization combined with climate change has increased the chance of flood risk in Ho Chi Minch city, Vietnam. Remote sensing data with a GIS tool are integrated with a hydrologic model to identify flood risk in the city. In this study, a Quick bird image was used to generate LULC maps; US Soil Conservation Service Technique Release 55 (SCS TR-55) is used for generating a rainfall–runoff model. The digital elevation data (DEM) and GIS were used to analyze tidal floods.

From the study it was observed that the flood depth of 2–10 cm is not a serious threat but a flood depth of 10–100 cm presents a significant issue because of the tidal effect. Apart from the tidal effect, increase in imperviousness, urbanization, and decrease in flow length have contributed greatly to flood risk. Finally, the study concludes with appropriate infrastructure and by improving the infiltration capacity of runoff by optimized drainage systems (Dang and Kumar 2017).

West Bank of Palestine

The most challenging task in the hydrology field is prediction and quantification of surface runoff. The CN is a key factor in determining surface runoff through the SCS-based hydrologic method. The traditional method of calculating CN is tedious and time consuming in hydrologic modeling. Therefore, the SCS method combined with GIS is used for calculating surface runoff. In this study, the hydrologic modeling flow of the West Bank catchment is analyzed by the SCS-CN method. The West Bank of Palestine is an arid to semiarid region with yearly rainfall of 100 mm to 700 mm from the Jordan River to the central part of the region. The calculated composite CN for the entire West Bank is assumed to be 50 for dry conditions in the basin. This study clearly proves that combining GIS with the SCS method is a powerful tool for estimating runoff volume in the basin (Shadeed and Almasri 2010).

Marand Basin, Iran

The LULC change and its impacts on peak runoff in Marand basin, Iran, were studied using remote sensing with GIS techniques. The runoff coefficient was estimated from the LULC extracted satellite images; slope map, hydrologic soil groups, rainfall intensity, and peak runoff for each subbasin were calculated. In this study by using linear function in fuzzy logic model and integration of two layers of peak runoff and elevation line, layers between 0 and 1 were transformed into fuzzy values. Then, by multiple overlap weights with these two layers, different classes of flood hazard map were developed. The flood hazard map is compared with participatory Geographic Information System (PGIS) and transferring the information to a confusing matrix; the accuracy of the flood map generated was 87.83%, these maps were compared with a LULC map, and the flood extent was determined (Mousavi et al. 2019).

South Carolina

In this study, the United States Geological Survey (USGS) hydrologic units of three different data sets were used. In the first data set, rainfall is observed at individual meteorological gauges, and the second dataset was from the National Centers for

Environmental Prediction (NCEP) Environmental Modeling Center (EMC): 4-km Gridded Radar-Estimated Precipitation (GRIB) Stage IV data are used. The third data set was from gauge data with the spatial coverage of the Parameter-elevation Relationships on Independent Slopes Model (PRISM) data. The two watersheds in South Carolina were examined for three different representations of heavy rainfall, using the HEC-HMS developed by the U.S. Army Corps of Engineers. From the analysis it was found that the latter two precipitation inputs that consider spatial representation of rainfall yielded similar performance and improved simulated streamflow as compared to simulations using rainfall observed at individual meteorological gauges. This method is used in locations that lack rain gauges (Gao et al. 2018).

European Union

In European Union countries such as Belgium, England, France, Netherlands, Poland, and Sweden, a legal and policy analysis of implementation of flood directives is provided. These directives are implemented at a national level for effective flood directives on increasing societal resilience. The study also shows a focus on transboundary river basin management and flood risk management. The studies lack a modeling approach, and flood risk management planning strategies are generally recommended (Priest et al. 2016).

Cities in Germany

German cities are frequently affected by flood. During years 2002 and 2013, floods caused a total damage cost of 6–8 billion Euro in Germany, the most expensive natural hazard event in Germany up to now. The event of 2002 recommends an integrated flood risk management system in Germany to cope with floods and to achieve effective and more integrated flood risk management. Flood risk management recommends a consideration of flood hazards in spatial planning and urban development, property-level mitigation and preparedness measures, more effective flood warnings, and improved coordination of disaster response. The continuous flood occurrence in these cities considers risk drivers, such as climate change, land use changes, and demographic change, to reduce flood risk assessment in these cities (Annegret et al. 2016).

Riyadh City

The surface runoff in the watershed of Riyadh city, Saudi Arabia, was estimated by the Natural Resource Conservation Service (NRCS) method. The remote sensing technique along with GIS is used to estimate SCS-CN for the watersheds of Riyadh city. The data used in the study are DEM, soil map, geology map, satellite images, and daily rainfall data. The HSG, LULC, and CN maps were developed for each basin in the study area. From the analysis it was noted that the study area was delineated into 40 watersheds and the weighted CN for the city is 92. The rainfall–runoff analysis shows that the area has a high and very high daily runoff of 35–50 and >50 mm, respectively (Radwan et al. 2018).

The major cities in India and in the world have experienced loss of life, property, disturbance in transport, and the incidence of epidemics as results of floods. Therefore, the lesson learnt from these major cities is decided to be to adopt some of the parameters for developing flood modeling in these medium size cities, using measures such as management of land use practices, climate risk, and proper maintenance of urban drainage structures. From the various literature reviews, it was observed that almost all flood modeling studies are done only for major and coastal urban cities globally and nationally. Very few limited studies were conducted in an integrated flood modeling approach and for medium size cities.

In this study, an integrated approach of hydrologic and hydraulic modeling with Arc GIS is adopted. The hydrologic and hydraulic modeling used in this study is HEC-HMS and HEC-RAS. The integrated flood modeling approach is adopted to develop a regional model for generating floodplain and flood hazard maps. The changes in LULC for different years and their impact on urban runoff are studied. Floodplain and flood hazard maps for different return periods are generated for various LULC changes.

The current study considers the development of integrated flood modeling for Koraiyar River basin passing through a medium size city, Tiruchirappalli city, in India. The study approaches the integrated flood modeling approach for the Koraiyar River basin by temporal and spatial data collection. Temporal data such as daily rainfall data are collected and analyzed by frequency distribution methods. The spatial data of remote sensing images of Landsat images classifies LULC changes and the DEM data address stream network classification and basin delineation. The LULC maps with the HSG map are imported to the hydrologic model HEC-HMS for generating peak discharge. The generated peak discharge is incorporated into the hydraulic model HEC-RAS to generate flood depths for different return periods.

Motivation of the Present Chapter

In recent years the severity of floods has increased in all cities in India. The most prominent cities that are frequently affected by floods are Mumbai, Bengaluru, Hyderabad, and Chennai. After the 2015 floods in Chennai, it is necessary to develop flood modeling for all developing cities in India to provide sufficient warning and flood forecasting systems. In India, flood modeling studies so far have focused on megacities and minimal studies have been done for a basin that passes through a medium size city. No integrated flood modeling studies have yet been done for the Koraiyar basin selected for this study.

Objectives of the Study

The present study aims to analyze the impacts of LULC changes spatially and temporally. It also addresses the influence of LULC in the runoff of a medium size urban catchment area, in India, for different rainfall return periods of an ungauged basin. The objectives of the study are as follows:

1. To estimate the rainfall intensity for a medium size urban catchment area using statistical techniques
2. To study the changes in LULC of a medium size urban catchment area and the impacts on urban runoff
3. To develop floodplain and hazard maps for different return periods
4. To recommend suitable planning and designing of the drainage system in the catchment area

Scope of the Study

The work is to be carried for the Koraiyar basin passing through Tiruchirappalli city, a medium size city. Flood studies for a 100 years return period have not been done. An integrated flood modeling approach and LULC analysis have not yet been attempted in the Koraiyar Basin. The modeling approach adopted for this study can be adopted to basins of similar land use and climatic conditions.

Study Area

The Koraiyar River and its catchment lie between latitude 10°32′40.24″N and 10°48′16.81″N and longitude of 78°32′23.94″E and 78°39′48.58″E in Tiruchirappalli city, South India. The river originates from the Othakkadai Karupur Redipatti hills in Manaparai taluk of Tiruchirappalli district and finally flows into the Uyyokondan channel in the center of Tiruchirappalli city, Tamil Nadu (Fig. 3.1). The river basin has a subtropical climate. There is no significant temperature difference between summer and winter, with summer (March–May) having an extreme temperature of 41 °C and a least of 36 °C; winter (December–February) generally being warm but pleasant with temperatures ranging from 19 °C to 22 °C. Hot, humid, dry summers and mild winters are the main climatic features of the basin. The rainy season that falls between October and December brings rain mostly from the northeast monsoon. The river flows through Manaparai, Thuvarankurichi, and Viralimalai, and the total length of the main river is 75 km, with a catchment area of 1498 km^2. The average annual rainfall for the Koraiyar urban catchment is about 757.40 mm to 866.70 mm. The surplus water flows through Puthur weir outlet in the left bank of the Uyyakondan River and traverses the Kodamurutty River for 6 km

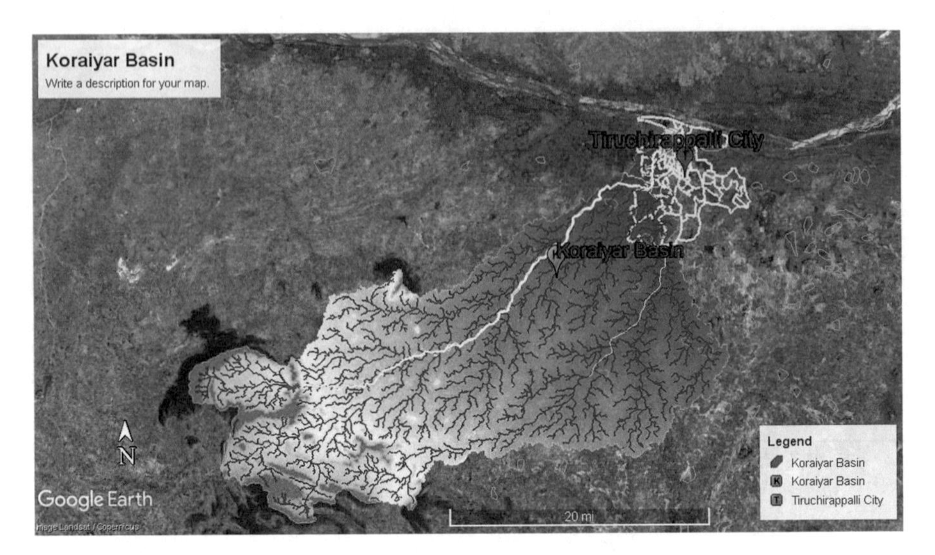

Fig. 3.1 Selected study area for flood modeling of Koraiyar River basin

before finally flowing into the Bay of Bengal. The rain gauge station for the basin is located at Trichy Airport.

Methodology Adopted for the Study

The methodology adopted for the present study is explained next.

Data Collection and Processing

The spatial and temporal data required for the study are satellite images and rainfall data. The satellite images of 30 m resolution Landsat ETM 7+ were collected from the USGS earth explorer website (https://earthexplorer.usgs.gov/) for the years 1986–2016 at 10-year intervals for LULC analysis. The soil map was acquired from the Food Agricultural Organization harmonized world soil database (http://www.fao.org/soils-portal/harmonized-world-soil-database-v12/en/.). The DEM data are downloaded from ALOS World 3D, a 30 m resolution digital surface model (DSM) constructed by the Japan Aerospace Exploration Agency used in HEC-Geo HMS for watershed delineation and subbasin development. The daily rainfall data are collected from the state surface and groundwater data center, Chennai, from 1977 to 2016.

Rainfall Analysis

The daily rainfall data are collected and converted into hourly rainfall data by the IMD method. In this study an extreme rainfall event that occurred in one day between 1976 and 2016 was selected for the study. In these 40 years, the highest extreme event, which occurred on 22 November 1999, was selected for the study. The IDF curves are generated based upon the suitable frequency distribution method. In this study the IDF curves are generated from Gumbel distribution methods for different return periods (Fig. 3.2).

Image Classification from LULC

The satellite data collected for the study are shown in Table 3.2.

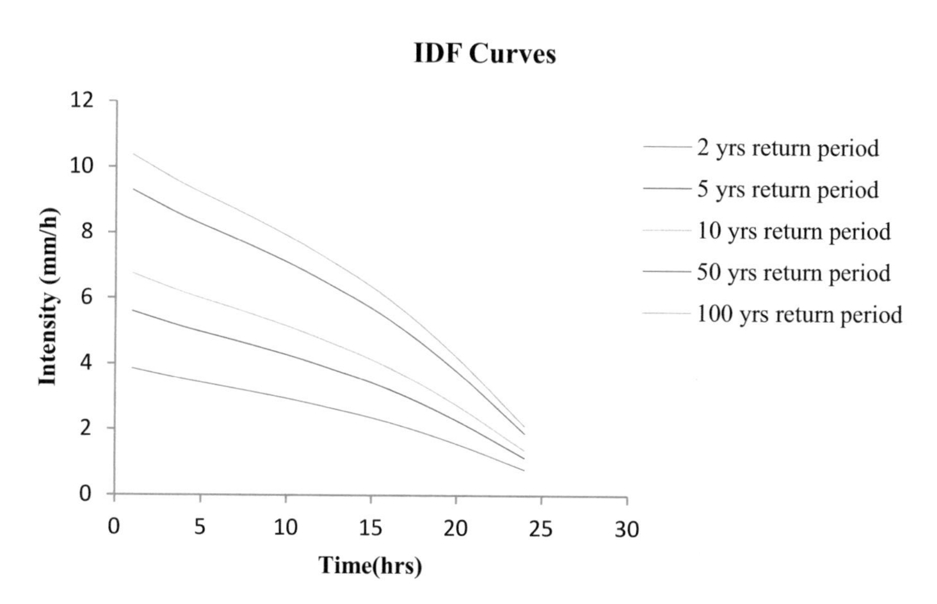

Fig. 3.2 Developed intensity duration frequency (IDF) curves for Koraiyar basin

Table 3.2 Satellite data used in the study

Sample no.	Data product	Imagery date	Resolution (m)	Path/Row
1	Landsat 4–5 TM C1 Level 1	26-03-1986	30	15/191
2	Landsat 4–5 TM C1 Level 1	15-12-1996	30	143/53
3	Landsat 7 ETM+ C1 Level 1	6-05-2006	30	143/53
4	Landsat 8 OLI/TIRS C1 Level 1	24-11-2016	30	143/ 53

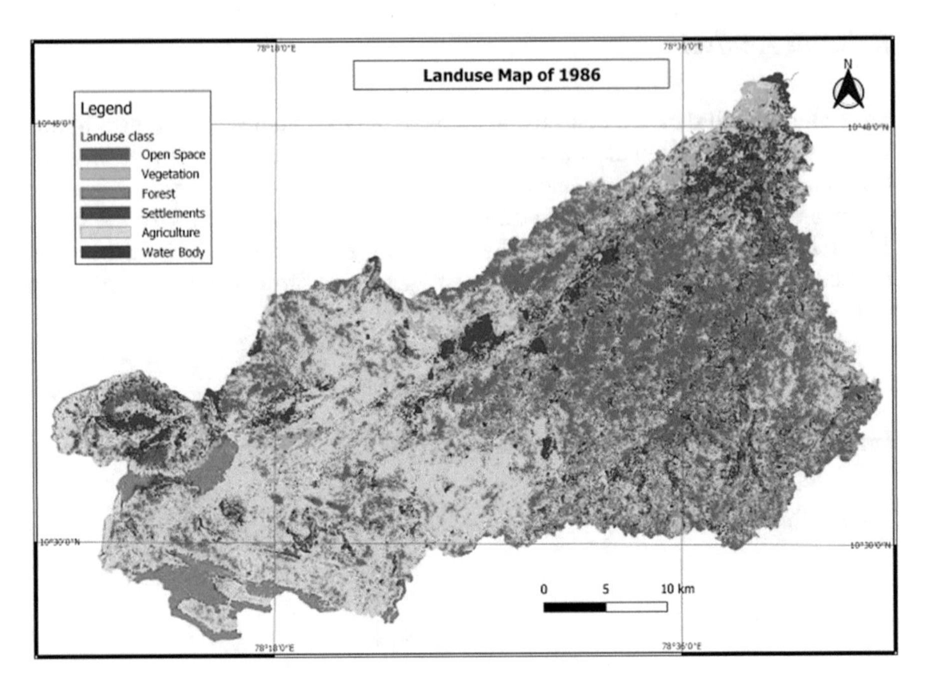

Fig. 3.3 Generated LULC map for year 1986

To analyze the LULC classification of the Koraiyar basin image, registration is the first step, which is applied in satellite images for the image classification analysis and change detection processes. The ArcGIS software version 10.2.2 is used in this study, and the base map of the Koraiyar basin is prepared from the DEM by fixing the outlet point. The map to image registration for the base map is prepared by using SOI (Survey of India Toposheet) 58 J/9, J/10, J/13, J/14 of the year 1970, and image registration is done by suitable ground control points (GCPs) with permanent features such as bridges in the Koraiyar basin. The remote sensing image was then geo-rectified by using the selected GCPs extracted from the geo-referenced toposheets. In the image to image registration, the base reference of the study area is required. The Landsat 4–5 TM image of the year 1986 is used as the base image for the co-registration of the images of years 1996, 2006, and 2016, so that all the images would have a uniform projection.The maximum likelihood technique is used for generating LULC maps (Fig. 3.3).

LULC Pattern Analysis from 1986 to 2016

The LULC classification of various land classes from 1986 to 2016 is shown in Table 3.3. From the analysis, an increase in open land of 4.63% in this time period is noted. This increase in open land results from the conversion of agriculture, forest,

Table 3.3 Percentage of LULC change in the Koraiyar basin between 1986 and 2016

Sample no.	LULC classes	1986–1996	1996–2006	2006–2016	1986–2016 (%)
1	Open land	−27.96	−8.85	59.34	4.63
2	Vegetation	−26.59	1440.53	−85.49	64.09
3	Forest	71.72	12.95	−70.56	−42.89
4	Settlement	45.11	15.88	21.76	104.74
5	Agriculture	15.84	−35.09	16.65	−12.29
6	Water body	−60.49	−57.60	81.08	−11.45

and water bodies to open land. The vegetation also shows a markable increase of 64.09% in these periods. During this period the settlement area increased by 1.74% of the whole basin, and the agriculture land and forest areas decreased by 12.29% and 42.89%, respectively. In agricultural land, increase in settlement areas such as residential, commercial, and industrial built-up areas is noted during these periods. The water bodies present in the basin decreased by about 11.45% during these times.

The overall analysis of the different LULC classes from 1986 to 2016 shows there is a marginal increase in the settlement area resulting from rapid urbanization in the basin. There is also a decrease in open land from 1986 to 1996 and from 1996 to 2006, but the notable increase in open land is observed during 2006–2016. From the overall analysis, it is concluded that there is an increase in open land.

When combining the change percentage of agriculture, open land, vegetation, and water bodies with change percentage of settlement area, growth of the settlement is seen to occur from the utilization of agriculture, vegetation, open land, and considerable forest area. The accuracy of the LULC classification was assessed by two methods, namely, producer accuracy and user accuracy.

Forecasting Future Land Use/Land Cover Analysis from 2026 to 2036

The changes in LULC classes in the basin were predicted for year 2036 based on the transition that occurred in the basin during 1996–2016. This duration is considered as a base period for LULC forecasting because rapid changes occurred in the basin during these years. For predicting the future LULC map of the basin, the Markov model was used to simulate the changes that occur in the future by multiplying with the primary matrix shown in Eq. (3.3):

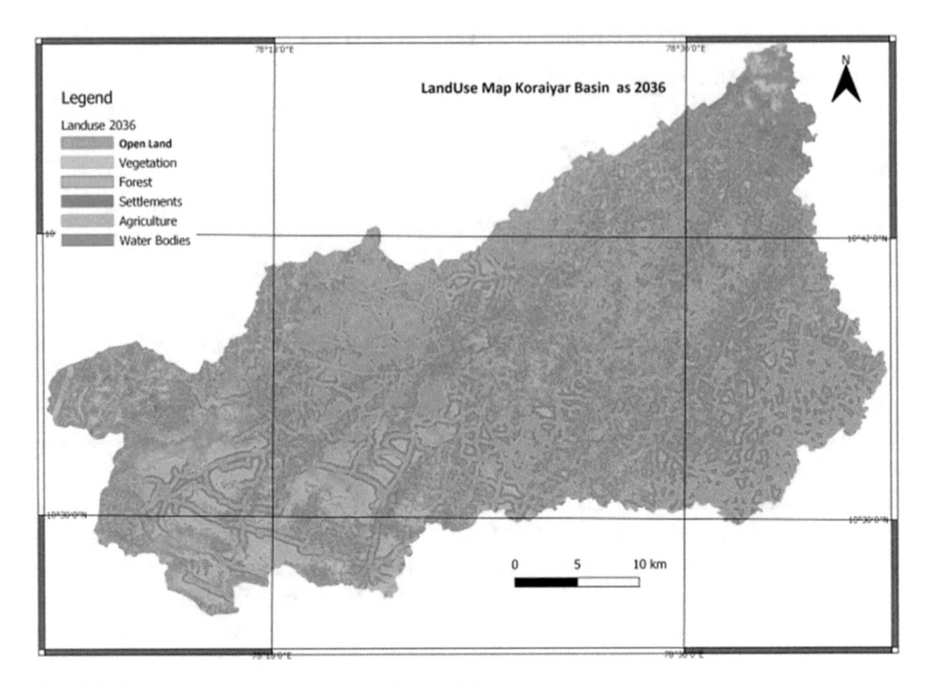

Fig. 3.4 Simulated future LULC map of year 2036

$$
P = P_{ij} = \begin{vmatrix} P_{11} & P_{12} & \cdots & P_{1m} \\ P_{21} & P_{22} & \cdots & P_{2m} \\ \cdots & \cdots & \cdots & \cdots \\ P_{m1} & P_{m2} & \cdots & P_{mm} \end{vmatrix} \tag{3.3}
$$

where P stands for probability conversion from state i to state j (Jianping et al. 2005).

The transition probability matrix of the years 1986–1996 and the primary matrix $P_{(0)}$ were used to derive the forecasted LULC for the year 2006. Similarly, the forecasted LULC for the year 2036 are derived from the primary matrix. The simulated LULC map of the Koraiyar basin for the future year 2036 is shown in Fig. 3.4. The results indicate the percentage of settlement area in the Koraiyar basin has increased from 45.11% to 104.74% during 1986–2016. For the future year of 2036, an increase of 3.22% is projected. Rapid settlements are observed in the northern part of the basin area because of major developmental activities such as residential, commercial, and industrial areas in the basin. The accuracy of the predicted LULC maps generated by the Markov technique was validated using observed and predicted LULC maps of the Koraiyar basin.

SCS-CN Method

Development of Hydrologic Soil Map

The soil map for the Koraiyar basin is extracted from the Harmonized World Soil Database. The soil map is rectified and mosaicked so that the study area was extracted from the harmonized world soil database by subsetting it from the full map. The soil in the basin is classified according to HSG A, B, C, and D as per the US Natural Resource Conservation Service (NRCS) (USDA 1974). The A-type soil covers 16.05% with alluvium sandy loam and loamy soil. The spread of B-type soil is noted in the basin, which has a moderate runoff rate in the basin. The C-type soil with subtle texture, grey soil, and black soil of 30.93% is distributed throughout the basin. The D-type of soil such as clay and red ferruginous soils are also observed in the basin. The characteristics of soil A have low runoff potential and a high water transmission rate. The soil group C indicates moderate infiltration, and D-type soil has high runoff potential and low water rate transmission.

SCS-CN Approach for Runoff Calculations

In the early 1950s, the United States Department of Agriculture, Natural Resources Conservation Service, named Soil Conservation Service (SCS) developed a method for assessing the volume of direct runoff from rainfall (Pancholi et al. 2015). This method is often referred to as the CN method empirically developed for the basins. The standard SCS-CN method is based on the following relationship between rainfall, P (mm), and runoff, Q (mm). The CN factor is a conversion of potential maximum retention (Chow et al. 1988), where Q is denoted as collected runoff (in mm) (Eq. (3.4)). The I_a is the initial abstraction (from infiltration, evaporation, and interception) (mm) as shown in Eq. (3.5). The I_a is substituted in Eq. (3.6) to give Eq. (3.7).

$$Q = \frac{(P - I_a)^2}{P - I_a + S} \tag{3.4}$$

$$I_a = 0.2 * S \tag{3.5}$$

$$S = \frac{254000}{CN} - 254 \tag{3.6}$$

where S ispotential maximum retention in (mm); CN is known as a curve number value

$$CN = \frac{(A_1CN_1 + A_2CN_2 + A_3CN_3 + \ldots\ldots A_nCN_n)}{(A_1 + A_2 + A_3 + \ldots\ldots. A_n)} \quad (3.7)$$

The whole curve number (CN) (A_1, A_2, A_3... A_n) represents the overall area of all LULC existing in the basin. A_1 to A_n is the area of different LULC classes, and CN_1 to CN_n represent the corresponding CN of LULC classes, where CN is the weighted curve number; CN_1 is the curve number from 1 to any number N; A_1 is the area with curve number CN_1; and A represents the entire basin. To compute the surface runoff depth in the basin, the hydrologic equations (Eqs. 3.4, 3.5, 3.6, 3.7) are used to determine the rate of rainfall.

Assessment of Rainfall–Runoff Using the SCS CN Method

Surface runoff is computed by selecting an extreme single rainfall event that occurred in the basin. The entire Koraiyar basin is divided into six land use/land cover classes of open land, vegetation, forest, settlement, agriculture, and water bodies. The CN is generated for these classes by the reclassify method shown in Fig. 3.5. The future CN for the basin is also simulated (Fig. 3.6). The surface runoff is computed from LULC classes by the SCS-CN method.

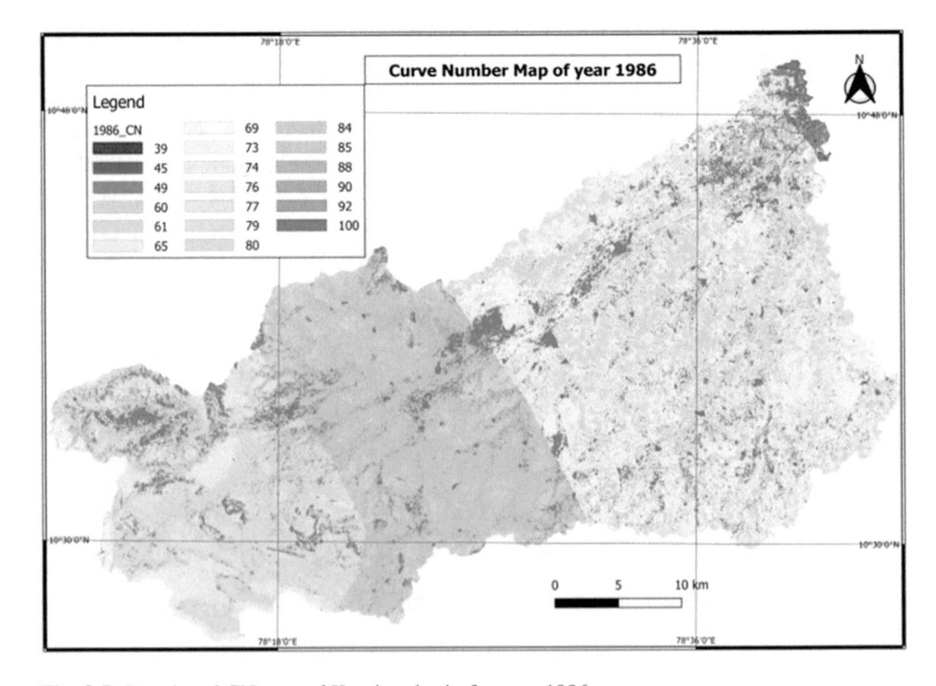

Fig. 3.5 Developed CN map of Koraiyar basin for year 1986

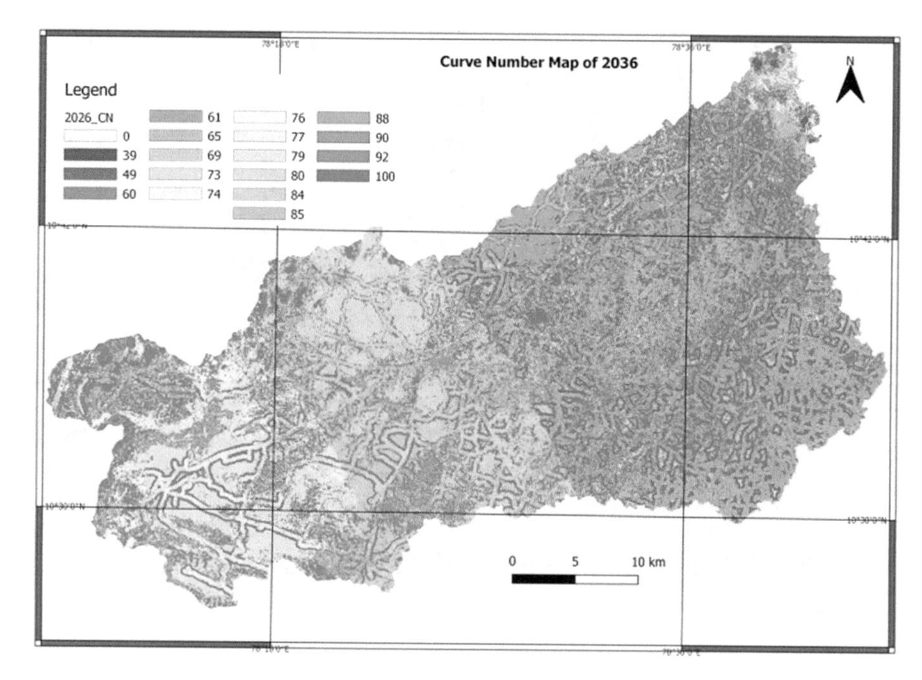

Fig. 3.6 Predicted CN map of Koraiyar basin for year 1986

Results and Discussion

Estimation of Runoff Curve Numbers from Indian Conditions

The Hydrologic Soil Group (HSG) divides the soil into types A, B, C, and D. The weighted existing and predicted curve numbers for the Koraiyar basin are shown in Table 3.4.

The CN model developed by the SCS CN approach shows that the overall generated curve number increases or decreases for the various land cover classes in the basin from 1986 to 2016, and the predicted CN is also shown in Table 3.5. In the year 1986, the curve number generated is 72, but a decrease in CN is noted in the year 1996, a decrease of 0.01% of the total number produced. The CN for the years 2006 and 2016 shows an increase from 72.83 to 73.20, with an increase of 0.005%. In the past 30 years, the increase in CN is 0.001% in the whole basin. The reason for the increasing and decreasing CN in the basin is the LULC transition process. It occurs in the form of an increase or decrease of open land and conversion of agriculture land into open land or change of vegetation into open land. The increase in CN is the result of reduction in agricultural land. The future simulation of curve number for the years 2026 and 2036 shows an increase. From the analysis, it is noted that that curve number is gradually increasing for both present and future LULC conditions. The increase in curve number reflects the increased runoff in the basin.

Table 3.4 Weighted curve number for Koraiyar basin

Sample no.	Land use	Runoff curve numbers for HSG			
		A	B	C	D
1	Open land	39	61	74	80
2	Vegetation	65	76	84	88
3	Forest	45	60	73	79
4	Settlement	77	85	90	92
5	Agriculture	49	69	79	84
6	Water body	100	100	100	100

Table 3.5 Generated curve number for Koraiyar basin

Sample no.	Calculated CN	
1	1986	2036
	72	74.70

Analysis of Runoff for Change in Different Years of Land Use Conditions

The extreme rainfall event that occurred on 22 November 1999 is used for generating peak discharge through the SCS method. The generated peak discharges for different return periods are shown in Table 3.6. From the analysis, the 2 years return period has less runoff compared to the 100 years return period. The peak discharge increases because of an increase in CN from 1986 to 2016. In the future simulation for the years 2026 and 2036, the runoff increases because of an increase in CN. The LULC change and its influence on surface runoff are analyzed from Eq. (3.8).

$$F = \frac{\sum R_2}{\sum R_1} \tag{3.8}$$

The calculated rainfall and the LULC data of 1986 runoff of the year are estimated for the Koraiyar basin. The runoff constitutes the function of rainfall of the year 1999 and the land use data of 1986–2016. From the analysis, it is noted that the total runoff of the year 1986 is $\sum R_1$, 43.86 m^3/s for the 2 years return period, and the total runoff of the year 1996 is $\sum R_2$, is 42.69 m^3/s. The impact factor for land use and land cover and its effects on the surface runoff on the Koraiyar basin for the 2 years return period is 0.973. It is a deciding factor for measuring the effects of LULC in the surface runoff in the basin.

Table 3.6 Peak discharge for different flow and land use conditions

Peak discharge for different flow and land use conditions

Sample no.		**1**	**2**	**3**	**4**	**5**
Maximum RF in mm		**310.5**				
Return period storm event		**2**	**5**	**10**	**50**	**100**
Peak discharge m^3/S at outlet	LULC 1986	43.86	127.22	188.10	327.66	387.99
	LULC 1996	42.69	125.30	185.87	325.01	385.23
	LULC 2006	45.41	129.72	190.98	331.06	391.54
	LULC 2016	46.10	130.83	192.26	332.56	393.11
	LULC 2026	49.09	135.54	197.64	338.84	399.63
	LULC 2036	48.97	135.35	197.43	338.60	399.39
% change in peak discharge	1986–1996	−2.68	−1.51	−1.18	−0.81	−0.71
	1996–2006	5.99	3.41	2.68	1.83	1.61
	2006–2016	1.51	0.85	0.67	0.45	0.40
	2016–2026	6.09	3.47	2.72	1.85	1.63
	2026–2036	−0.24	−0.13	−0.10	−0.07	−0.06
	1986–2036	10.44	6.01	4.73	3.23	2.85

Development of Hydrologic Modeling for Koraiyar Basin

The hydrologic model developed by integrating with GIS, known as HEC-Geo HMS, is used to develop the model for the basin. The runoff from the rainfall is generated through HEC-HMS.

Digital Elevation Model of Study Area

The toposheet does not contain the elevation contours of the basin, so the elevation of the study area could not be done by using the topo sheet alone. Therefore, the Advanced Land Observing Satellite (ALOS) from Japan Aerospace Exploration Agency (JAXA) of 30 m resolution is adopted. The basin is delineated into sub-basins, and a total of 27 subbasins are delineated in the Koraiyar basin (Fig. 3.7). The total area of the basin is 1498 km^2.

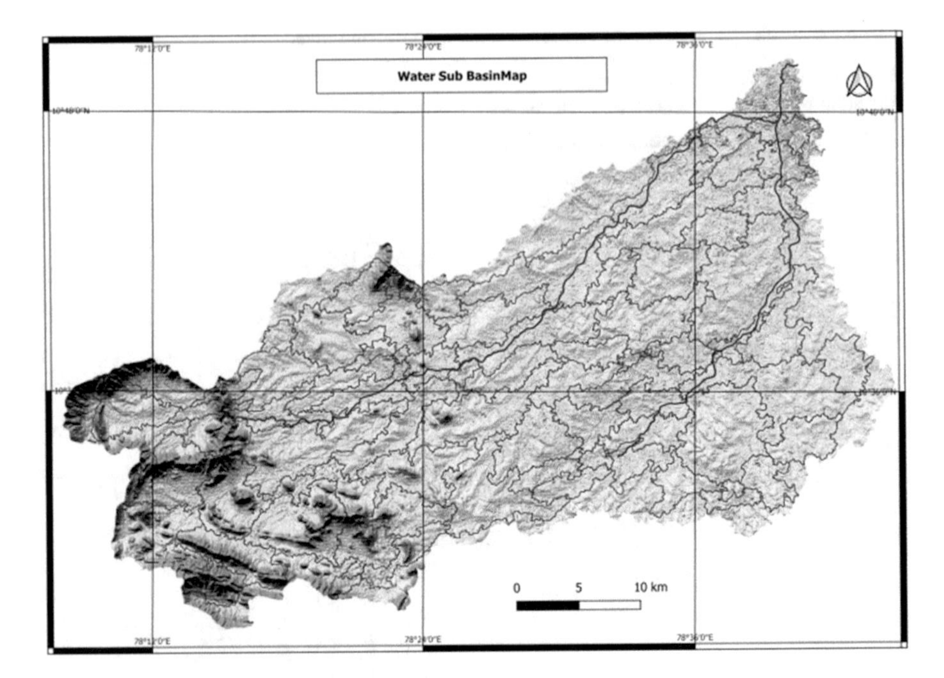

Fig. 3.7 Subbasin map of Koraiyar River Basin

Basin Model

In HEC-HMS in the basin model, the required inputs are LULC information, hydrologic soil groups, curve number (CN), and rainfall event. The land use and hydrologic soil group are processed in HEC-GeoHMS to generate CN for the basin. The other physical characteristics of the basin, stream length, slope, location of subbasin centroid, elevation, the longest flow path for each subbasin, and the length along the stream path, are extracted from DEM.

Slope Classes of Koraiyar Basin

The slope is an essential measure for runoff in the basin. The slope map generated is shown in Fig. 3.8. Slope less than 3° is considered as nearly level, and slope greater than 15° was given importance because of runoff. In the Koraiyar basin, slope varies from 0.5° to 53°. The SCS curve number is adopted in this study because of its simplicity and requires only one parameter for each subbasin is the lag time. The lag time is 0.6 times of concentration for the ungauged basin as per USGS (2012) and

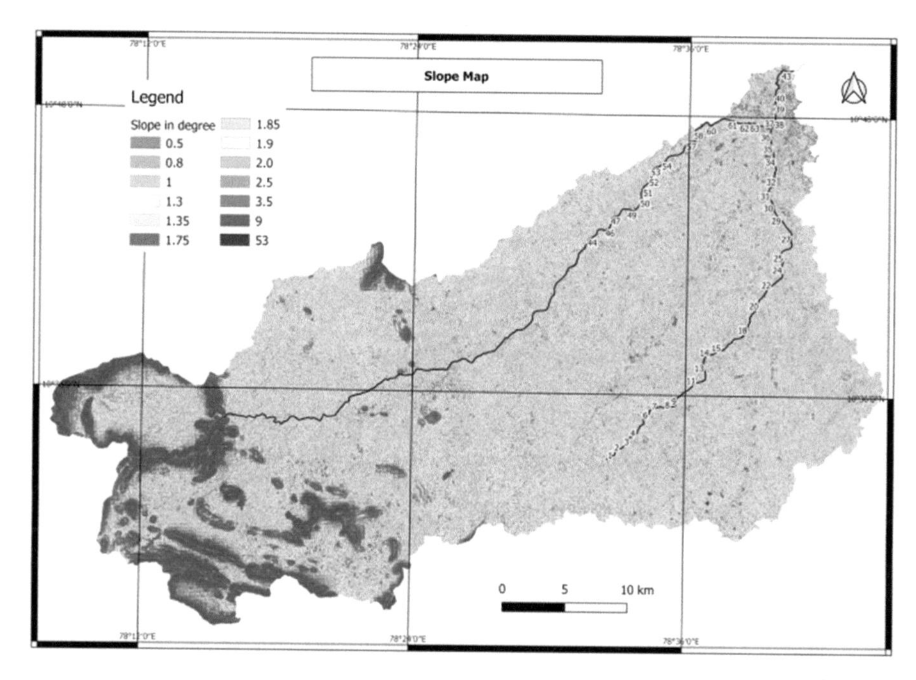

Fig. 3.8 Slope map

Table 3.7 Time of concentration and lag time for basin

Subbasin	Tc	Tlag = 0.6*tc	Subbasin	Tc	Tlag = 0.6*tc
W280	1.19	0.71	W410	1.59	0.95
W300	0.92	0.55	W460	1.22	0.73
W310	0.12	0.07	W430	2.06	1.23
W300	4.28	2.57	W430	0.37	0.22
W370	0.78	0.47	W510	1.08	0.65
W370	0.65	0.39	W430	0.54	0.33
W320	3.51	2.11	W470	7.50	4.50
W330	2.82	1.69	W360	5.88	3.53
W340	2.88	1.73	W450	2.09	1.25
W400	2.99	1.79	W440	1.87	1.12
W370	3.74	2.25	W480	4.52	2.71
W390	2.41	1.45	W530	5.35	3.21
W420	2.30	1.38	W540	13.21	7.92

shown in Eq. (3.9). The time of concentration is calculated by Giandotti's formula, as shown in Eq. (3.10) Giandotti (1934). The calculated time of concentration and lag time for the basin are shown in Table 3.7.

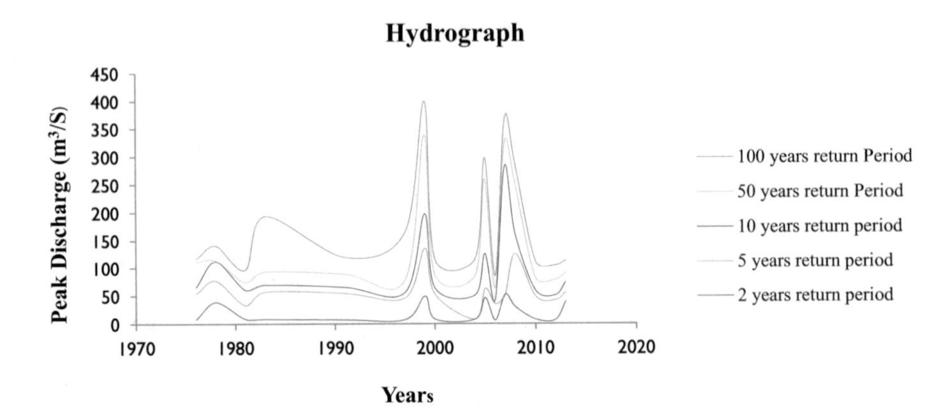

Fig. 3.9 Simulated hydrograph for extreme event in Koraiyar Basin

Table 3.8 Simulated and observed peak flow for single extreme event occurring in Koraiyar basin

Single Extreme Event Date & Rainfall in mm	Return period (years)	Peak flow (m^3/s)		Volume (mm^3)	
22nd November 1999 & 310.55		Simulated	Observed	Simulated	Observed
	100	576.3	484.9	11.84	10.97
	50	505.4	422.5	10.58	9.54
	10	326.6	275	6.72	6.17
	5	119.2	207.9	3.16	4.64
	2	101.2	106.7	2.7	2.33

$$T_{lag} = 0.6 * T_c \tag{3.9}$$

$$T_c = \frac{L^{0.8}(S+1)^{0.7}}{1900\sqrt{Y}} \tag{3.10}$$

The frequency storm data are used in the meteorological model. The frequency storm method is used for producing rainfall depth for different return periods. The selected extreme event rainfall is given as input to the HEC-HMS model, and control specifications are used to define the time interval for model simulation. In this study, an extreme rainfall event that occurred on 22 November 1999 is used for simulating rainfall runoff in the present land use conditions to find the effect of LULC changes and the impact on surface runoff. The Muskingum routing method is adopted to route the water movement in the reach. The expected output from the model is the peak discharge from the basin outlet. The hydrographs generated during simulation are shown in Fig. 3.9. The simulated peak discharge for the basin is shown in Table 3.8. which is inputted to the hydraulic model HEC-RAS for developing flood depth and spread of the basin.

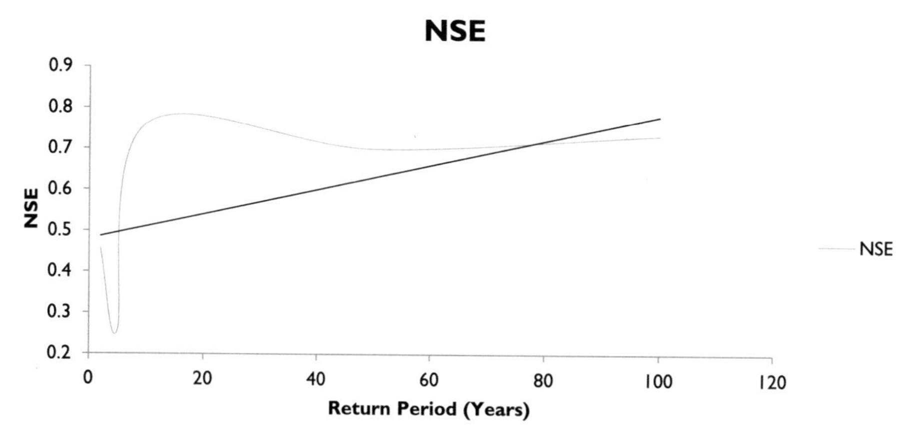

Fig. 3.10 Simulated NSE values for extreme events

Calibration and Validation of HEC-HMS for Simulation of Flood Events

The calibrated parameter values of all the subbasins for runoff simulation show that the average value of Muskingum travel time is between 0.5 and 1. The NSE values are mostly 0.5–0.6, showing the model performance is satisfactory (Fig. 3.10). The changes in curve number and initial abstraction have a prominent role in this basin and are optimized by several trial and error methods by 100 iterations. In the future, the parameters such as CN are manually adjusted, and the other most influential parameters are also identified.

Hydraulic Model for Koraiyar Basin

For developing hydraulic modeling, the outputs from the hydrologic modeling are given as input to hydraulic modeling for generating the flood spread. The other parameters required for generating hydraulic models are cross sections for the river basin is roughness coefficient (Manning's constant '*n*' contraction, and expansion constants). An aggregate of 60 cross sections is taken for the Koraiyar River basin (Fig. 3.11). The cross sections are used as one of the parameters for generating floodplain maps. The roughness coefficient for each cross section of the basin is assigned based on different LULC classes.

The roughness coefficient is estimated combining land use data with Manning's constant. The HEC-RAS version 5.0.7 is used for calculating steady flow water surface profiles. The other tools, such as inundation mapping tools, are also present in this hydraulic model. The flood peak for different return periods of 2, 5, 10, 50,

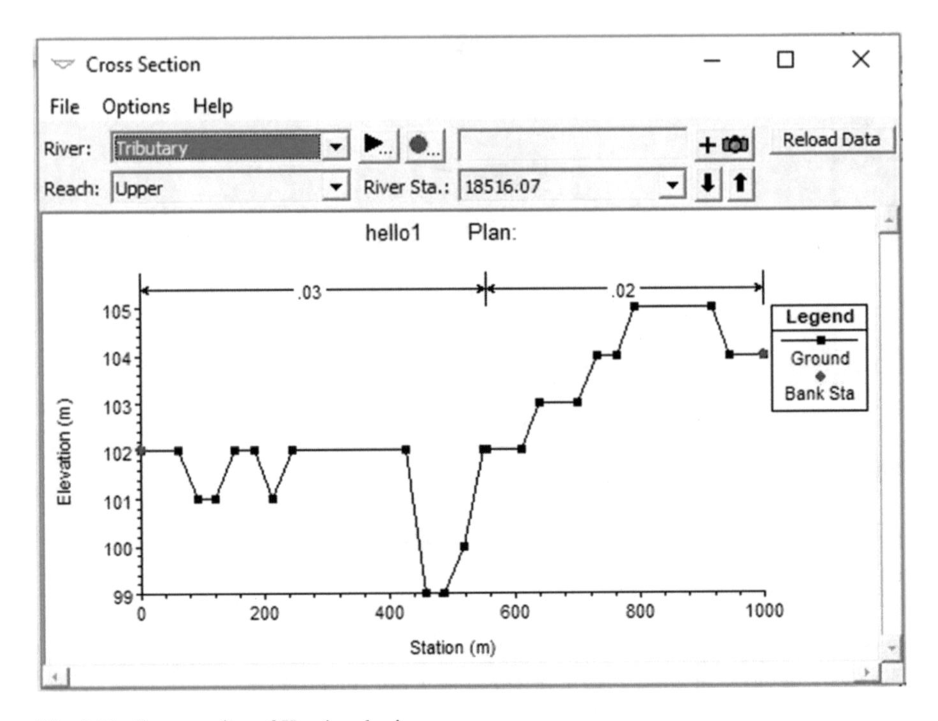

Fig. 3.11 Cross section of Koraiyar basin

and 100 years is considered. The discharges of the different return periods are calculated.

Processing of Hydraulic Modeling

Hydraulic modeling is used to obtain floodplain and flood hazard maps for the Koraiyar basin in Tiruchirappalli city. The flood portion is defined through 60 cross sections delineated from the basin. The distance between each cross section of the basin is 100 m. The maximum and minimum widths of the river basin are 400 and 110 m, respectively. The roughness coefficients used in the study area are 0.060 for vegetation, 0.150 for forest areas, 0.020 for settlement, 0.035 for water bodies, 0.05 for agricultural land, and 0.040 for open land. A mixed condition of flow is chosen for the HEC-RAS. The upstream boundary condition is set as a critical depth, whereas the downstream end is considered as normal depth. Once all the input data and boundary conditions in HEC-RAS are set, the simulation is carried out as a steady-state analysis. The flood hazard maps are generated by keeping the extreme event of rainfall as a constant factor for different LULC for the years 1986 and 2036.

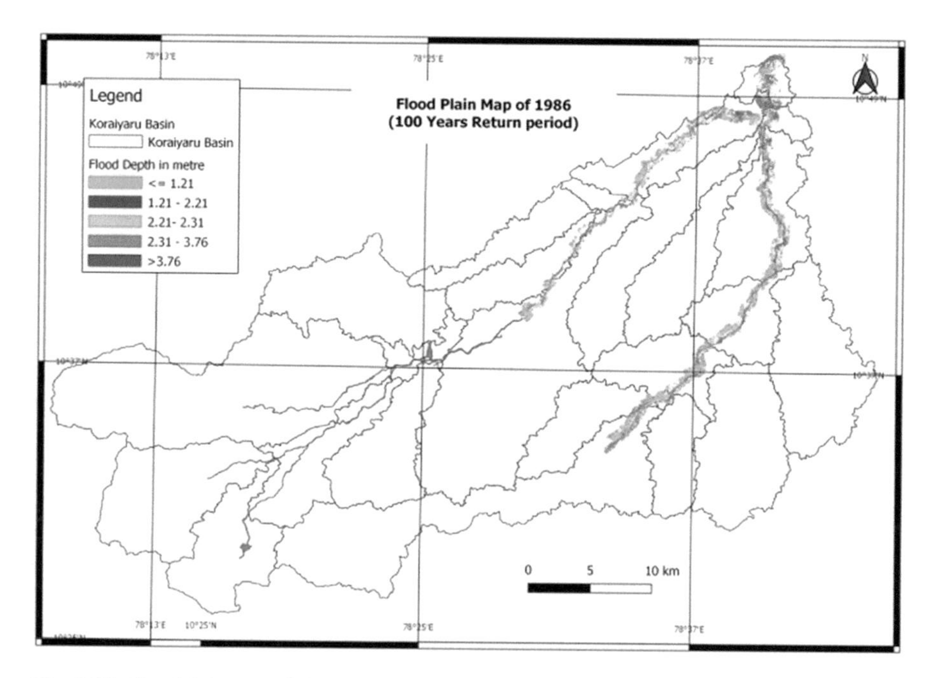

Fig. 3.12 Floodplain map of the Koraiyar basin for LULC of the year 1986 with 100 years return period

Generation of Floodplain Maps

Floodplain maps of the study area are developed for the years 1986 and 2036 for various LULC classes. The floodplain maps for the basin are obtained through modeling the land use conditions for 2-, 5-, 10-, 50-, and 100-year return period rainfall depths (Figs. 3.12 and 3.13). The areal extent of the floodwater spread for the basin area is shown in Table 3.9.

Delineation of Flood Hazard Maps

For efficient flood forecasting and warning, flood hazard maps are required in the initial stages of urban developmental activities in the basin. The flood hazard maps in this study are generated for an extreme rainfall event in the Koraiyar basin for various LULC changes between 1986 and 2036 (Figs. 3.14 and 3.15).

The maps obtained show that the flood hazard of the basin increases between the years 1986 and 2036. In this study, the flood hazard maps of maximum flood extent for different return periods are prepared from the maximum peak discharge, slope, and settlement density of the basin by using Arc GIS software 10.2.2 to generate flood hazard maps.

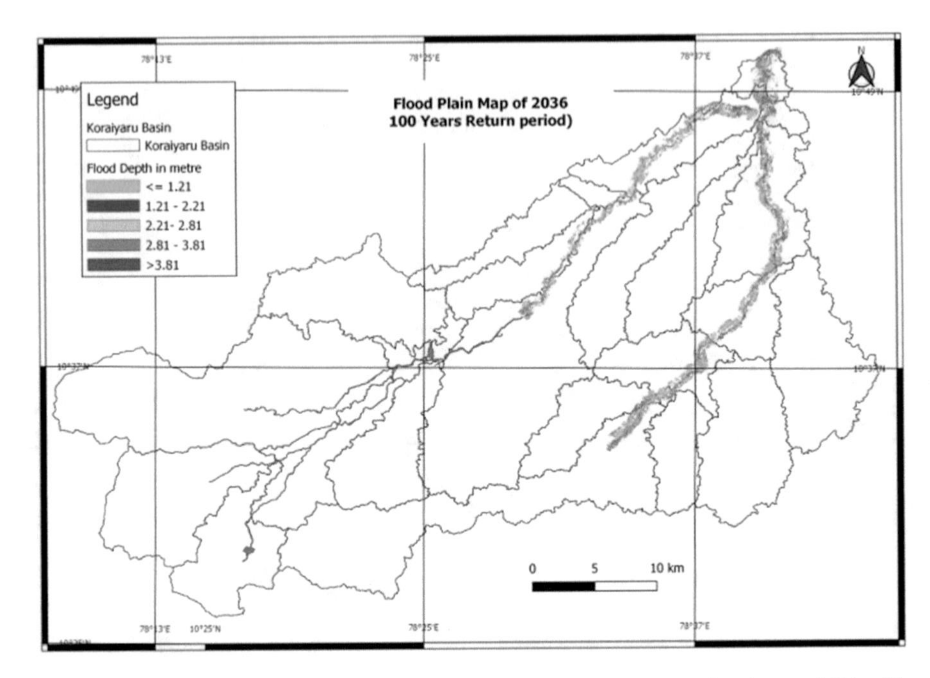

Fig. 3.13 Floodplain map of the Koraiyar basin for LULC of the predicted year 2036 with 100 years return period

Table 3.9 Floodplain area for different LULC and return periods

Return period in years	Floodplain area (in km^2)	
	LULC 1986	LULC 2036
2	20.09	20.87
5	21.35	21.93
10	32.10	33.64
50	39.41	40.92
100	41.16	42.88

The flood hazard maps are prepared in the raster format to generate the flood depth for various return periods. The generated raster formats are converted into vector format, and the extreme flood spread area polygons are clipped for the different return periods for different LULC conditions. The clipped raster maps are made uniform by using the slice tool in Arc GIS. From the allocated weights, the flood hazard map is categorized into seven classes, namely, no flood hazard, very very low, very low, low, moderate, high, and very high, by the standard deviation method of classification. The class-wise susceptibility areas of flood hazard for different LULC of the years 1986–2036 are shown in Table 3.10.

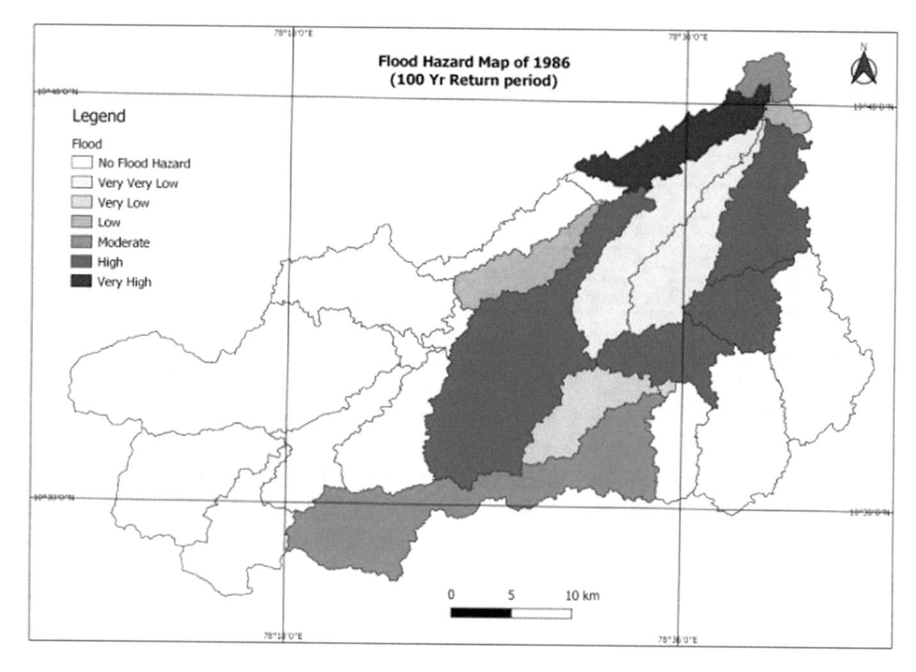

Fig. 3.14 Flood hazard maps of Koraiyar basin for LULC of the year 1986 for 100 years return period

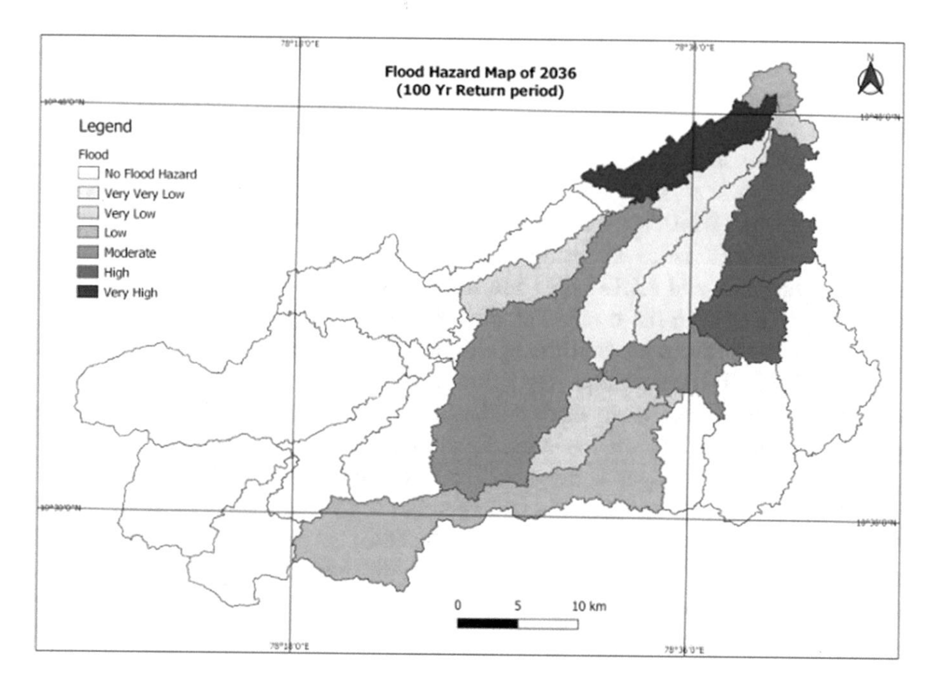

Fig. 3.15 Flood hazard maps of Koraiyar basin for LULC of predicted year 2036 for 100 years return period

Table 3.10 Flood hazard areas for different LULC and return periods

Sample no.	Classification of flood	Flood hazard area (in km^2)	
		LULC 1986	LULC 2036
1	Very very low	13.63	13.75
2	Very low	22.21	22.68
3	Low	29.58	31.42
4	Moderate	75.73	78.49
5	High	485.72	500.49
6	Very high	192.26	198.85

Conclusions

In this study, the GIS technique is used to delineate the LULC of various land classes and to examine their dynamics. From the study, an increase and decrease of various LULC classes of the Koraiyar basin from 1986 to 2016 were observed. During this period, it is noted that the settlement area increased by 104.74% and water bodies decreased by 11.45%. A decrease in agricultural land of about 12.29% is observed for the past 30 years, whereas open denuded land has increased. This increase in open land is the result of noncultivation of agricultural lands during the non-monsoon season and the practice of cultivation for more extended periods at other times, which is verified by field observations. The Markov model is used for predicting the future trend of LULC dynamics. The model not only accurately describes the past and present changes but also predicts probable future land use changes. The simulation trend shows that the overall increase in a built-up area is predicted to be 7.17% from 1986 to 2036. The CN-generated results show a continued increase in the CN values from the past to the present, and this increase proves the associated increase in peak runoff rates and total runoff volume. The 30-m resolution DEM data were used for delineation of the Koraiyar basin and its catchment characteristics by using HEC-GeoHMS, an Arc Hydro extension in ArcGIS. The soil and LULC data are used to generate the CN and analyze its influence on runoff in the basin. The HEC-HMS hydrologic modeling software is adopted in the basin to predict flood peak runoff and volume. In the metrological model of HEC-HMS, the frequency storm method is adopted for selected extreme events. From the LULC analysis of the basin, it is noted that a 1.4% increase in the settlement area of the whole basin led to an increase of 3% in the runoff rate. The analysis of LULC changes shows that the basin has only a marginal increase in settlement, and open land results in a marginal increase in the runoff and volume of the basin. Flood plain and hazard maps for the basin are generated for various land use conditions from 1986 through 2036 for the different return periods. The minimum and maximum flood depth range obtained during this period of study is 1.21–4.71 m. The flood hazard map obtained shows that a maximum of 818–852 km^{2}of the basin area is susceptible to floods. It is also observed that the very low hazard zone occupies more area in the basin in the present and the future prediction of flood hazard. Finally, it is concluded that this model can be adopted in the areas of limited data and an ungauged basin. The model also proves to be an accessible tool for real-time solutions and immediate actions to be implemented. The flood risk map developed in the basin is used for providing public awareness and for early warning of flood forecasting systems. The developed maps are useful to Tiruchirappalli City Corporation for early flood warnings and also as a valuable tool for proper disaster and mitigation measures.

References

Ahamad I, Deshang T, TianFang W, Bakhtawar W (2015) Precipitation trends over time using Mann–Kendall and Spearman's rho tests in Swat River Basin, Pakistan. Adv Meteorol 10:15

Ali H, Shui LT, Mirzaei M, Memarian H (2012) Incorporation of GIS based program into hydraulic model for water level modeling on river basin. J Water Resour Prot 4:25–31. https://doi.org/10.4236/jwarp.2012.41004

Aliani H, Malmir M, Sourodi M, Kafaky SB (2019) Change detection and prediction of urban land use changes by CA–Markov model (case study: Talesh County). Environ Earth Sci 78:546. https://doi.org/10.1007/s12665-019-8557-9

An Thi Ngoc Dang, Kumar L (2017) Application of remote sensing and GIS-based hydrological modelling for flood risk analysis: a case study of District 8, Ho Chi Minh city, Vietnam. Geomat Nat Haz Risk 8(2):1792–1811

Annegret HT et al (2016) Review of the flood risk management system in Germany after the major flood in 2013. Ecol Soc 21(2):51. https://doi.org/10.5751/ES-08547-210251

Chow VT, Maidment DR, Mays LW (1988) Applied hydrology. McGraw-Hill International Editions, Singapore

Chowdhury RK, Beecham S (2010) Australian rainfall trends and their relation to the southern oscillation index. Hydrol Process 24(4):504–514

Clarke KC, Gaydos LJ (1998) Loose-coupling a cellular automaton model and GIS: long-term urban growth prediction for San Francisco and Washington/Baltimore. Int J Geogr Inf Sci 12:699–714

Dewan AM, Makoto TK (2006) Flood hazard delineation in greater Dhaka, Bangladesh using an integrated GIS and remote sensing approach. Geocarto Int 21:2

Dewan AM, Yamaguchi Y (2009) Land use and land cover change in Greater Dhaka, Bangladesh: using remote sensing to promote sustainable urbanization. Appl Geogr 29:390–401

Fox DM, Witz E, Blanc V, Soulié C, Penalver-Navarro M, Dervieux A (2012) A case study of land-cover change (1950–2003) and runoff in a Mediterranean catchment. Appl Geogr 32 (2):810–821

Gao P, Carbone GJ, Lu J (2018) Flood simulation in South Carolina watersheds using different precipitation inputs. Adv Meteorol Article ID 4085463:1–10. https://doi.org/10.1155/2018/4085463

Ghosh S, Luniya V, Gupta A (2009) Trend analysis of Indian summer monsoon rainfall at different spatial scales. Atmos Sci Lett 10:285–290

Giandotti M (1934) Previsione delle piene e delle magre dei corsi d'acqua [Forecast of floods and lean waters]. Istituto Poligrafico dello Stato 8:107–117

Haan CT (1977) Statistical methods in hydrology. Iowa State University Press, Ames

HEC (Hydrologic Engineering Center) (2010) HEC-RAS Hydraulic Reference Manual, Version 4.1. U.S. Army Corps of Engineers Hydrologic Engineering Centre, Davis

Hsu HM, Chen SH, Chang TJ (2000) Inundation simulation for urban drainage basin with storm sewer system. J Hydrol 234:21–37

Hu H, Walton R (2008) Advanced guidance on use of steady HEC-RAS. World Environmental and Water Resources, Congress 2008, pp 1–10

Hunter NM, Bates PD, Horritt MS, Wilson MD (2007) Simple spatially-distributed models for predicting flood inundation: a review. Geomorphology 4(90):208–225. https://doi.org/10.1016/j.geomorph.2006.10.021

Jianping LI, Bai Z, Feng G (2005) RS-and-GIS-supported forecast of grassland degradation in southwest Songnen plain by Markov model. Geo-spatial Inf Sci 8:104–106

Kothyari UC, Garde RJ (1992) Rainfall intensity-duration-frequency formula for India. J Hydraul Eng 118:323–336

Lambin EF (1997) Modelling and monitoring land-cover change processes in tropical regions. Prog Phys Geogr 21:375–393

Linsley RK, Franzini JB, Freyberg DL, Tchobanoglous G (1992) Water resources engineering. McGraw-Hill, New York

Logsdon MG, Bell JE, Westerlund VF (1996) Probability mapping of land use change: a GIS interface for visualizing transition probability. Comput Environ Urban Syst 20:389–398

Majumdar PP (2012) Stochastic hydrology. Nptel lectures

Mark O, Weesakul S, Apirumanekul C, Aroonnet SB, Djordjevic S (2004) Potential and limitations of 1D modelling of urban flooding. J Hydrol 299(3–4):284–299

Mason DC, Horritt MS, Hunter NM, Bates PD (2007) Use of fused airborne scanning laser altimetry and digital map data for urban flood modelling. J Hydrol Process 21:1436–1447

Mays LW, Tung YK (1992) Hydrosystems engineering and management. McGraw-Hill, New York

Mignot E, Paquier A, Haider S (2006) Modelling floods in a dense urban area using 2D shallow water equations. J Hydrol 327(1):186–199

Mousavi SM, Roostaei S, Rostamzadeh H (2019) Estimation of flood land use/land cover mapping by regional modelling of flood hazard at sub-basin level case study: Marand basin. Geomat Nat Haz Risk 10(1):1155–1175. https://doi.org/10.1080/19475705.2018.1549112

Ojha CSP, Berndtsson R, Bhunya P (2008) Engineering hydrology. Oxford University Press, New Delhi

Pancholi VH, Shah MK, Motiani AT, Indraprakash (2015) Estimation of runoff and soil Erosion for Vishwamitri River watershed, Western India using RS and GIS.American. J Water Sci Eng 1(2):7–14

Priest SJ, Suykens C, Van Rijswick HFMW, Schellenberger T, Goytia SB, Kundzewicz ZW, Van Doorn-Hoekveld WJ, Beyers JC, Homewood S (2016) The European Union approach to flood risk management and improving societal resilience: lessons from the implementation of the Floods Directive in six European countries. Ecol Soc 21(4):50

Radwan F, Alazbal AA, Mossad A (2018) Estimating potential direct runoff for ungauged urban watersheds based on RST and GIS. Arab J Geosci 11:748. https://doi.org/10.1007/s12517-018-4067-4

Rama Prasad NN, Narayanan P (2016) Vulnerability assessment of flood-affected locations of Bangalore by using multi-criteria evaluation. Ann GIS 22(2):151–162. https://doi.org/10.1080/19475683.2016.1144649

Rangari VA, Sridhar V, Umamahesh NV, Patel AK (2018) Floodplain mapping and management of urban catchment using HEC-RAS: a case study of Hyderabad city. In: J Inst Eng India Ser A. https://doi.org/10.1007/s40030-018-0345-0

Reddy AM (2002) Text book of remote sensing and geographical information systems. B.S. Publications, Ottawa

Sahoo SN, Sreeja P (2017) Development of flood inundation maps (FIM) and quantification of flood risk in an urban catchment of Brahmaputra River. J Risk Uncertain Eng Syst A 3(1): A4015001

Sarma PK, Lahkar BP, Ghosh S, Rabha A, Das JP, Nath NK, Dey S, Brahma N (2008) Land use and land-cover change and future implication analysis in Manas National Park, India using multi-temporal satellite data. Curr Sci 95(2):25

Schmitt TG, Thomas M, Ettrich N (2004) Analysis and modeling of flooding in urban drainage systems. J Hydrol 299:300–311

Shadeed S, Almasri M (2010) Application of GIS-based SCS-CN method in West Bank catchments, Palestine. Water Sci Eng 3(1):1–13. https://doi.org/10.3882/j.issn.1674-2370.2010.01.001

Siegel S, Castellan N (1998) Non-parametric statistics for the behavioral sciences. McGraw-Hill, New York

Subyani AM, Al-Amri NS (2015) IDF curves and daily rainfall generation for Al-Madinah city, western Saudi Arabia. Arab J Geosci. https://doi.org/10.1007/s12517-015-1999-9

Suriya S, Mudgal BV (2012) Impact of urbanization on flooding. Thirusoolam subwatershed: a case study. J Hydrol 412:210–219

USACE (2000) Hydrologic modeling system HEC-HMS: applications guide. Hydrologic Engineering Center, Davis

USDA-SCS (1974) Soil survey of Travis County, Texas. Texas Agricultural Experiment Station/USDA Soil Conservation Service, College Station/Washington, DC

USGS (2012) Estimating basin lag time and hydrograph timing indexes used to characterize storm flows for runoff- quality analysis. Scientific Investigations Report. U.S. Geological Survey, Reston

Wheater HS (2006) Flood hazard and management: a UK perspective. Philos Trans R Soc A 364:2135–2145. https://doi.org/10.1098/rsta.2006.1817

Yang X, Lo C (2002) Using a time series of satellite imagery to detect land use and land cover changes in the Atlanta, Georgia metropolitan area. Int J Remote Sens 23(9):1775–1798

Yu D, Lane SN (2006a) Urban fluvial flood modelling using a two-dimensional diffusion-wave treatment. Part 1: mesh resolution effects. Hydrol Process 20(7):1541–1565

Yu D, Lane SN (2006b) Urban fluvial flood modelling using a two-dimensional diffusion-wave treatment. Part 2: development of a sub-grid-scale treatment. Hydrol Process 20(7):1567–1583

Zhang RC, Tang S, Ma H, Yuan L, Gao FW (2011) Using Markov chains to analyze changes in wetland trends in arid Yinchuan plain, China. Math Comput Model 54:924–930

Zope PE, Eldho TI, Jothiprakash V (2015) Impacts of urbanization on flooding of coastal urban catchment: a case study of Mumbai City, India. J Nat Haz 75:887–908

Zope PE, Eldho TI, Jothiprakash V (2016) Impacts of land use–land cover change and urbanization on flooding: a case study of Oshiwara River Basin in Mumbai, India. Catena 145:142–154

Chapter 4
Impact of Climate Change and Adaptation Strategies for Fruit Crops

Tanmoy Sarkar, Anirban Roy, Sanvar Mal Choudhary, and S. K. Sarkar

Abstract The long life cycles of fruit crops causes these crops to be more challenged under stress as compared to crops of shorter duration that have more adaptation potential toward ending the crop cycle as quickly as possible. The inherent potential of these plants, aside from environmental challenges, maintains a viable production chain involving various strategies for plant management, including plant breeding activities exploiting variations. Genetic control of various traits to keep crops thriving in adverse situations allows better plant variety selections. Exploitation of available natural spontaneous and induced variations and selection based on easy phenotyping protocols provide short-duration annual crop species such as climate-responsive fruit crop varieties. Because both biotic and abiotic stresses are becoming unpredictably more frequent, interventions using recently developed biotechnological information, such as genomic data, can be a boon for climate-smart fruit crop development.

Keywords Adaptation · Climate change · Fruits · Genetics

Introduction

The prediction for carbon dioxide in environmental concentrations is a 100% increase by the end of the present century because of emissions to the environment from industry to developmental activities, and the global temperature will increase

T. Sarkar · S. K. Sarkar
Department of Fruit Science, Bidhan Chandra Krishi Viswavidyalaya, Mohanpur, Nadia, West Bengal, India

A. Roy
Department of Genetics and Plant Breeding, Bidhan Chandra Krishi Viswavidyalaya, Mohanpur, Nadia, West Bengal, India

S. M. Choudhary (✉)
Department of Horticulture, Mahatma Phule Krishi Vidyapeeth, Rahuri, Maharashtra, India

M. N. Islam, A. van Amstel (eds.), *India: Climate Change Impacts, Mitigation and Adaptation in Developing Countries*, Springer Climate,
https://doi.org/10.1007/978-3-030-67865-4_4

by 6 °C (IPCC 2007). Climate change will affect tropical and subtropical cropping systems more, mostly small-scale farming communities (Calberto et al. 2015; Ranjitkar et al. 2016). The occurrence and severity of water stress has been predicted to be one of the challenging issues in the future (Feller and Vaseva 2014). The few regions suitable for growing mango, and the drastically changing patterns in fruiting and flowering in mango, continue to alarm us about climate change effects. Flowering, fruiting, and ultimately productivity are seen to be affected by variations of rainfall from the past few years (Sdoodee et al. 2010). A fruit crop is a good candidate for carbon sequestration, which also reduces the negative effect of climate change. At field conditions, various stresses are combined and affect the crop severely that cannot be quantified by imposing a single stress exposure, which invites several platforms such as various integrated phenotyping and genotyping platforms for an integrated breeding approach (Pereira 2016). Unexplored rootstock and scion, evolving genome sequencing technologies, genomic sequences, quick mapping traits, biotechnological tools for gene editing, faster phenotyping, and functional validation of putative genes as major candidates controlling development under stress have been integrated, keeping breeding for stress-tolerant fruit crops viable. Various new changes in fruit have been reported frequently, such as discolouration of peel, acid reduction, softening, abrupt increase in size, and quick degradation of fruit quality (Morinaga 2016). A wide range of environmental changes such as frosts, storms, uneven rainy seasons, early withdrawal of rain, and extreme temperatures are major causes that limit the expansion of underutilized climate-smart minor fruit crops because of lack of modeling, which can be best utilized under a climate change scenario to diversify cropping pattern. As this era is mostly utilizing genomic technologies with better genetic gain for crop improvement, the entire genomic platform is leading the way for modern fruit breeding under a climate change scenario, which was fully phenotype based previously (Crossa et al. 2017).

Optimum Climate for Major Fruit Crops

The optimum growth of most tropical and subtropical fruit crops occurs at a mean temperature range between 22 and 30 °C, whereas temperate fruits grow best at 10–20 °C. Some tropical fruit crops, such as mango and litchi, can tolerate more than 45 °C and the temperate fruits, such as apple, can also endure below −2 °C for a short period. Mango and litchi grow at 0 °C and as much higher than 45 °C; the ideal temperature for mango is 25–30 °C with precipitation of 900–1000 mm in a year, and with high humidity during the growing season. Litchi, being a subtropical to tropical fruit, can grow around the optimal temperatures of 25–35 °C with annual rainfall of 1200–2000 mm, frost free, and summer heat is essential. Rambutan grows well at about 20–30 °C, but below 10 °C growth and fruit set are drastically reduced. The guava can grow successfully in tropical and subtropical regions up to 1500 m mean sea level. The average temperature is 20–30 °C for optimum growth of the

Table 4.1 Optimal climatic conditions for some important tropical fruits

Fruits	Optimum temp. (°C)	Optimum annual rainfall (mm)	Lat. range (Distribution)	References
Mango	24–30	900–1150	30°N–27°S	Davanport (2009)
Guava	23–28	>1000	26°N–30°S	Verjeij and Coronel (1992)
Mangosteen	25–35	122–1800	12°N–10°S	Osman and Milan (2006)
Litchi	25–35	1200–2500	28°N–27°S	Tindall (1994)
Citrus	16–27	1400–1800	44°N–27°S	Verjeij and Coronel (1992)
Rambutan	24–32	1500–2500	18°N–17°S	Tindall (1994), UDPMS (2002) and Verjeij and Coronel (1992)
Jackfruit	16–28	1000–1500	28°N–30°S	Ghosh (2000) and Haq (2006)
Ber	22–36	700–1600	22°N–18°S	–
Papaya	18–27	800–1500	25°N–29°S	–

plant, but quality guava is obtained where low night temperature (10 °C) prevails during the winter season. Guava tolerates high temperature and drought conditions in north India in summers but it is susceptible to severe frost, which can kill the young plants (Table 4.1).

Citrus is a subtropical fruit that also grows well under tropical conditions; the optimal temperature is 16–20 °C, although it can be grown above 30 °C, but very high temperature is harmful for fruit quality. Some citrus groups such as mandarin require low temperatures for optimum growth. With proper irrigation facilities in the drought areas, good production of citrus is possible. At least 750 mm well-distributed rainfall is necessary for citrus. The custard apple can be grown at temperatures about 25–35 °C, but temperatures above 40 °C and less than 0 °C are harmful for custard apple production. The annual rainfall of 900 mm is adequate with high humidity to help maximize fruit set, but poor fruit set occurs with long durations of high temperature and dry conditions. Aonla grows under an average temperature about 28–35 °C, needing to be frost free during winter. It has adapted to a wide range of climatic conditions from sea level up to above 1800 m MSL (Bose et al. 1999). Grape is a most important subtropical fruit; it grows well under mild temperatures not exceeding 35 °C during summers (Table 4.2). A hot and dry climate is ideal for fruit production. Higher night temperatures above 25 °C during ripening hamper colour development; rainfall time and frequency are very important. Rains during flowering and berry ripening cause enormous damage to the berry.

Apple is undoubtedly the most important temperate fruit. It requires chilling temperature for the germination of seed, about 4 °C for 1000 h. It can tolerate temperatures below 0 °C and above 35 °C. The optimal temperature for growth of apple is 10–20 °C, but when temperature falls below −2.2 °C, apple is injured by the cold. It is generally grown at elevations from 1600 to 2300 m MSL. Strawberry is one of the most cultivated temperate fruits; it is grown at an optimal temperature about 15–25 °C with frost-free conditions during the growing season (Table 4.3). In open conditions at higher elevations where frost and snowfall are common factors, those areas are sensitive for strawberry production, so it is mostly cultivated under

Table 4.2 Optimal climatic conditions for some important subtropical fruits

Fruits	Optimum temp. (°C)	Optimum annual rainfall (mm)	Climatic description
Grape	25–35	900	Mild climate with moderate humidity
Pomegranate	24–32	700–1100	Hot dry summer
Fig	28–32	850–1000	High temp cause inferior fruit quality
Phalsa	25–30	950	Semiarid with mild winter
Bael	26–30	800–1100	Semiarid with moderate rainfall
Aonla	20–34	700–2000	Semiarid, moderate humidity
Mulberry	23–32	800–1000	Low temperature, cold hardy
Jamun	25–33	900–1000	Dry and humid
Avocado	24–30	1000–1800	Hot and humid

Table 4.3 Optimal climatic condition for some important temperate fruits

Fruits	Optimum temp. (°C)	Optimum annual rainfall (mm)	Chilling hours (4–7 °C)	Climatic description	References
Apple	21–24 °C (during active growth period)	100–1250	1200–1500	Winter and abundant sunshine for good colour development	Verma (2014a, b, c) and Jindal et al. (2000)
Pear	26–35 (in summer), 2 (in winter)	1000–1250	1200 (Bartlett needs about 1500)	Summers should be less humid	Verma (2014a, b, c)
Peach	23–26	1000–1200	100–100	30°N and 45°S latitude distribution	Hancock et al. (2014)
Strawberry	22–25 °C day and 7–13° C night temperature	700–800		Thermo and photo sensitive	Tyagi et al. (2015) and Sharma (2002)
Walnut	10–25	500–700	700–1500	Cool winters and dry summers	Verma (2014a, b, c) and Adem (2003)
Pecan nut	24–30 °C	600–1000	400–700	25°N and 35°S latitudes	Barrios (2010), Fronza et al. (2018) and Kuden et al. (2013)

protective structures. Higher temperatures during ripening of berry cause abnormal colour formation and also increases acidity in the fruit.

Impact of Climate Change on Fruit Crops

Climate change may have direct effect on increasing temperatures, which may have negative impact on fruit crops, temperate fruit crops as well as tropical and subtropical fruits. Elevated CO_2 and changed precipitation are also considered important, with direct effects on production technology, delay or early harvest, reduced available irrigation water, increased irrigation cost, increased insect pest attacks, increased physiological disorders, inferior fruit quality, and lack of suitable cultivars, and also impacts on soil from excess rainfall and changes in temperature that negatively impact fruit crops (Nath et al. 2019).

Optimum growth of plants needs a particular temperature for proper development: temperate fruits require low temperature whereas tropical and subtropical fruit require a medium temperature range. Because of climate change, a rise in a temperature above 1 °C may shift major potential areas for fruit crops (Nath et al. 2019). Temperature increases likely more prominently affect the reproductive biology of crops, with reduction in the population; hence, dormancy breaking will occur earlier. High temperature and less moisture increase cracking and sunburn in apples, apricot, and cherries, and fruit cracking in litchi (Kumar and Kumar 2007). The faster maturity and quick ripening may reduce the fruit availability period (Kumar 2015), altering important quality parameters such as synthesis of sugars, organic acids, antioxidant compounds, peel colour, and firmness (Moretti et al. 2010). Grapes have higher sugar content and lower levels of tartaric acid when grown under high temperatures (Kliewer and Lider 1970).

Mango (*Mangifera indica* L.)

Mango phenological patterns and indirectly the vegetative and reproductive processes have been influenced by climate change, leading to reduced quality and production. Rainfall and air temperature are significant in the vegetative and phenological phases in mango and are also the two important factors determining the suitability of an area climate for mango production (Makhmale et al. 2016). Climate changes have already brought wide changes in flowering and fruiting patterns. Mango is distributed in tropical and subtropical regions, so it can tolerate high temperature. Mango flowering starts from January onward; a favorable temperature regime of 27/13 °C (day/night) also increases flowers, whereas low temperature below 10 °C with high humidity (>80%) and cloudy weather delay flowering (Chadha 2015). With high temperature during the night before flowering, a reproductive bud may change to a vegetative bud. Elevated temperature reduces the flowering period and the number of days for effective pollination. Severe winters may cause mango malformation.

High rainfall and cloudy weather during flowering cause flower drop and low fruit set. High rainfall during full bloom may affect pollination by pollen washout and also reduced activity of pollinators (Rajan 2012). Rainfall and hail storms during the maturation of the fruit may affect fruit quality and increase fruit drop, ultimately resulting in reduced yield. High humidity and unseasonal rain during fruit maturity invites severe attacks by insects, pests, and diseases, fruit fly, anthracnose, and mango stone weevil (Makhmale et al. 2016). With temperature increase from 20 to 35 °C, attacks of fruit fly increased in mango cv. Chausa (Kumar and Shukla 2010). Spongy tissue, mostly occurring in cv. Alphonso mango, is induced by high temperature buildup within the fruits (Makhmale et al. 2016).

Grape (*Vitis vinifera*)

Climate change has significant effects on the grape and wine industry. Approximately more than 10,000 grape cultivars are known in the Vitiaceae alone, showing wider biological diversity (Mullins et al. 1992). Warmer temperatures extend the period of physiological maturity and hence augment metabolic rates and affect metabolite accumulation. Above 30 °C, berry size and weight are reduced and metabolic activities and accumulation of sugar may stop (Coombe 1987; Hale and Buttrose 1974), although elevated temperatures accelerate berry maturation, and also affect final sugar accumulation. Higher temperatures (30 °C) may lead to suspended solid concentrations, and also increase Brix levels more than 24–25 Brix, not for the photosynthesis and transportation of sugar concentration from leaves, but for evaporative loss (Keller 2009, 2010). Flavonoids have an important role for the quality of grape and wine. Temperature has been shown to have a direct and important role for their formation. Low temperatures (14/9 °C day/night) are not conducive to develop anthocyanin concentrations, and temperatures of 30 °C or more also hamper lower anthocyanin synthesis (Spayd et al. 2002; Tarara et al. 2008). Rainfall at the time of blooming and temperatures below 15 °C and above 32 °C severely pollen germination, fertilization, and fruit set. High humidity after pruning of 30–110 days favours fungal disease development such as powdery mildew or anthracnose in warm areas.

Other climate change factors likely to influence grape production are related to pest and disease attack as well as the new vectors responsible for disease spread. The distribution of glassy-winged sharpshooter is highly temperature dependent, and with rising winter temperature their infestation is shifted towards high (Daugherty et al. 2009). Climate change may modify the complex interrelationships between vine and pest development, such as the vector of phytoplasma grapevine disease, *Scaphoideus titanus* (Stock et al. 2005), which is at an advantage in warmer viticulture regions.

Banana (*Musa* spp.)

Bananas are mostly grown in the tropical and subtropical regions of the world. The banana is also ubiquitous in availability in nonproducing regions through national as well as international trade, which accounts for 15% of global production (Turner et al. 2007). Quantifying the optimal climatic condition of banana climate sensitivity predicts the wide impacts of climate change on production systems. In our global model, banana growth is greatest at the optimum annual temperature of 27 °C (±0.04 °C) (Calberto et al. 2015; Ramirez et al. 2011), from 20.1 °C (95% CI ± 0.1 °C) for Brazil to 30.4 °C (95% CI ± 0.1 °C) for Africa (Fig. 4.1).

The main factors of banana growth that hamper production, apart from disease and pests, are long dry seasons (Van Asten et al. 2011), cold front events, and also cyclonic winds. Most of the cultivars below 15 °C have significantly slow growth and below 10 °C the growth of new leaves is restricted. Banana requires about 1000–2200 mm water/year for adequate growth (Ramirez et al. 2011), so decreases in precipitation may affect the maturity of plants and fruit. Warmer night temperature deteriorates fruit quality (Chen 2012). Crop duration has been reduced by dry spells during flower emergence.

Consistent increases in temperature in the past few years have direct impact on yield. Regional model-based forecasting recorded that by 2050 (Fig. 4.2), past positive effects of changes of climate on yield of banana are likely to continue at a lower magnitude. Warmer countries having more optimal temperatures have seen increased productivity, whereas exceeding the regional optimum temperature shows declines in production (Varma and Bebber 2019).

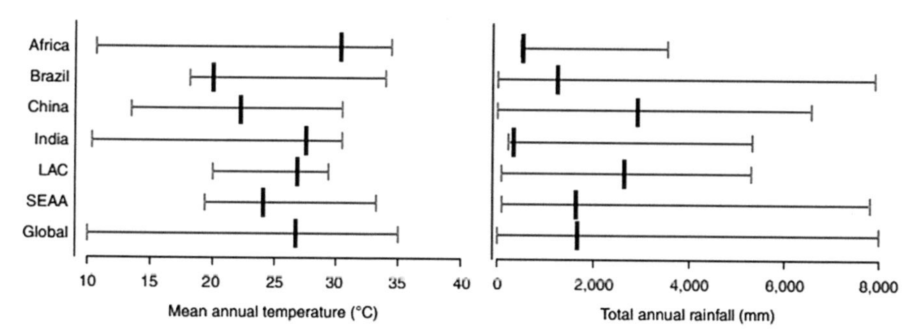

Fig. 4.1 Cardinal value of mean annual temperature and total annual rainfall for banana yields (Varma and Bebber 2019)

Practical of Horticulture

1. Identification of horticultural crops
2. Different propagation techniques of horticultural crops
3. Given them to make Herbarium sheet of horticultural crops

Practical of Forestry

1. Identification of forest trees
4. Given them to make Herbarium sheet of Forest trees
2. Planning of making home garden/multi-stories Farming

Ongoing - How to take measurement of falling trees

Fig. 4.2 Assessment of future climate risk for major banana-producing countries (by 2050). This categorization was carried out by combining changes in predicted yield under the RCP 8.5 climate change scenario and the effect of cultivation efficiency (technology trend) on past yields. Countries classified as 'at risk' are those where yields are predicted to decline because of climate change and that have shown a negative technology trend on past yields. 'Adaptable' countries could see future climate-driven yield declines but mitigation could be possible given past positive technology trends. Countries at an 'advantage' are those that are predicted to experience climate-driven increases in yield by 2050 regardless of past technology trends. Countries are grouped by region. Brazil, China, and India are considered to be stand-alone regions (Varma and Bebber 2019)

Citrus (*Citrus* spp.)

Citrus is one of the major fruit crops after mango, banana, and apple. More than 100 varieties are grown commercially around the world, so it has great economic return in international trade. Citrus is widely grown in different climatic zones from tropical to subtropical to semiarid zones (Abobatta 2019). Temperature is a critical factor for successful citrus production; it is grown at an optimum range of temperatures between 16 and 27 °C. High temperature reduces fruit growth by decreasing net CO_2 assimilation and also increases fruit drop. Adverse effects of high temperature and moisture stress can delay fruit maturity, flatten peel colour, and increase fruit splitting and creasing; low temperature (frosts and freezing) damages the fruits or, if it persists for a long time, may kill the plant. In navel oranges the content of TSS (Total Soluble Solids) was affected by increasing acidity. Among other important climatic factors, the rainfall in September and October has great effect on fruit quality in terms of TSS, and less rainfall increases the TSS content in fruits (Mitra 2018). It changes climate affect also in fruit enlargement, peel puffing, and reduction of acid concentration in Satsuma mandarins (Sugiura 2019). Elevated temperature also increases the water requirement in tropical regions and reduced flowering in subtropical regions. Warming also favours scale and mite attacks. Increasing temperature above 30 °C promotes vegetative flush instead of flowering flush (Nath et al. 2019). High temperature increases fruit cracking and sunburn (Figs. 4.3 and 4.4) in 'Washington' sweet orange and mandarin hybrids such as 'Murcott' and 'Nova' (Gracia-Luis et al. 1994).

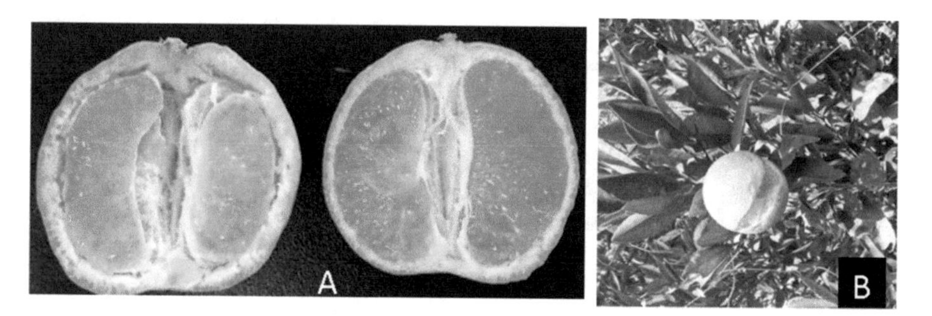

Fig. 4.3 (**a**) Peel puffing in satsuma mandarin (left); normal fruit (right) (Sugiura 2019). (**b**) Fruit cracking in Valencia fruits (Abobatta 2019)

Guava (*Psidium guajava* L.)

Guava has wider adoptability in tropical and subtropical regions of the world. In India, most of the cultivation takes place under a tropical and subtropical climate, so diverse germplasm has been recorded. Guava can tolerate temperatures below 15° to above 35 °C, but the optimal temperature range is 23–28 °C (Nath et al. 2019). The plants grow best with an annual rainfall about 1000 mm. The guava plant is very sensitive to waterlogged conditions, so good drainage is necessary for young plants. The guava flowers throughout the year. Good quality of fruit is produced in the winter season, whereas in the rainy season fruits are inferior in quality because of the high water content, insipid in taste and with incidences of pests and disease. The guava plant is tolerant to environmental stresses but young plants are very susceptible to drought and cold. Temperature increases may affect the red colour in guava production. Development of red colour on the peel of guava requires mild days with a cool night temperature. Red dot skin develops in the Apple Color guava under subtropical conditions in North India because of the warmer night conditions. Production of red colour guava recorded negative correlation with TSS, dry mass percentage, and fruit firmness with temperature during fruit growth (Rajan 2008). The mean temperature of the coldest quarter will be 3.2 °C higher than normal temperature, resulting in less red colour development on fruits (Rajan 2008).

Apple (*Malus* spp.)

The apple is the most important temperate-zone fruit in the world; it is also known as the 'king of temperate fruit.' The optimal temperature is 21–24 °C (during the active growth period) with annual precipitation about 100–125 cm (Verma 2014a, b, c). The plant requires chilling temperatures at 4–7 °C for 1250–1500 h (Jindal et al. 2000). Breeders have also developed a low chilling variety of apple but those varieties grow well in subtropical as well as tropical areas. The quality of the apples

Fig. 4.4 Sunburn in apple and citrus (Sugiura 2019)

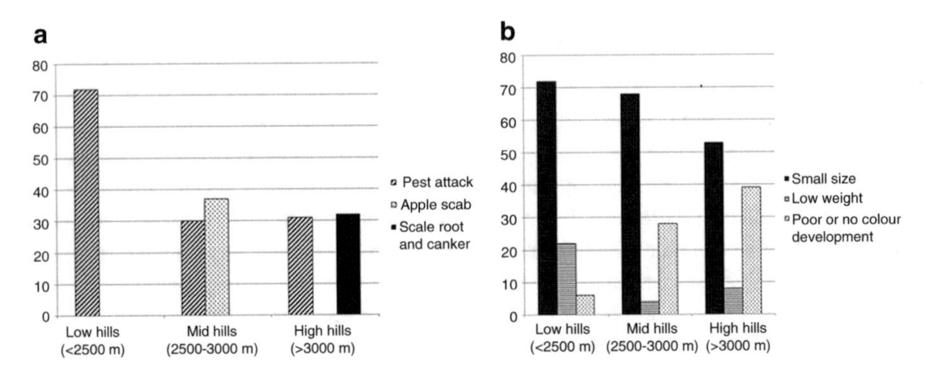

Fig. 4.5 Perceptions of respondents along the altitudinal gradient in the Kinnaur district of Himachal Pradesh on the impact on fruit quality (**a**) and insect pest attacks (**b**) caused by climate change (Basannagari and Kala 2013)

grown in these areas is very inferior compared to a temperate zone, and infestations of disease and pests are also greater.

Decline in moisture and increasingly frequent droughts may raise atmospheric temperature in temperate zones. Winter temperatures and rainfall in the form of snow are required for apple plant dormancy, and there is a chilling requirement for bud break to ensure the flowering (Jindal et al. 2000). So, delay in cold has effects on chilling requirements in December–January. However, extremely low temperature also affects plant growth and fruits by chilling injury, resulting in poor yields. Summer temperatures higher than 26 °C and lower than 15 °C at the blooming period reduce the apple crop. The increases in atmospheric temperature at different hill elevations ultimately affect the quality of the fruits, increase infestations of disease and pests, and also cause sunburn (Figs. 4.4 and 4.5) (Basannagari and Kala 2013). Apple cultivation has shifted from low elevation to high with increasing temperature. The harvesting period was also delayed for a week. In Japan and Korea, the areas suitable for farming of apple had an average annual temperature about 7–13 °C, and farming gradually shifted north because of heat injury (Kim et al. 2010; Fujisawa and Kohayashi 2013).

Other Temperate Fruit Crops

In India the temperate zone is near the Himalayan regions: Jammu Kashmir, Himachal Pradesh, Uttarakhand, Sikkim, Arunachal Pradesh, and some parts of Nagaland. Monitoring by the India Meteorological Department (IMD) reveals that temperatures are increasing significantly, 0.05 °C per year (Lal et al. 2018). In Kashmir valley the average temperature has risen by 1.45 °C in the past 28 years, but in Jammu regions it has increased by 2.32 °C (Ahmed and Lal 2015). Elevating temperatures shift the temperate fruit crops towards upper elevations. In lower elevations, for increasing temperature as well as precipitation, and unusual hailstorms in summer, plum, peach, and apricot fruits are heavily damaged (Choudhary et al. 2015).

In pome and stone fruits, increasing temperature negatively affects key phases such as erratic bud break, bud dormancy, and plant yield (Legave et al. 2013). In most temperate fruits, when the chilling requirement is fulfilled, the bud is ready for the flower bud burst (Hovarth et al. 2003). In apricot (*Prunus armeniaca* L.), environmental factors strongly influence floral biology patterns of flower bud anomalies, drops, and oenological processes (Oukabali and Mahhou 2007; Legave et al. 2006). The warmer temperature results indeed in a trending of earlier harvest dates in pears and peaches. In the Japanese persimmon, pigment development is important for maturity indices because high temperatures impede coloration.

Strategies to Cope with Climate Change and Adaptation

To reduce large-scale losses, diversification of cropping pattern, choices of crops and environment, and diversified farming can lead to sustainable adaptation strategies of any cropping system under a changing climate scenario. Various functional models for isolating early drought tolerance ability in fruit crops based on physiological traits, including stomatal conductance, and actual transpiration have been deduced (Kullaj et al. 2017). It has been established that the reproductive potential of a crop depends on a higher rate of photosynthesis, which increases with various stresses and accumulates photosynthate on a small leaf (Boominathan et al. 2004). Cellular defense mechanisms such as epicuticular wax increase during stress conditions in plants (Fich et al. 2016). In some crops, such as peach, a strategy including the most promising traits such as flowering and adaptation of a specific combination of scion and stock under climatic variations has been studied. As a plant changes its methylation pattern upon stress exposure, similarly one DNA methylated gene having conversion potential from dormant to active by methylation has been identified in bin almonds (Prudencio et al. 2018).

(a) Adaptation through breeding approaches

I. Selection methodology

Natural selection of a particular variety of fruit crop depends on long-term natural selection based on variations evolved with time. To induce or to find the most promising genotype suitable for particular ecological conditions, it needs to be selected based on early screening of a large set of rootstocks or available germplasm by imposing available artificial selection pressure. By screening of a set of germplasm in a variety of environments, although the concerned breeding target is for a particular ecological condition, the most widely accepted and stable genotype and its potential can be exploited by using the selected genotype. Selection of resistant cultivars based on various abiotic stresses can provide wide protection to variable climate change. Deeper-rooting genotypes are selected for drought tolerance, as in banana, which explains a considerable variation (Uga et al. 2013). In a collaborative approach involving 1576 varieties of grapevine including landraces, wild accessions originating from various countries were genotyped using an Infinium single nucleotide polymorphism (SNP) chip, which resulted in classification of genotypes in groups and identified one set with domestication traits most appropriate for developing a variety under climate change (Micheletti et al. 2015).

II. Maintenance of genetic resources

In past decades our most important target was to maintain genetic variations for the future to cope with unpredicted climatic variation. Continuous use of similar rootstocks or genotypes retains the common background and ultimately makes plants vulnerable to various environmental threats. Without duplicating the genetic stock, minimizing duplication using molecular markers can be a viable option for maintaining future genetic stock.

III. Natural variations

A great diversity of fruit crops is present in tropical areas, about 2700 species including commercially grown and exploited fruit crops (Paull and Duarte 2011). One potential area for climate-resilient variety development in recent times are induced alleles by irradiation, propagating material and maintaining a large set of variation followed by selection of desirable stocks. Major breeding strategies under low-cost investment utilize exploitation of the available primary gene pool to wild accessions: phenotyping based on climate-smart traits such as flowering, vigour, maturity, response of genotypes to elevated carbon dioxide and temperature, and maintaining frequent changes including wild sources and assembling various genetic loci within the same potential background is needed.

IV. Genes and quantitative trait loci (QTLs)

Some soils have inherently low phosphate content, and this issue can be resolved by a genotype having a low phosphate-responsive gene or can be transferred to recipient parent ortholog genes involved in the response to low P content in *Arabidopsis*, as AtLPR1 (low phosphate-resistant root) was reported within the confidence interval of QTLs for the number of roots of different diameters in grapevine (Tandonnet et al. 2018). Although the calcium-dependent protein kinase (CDPK) family of genes involves various biotic stress, still this group of transcripts has been characterized for drought, cold, and heat stress, identified in wild *Vitis* germplasm (Dubrovina et al. 2018). One gene of the WRKY family has been characterized for water and salt tolerance and has been overexpressed in tobacco plants as VvWRKY2 (He et al. 2018). In a recent study in peach, GWAS was used considering cultivated and wild peach genotypes to decipher the evolutionary history and adaptive traits related to agronomical properties (Li et al. 2019).

V. Genomics information for climate-smart fruit breeding

The recent initiative to explore the genomic sequence of *Prunus* species will lead to easily available molecular markers that can be used for selection of genotypes suitable for adverse climatic factors such as drought, heat, and salinity (Verde et al. 2013; Shirasawa et al. 2017); other biotechnological experiments can be designed involving genomic sequence and characterization. Initially, the derived simple sequence repeat (SSR) was based on preliminary genomic information (Thomas et al. 1994; Laucou et al. 2011), NGS-derived EST-SSR, SNP information, indel polymorphism, and other larger structural genomic variants in various fruit crops such as vine grapes (Cardone et al. 2016; Royo et al. 2018). Genomic selection to attain genetic gains exploited in various major food crops can be utilized in fruit crop breeding (Crossa et al. 2017).

VI. Genome editing technology

As dormancy can be one of the adoptive strategies for plants to withstand various stress, its activation and genetic control mechanism and controlling constitutive expression by transgenic approach can be used; in this approach a few recently identified dormancy genes can be exploited, ParCBF1 (C-repeat binding factor), *ParDAM5* (dormancy-associated MADS-BOX), and *ParDAM6* genes (Balogh et al. 2019) in *Prunus*. Cas9-directed genome editing involves complementarity to the target site: it generates a double-stand break within that specific region that allows targeted mutagenesis, and this technology was widely and very quickly developed for genome editing (Puchta and Fauser 2014). 'Neo Muscat' is an example of use in genome editing technology, wherein a construct that targets the phytoene desaturase gene has been edited and an albino plant developed (Nakajima et al. 2017). One derivative of Thompson seedless grapes with altered WRKY52 has been derived by genome editing (Wang et al. 2018).

VII. Flowering time

As stress response is correlated with flowering, the genetics of flowering and its temporal variations can be utilized in a breeding fruit crop. The gene MdTFL1 targeting the same phenotype has been used for a silencing study, and its role for controlling flowering has been discovered in apple (Flachowsky et al. 2012). Even from a major class of genes, few putative genes have been characterized that express sufficiently the control miRNA responsible for the transition state, that is, vegetative to floral initiation (Jia et al. 2017).

VIII. Drought stress

Longer drought spells occurring frequently in the subtropical environment before flowering hamper the flower to fruit conversion and ultimate production potential. The genetic potential of drought-tolerant crops can be exploited to develop tolerant cultivars in the crop of interest. Under drought conditions a plant always finds a strategy to avoid stress. It may change its physiology accordingly by reducing stomatal conductance and changes in root system architecture, reducing life cycle and early flowering in this circumstance and also maintaining some compatible solutes, that is, osmotic adjustment strategies (Blum 2017) in the vacuole. Maintaining physiological stasis under the affected water potential of the plant and carbon gain measured through photosynthesis activity can be considered for breeding climate-smart fruit crops (Jimenez et al. 2013). MdMYB88 and MdMYB124 genes have been reported to be associated with adaptive drought tolerance in apple that maintains the plant water level under stress conditions (Geng et al. 2018). Various major transcription factors related to drought tolerance mechanisms such as MYB, MYC, NAC, NAM, bZIP, and DREB are reported for drought tolerance-based differential regulation (Argarwal and Jha 2010).

IX. Salinity stress

Salinity stress is one of the problems of major arid and semiarid regions. Various frequent flashes of rain and subsequent inundation at the coastal belt cause a stress that impacts the crop and is also withstood by the crop plant by channelizing extra salt at the vacuoles. Genes such as LEA5, glutathione S-transferases, and SOD have been identified to be associated with salt-induced oxidative stress (Beeor-Tzahar et al. 1995).

(b) Adaptation through management practices

I. Use of rootstock

Rootstocks have an important role in successful establishment of plants into wide environmental and soil conditions. Most fruit crops are propagated vegetatively using rootstock tolerant to different abiotic stresses such as salinity, drought, iron chlorosis, and flooding (Irenaeus and Mitra 2014). In grapes, Dogridge is used for drought and salinity tolerance. Sharad Seedless grafted on Dogridge rootstock showed maximum water use efficiency (Satisha and Prakash 2006). *Juglans hindsii* rootstock is used in walnut for tolerance to waterlogged soils.

II. Changes in microclimate

The microclimate around plants can be modified to minimize stress by heat and cold. Some irrigation methods should be followed to modify the microclimate, such as an overhead sprinkler system, a mid-canopy sprinkler, shade net, canopy management, and water channels.

III. Water management

In drought-prone areas, most orchards depend on rainfed cultivation, which should be followed by drip irrigation to increase water use efficiency. More favorable soil moisture coupled with increased CO_2 may increase biomass and soil carbon. Surface water harvesting in suitable structures and preventing evaporation loss are important adaptations against drought conditions.

IV. Soil moisture conservation

Conservation of soil moisture in all fruit crops increases soil microbial activity and soil microclimate and also reduces the incidence of abiotic as well as biotic stresses. The use of plastic mulch is effective on a long-term basis compared to organic mulches. The use of plastic mulch in fruit crops increases yield 45.23% in mango, 14.63% in pineapple, and 12.61% in litchi compared to non-mulched plants (Patil et al. 2013).

V. Protected cultivation

In protected cultivation, the initial expenditure is greater, but for long-term production the cultivation cost has to be covered. In fruit crops, generally less production takes place under protected cultivation, but now several high-density plantings using dwarf rootstock are possible to cultivate. Protected cultivation has an advantage in terms of modified microclimate and is free from abiotic stresses such as hail storms, frost, drought, and excess precipitation, with reduced attacks by insects and diseases. To prevent early frosts and create a microclimate, an overhead net may be used (Fig. 4.6).

Fig. 4.6 Apple orchard covered with overhead net to avoid frost

Conclusion

To tap the pace of climate change and the development of climate variation, responsive fruit crop varieties need complete genomic information and more biotechnological interventions. Stringent selection strategies, with wide variations to be exploited in breeding populations, marker trait associations, a wide array of robust markers, cross-species transferability, and a candidate gene-based approach followed by gene targeting studies, are more and more needed. As the outcome of fruit breeding is a time-consuming approach, a multidimensional approach involving multiple backgrounds of genotype and continuous preplanned efforts must be activated for development programmes. Sometimes a poor transformation can be addressed by a system biology approach for the detection of genes and networks for adoption of various fruit crops. Utilization of integrative platforms involving genotype and phenotype for a particular crop and its interactions can be used simultaneously.

References

Abobatta WF (2019) Potential impacts of global climate change on citrus cultivation. MOJ Ecol Environ Sci 4(6):308–312

Adem HH (2003) Walnuts. Australian walnut Industry. https://walnut-net.wildapricot.org/Resources/Documents/Research%20Programs/Walnut%20production%20_H_Adem.pdf

Ahmed N, Lal S (2015) Climate change: impact on productivity and quality of temperate fruits. In: Climate dynamics in horticultural science, vol. I. Apple Academic Press, Toronto

Argarwal PK, Jha B (2010) Transcription factors in plants and ABA dependent and independent abiotic stress signaling. Biol Plant 54:201–212

Balogh E, Halasz J, Solteszt A, Galiba G, Hegedus A (2019) Identification, structural and functional characterization of dormancy regulator genes in apricot (*Prunus armeniaca* L.). Front Plant Sci 10:402

Barrios DO (2010) SWOT analysis and perspectives of pecan production in Chihuahua-Mexico. Revista Mexicana de Agronegocios 14(27):348–359. Available from http://www.redalyc.org/articulo.oa?id=14114743006. Accessed 12 Dec 2017

Basannagari B, Kala CP (2013) Climate change and apple farming in Indian Himalayas: a study of local perceptions and responses. PLoS One 8(10):77976

Beeor-Tzahar T, Ben-Hayyim G, Holland D, Faltin Z, Eshdat Y (1995) A stress-associated citrus protein is a distinct plant phospholipid hydroperoxide glutathione peroxidase. FEBS Lett 366:151–155

Blum A (2017) Osmotic adjustment is a prime drought stress adaptive engine in support of plant production. Plant Cell Environ 40:4–10

Boominathan P, Shukla R, Kumar A, Manna D, Negi D (2004) Long term transcript accumulation during the development of dehydration adaptation in *Cicer arietinum*. Plant Physiol 135:1608–1620

Bose TK, Mitra SK, Farooq AA, Sadhu MK (1999) Tropical fruits. Nayaprakash, Bidhan Sarani

Calberto G, Staver C, Siles P (2015) Climate change and food systems: global assessments and implications for food security and trade (Albehri A ed) Ch. 9. FAO, Rome

Cardone MF, D'Addabbo P, Alkan C, Bergamini C, Catacchio CR, Anaclerio F, Chiatante G, Marra A, Giannuzzi G, Perniola R, Ventura M, Antonacci D (2016) Inter-varietal structural variation in grapevine genomes. Plant J 88:648–661

Chadha KL (2015) Global climate change and Indian horticulture, climate dynamics in horticultural science. In: Chaudhary ML, Patel VB, Siddiqui MW, Verma RB (eds) Impact, adaptation and mitigation, vol. 2, vol 2. Apple Academic Press, Hoboken, p 351

Chen Q (2012) Adaptation and mitigation of impact of climate change on tropical fruit industry in China. Acta Hortic 928:101–104

Choudhary ML, Patel VB, Siddiqui MW, Mahsi SS (2015) Climate dynamics in horticultural science, vol. I: principles and applications. Apple Academic Press, Toronto

Coombe B (1987) Influence of temperature on composition and quality of grapes. Proceedings of the international symposium on grapevine canopy and vigor management. Acta Hortiic 206 (22):23–35

Crossa J, Pérez-Rodríguez P, Cuevas J, Montesinos-López O, Jarquín D (2017) Genomic selection in plant breeding: methods, models, and perspectives. Trends Plant Sci 22:961–975

Daugherty MP, Bosco D, Almeida RPP (2009) Temperature mediates vector transmission efficiency: inoculum supply and plant infection dynamics. Ann Appl Biol 155:361–369

Davanport TL (2009) Reproductive physiology. In: Litz RE (ed) The mango: botany, production and uses, 2nd edn. CABI, Wallingford

Dubrovina AS, Aleynova OA, Manyakhin AY, Kiselev KV (2018) The role of calcium dependent protein kinase genes CPK16, CPK25, CPK30 and CPK32 in stilbene biosynthesis and the stress resistance of grapevine *Vitis amurensis* Rupr. Appl Biochem Microbiol 54:410–417

Feller U, Vaseva II (2014) Extreme climatic events: impacts of drought and high temperature on physiological processes in agronomically important plants. Front Environ Sci 2:39

Fich EA, Segerson NA, Rose JKC (2016) The plant polyester cutin: biosynthesis, structure, and biological roles. Annu Rev Plant Biol 67:207–233

Flachowsky H, Szankowski I, Waidmann S, Peil A, Tränkner C (2012) The *MdTFL1* gene of apple (*Malus* × *domestica* Borkh.) reduces vegetative growth and generation time. Tree Physiol 32:1288–1301

Fronza D, Jonas JH, Both V, Rogério de Oliveira A, Meyer EA (2018) Pecan cultivation: general aspects. Ciênc Rural Santa Maria 48(2):1–9. https://doi.org/10.1590/0103-8478cr20170179

Fujisawa M, Kobayashi K (2013) Shifting from apple to peach farming in Kazuno, northern Japan: perceptions of and responses to climatic and nonclimatic impacts. Reg Environ Chang. https://doi.org/10.1007/s10113-0130434-6

García-Luis A, Duarte AMM, Porras I (1994) Fruit splitting in 'Nova' hybrid mandarin in relation to the anatomy of the fruit and fruit set treatments. Sci Hortic 57(3):215–231

Geng D, Chen P, Shen X, Zhang Y, Li X (2018) *MdMYB88* and *MdMYB124* enhance drought tolerance by modulating root vessels and cell walls in apple. Plant Physiol 178:1296–1309

Ghosh SP (2000) In: Govindasamy B, Duffy B, Coquard J (eds) Status report on genetic resources of jackfruit in India and SE Asia. IPGRI, New Delhi

Hale CR, Buttrose MS (1974) Effect of temperature on ontogeny of berries of *Vitis vinifera* L. cv. Cabernet Sauvignon. J Am Soc Hortic Sci 99:390–394

Hancock JF, Scorza R, Lobos GA (2014) Peaches. In: Temperate fruit crop breeding: germplasm to genomics. Springer/Kluwer, Dordrecht

Haq N (2006) Fruits for the future 10. Jackfruit (*Artocarpus heterophyllus*) Southampton Centre for Underutilised Crops, pp 192

He R, Zhuang Y, Cai Y, Agüero CB, Liu S, Wu J, Deng S, Walker MA, Lu J, Zhang Y (2018) Overexpression of 9-*cis*-epoxycarotenoid dioxygenase cisgene in grapevine increases drought tolerance and results in pleiotropic effects. Front Plant Sci 9:970

Horvath DP, Anderson JV, Chao WS, Foley ME (2003) Knowing when to grow: signals regulating bud dormancy. Trends Plant Sci 8:534–540

IPCC (2007) Climate change 2007. The physical science basis. In: Contribution of Working Group I to the fourth assessment report of the Intergovernmental Panel on Climate Change. Cambridge University Press, Cambridge/New York

Irenaeus TKS, Mitra SK (2014) Understanding the pollen and ovule characters and fruit set of fruit crops in relation to temperature and genotype: a review. J Appl Bot Food Qual 87:157–167

Jia XL, Chen YK, Xu XZ, Shen F, Zheng QB (2017) miR156 switches on vegetative phase change under the regulation of redox signals in apple seedlings. Sci Rep 7:1422–1423

Jiménez S, Dridi J, Gutiérrez D, Moret D, Moreno MA, Gorgocena Y (2013) Physiological and molecular responses in four *Prunus* submitted to drought stress. Tree Physiol 33:1061–1075

Jindal KK, Chauhan PS, Mankotia MS (2000) Apple productivity in relation to environmental components. In: Jindal KK, Gautam DR (eds) Productivity of temperate fruits. Dr YS Parmar University of Horticulture and Forestry, Solan, pp 12–20

Keller M (2009) Managing grapevines to optimise fruit development in a challenging environment: a climate change primer for viticulturists. Aust J Grape Wine Res 9(145):1–14

Keller M (2010) The science of grapevines: anatomy and physiology. Academic Press, New York

Kim C, Lee S, Jeong H, Jang J, Kim Y, Lee C (2010) Impacts of climate change on Korean agriculture and its counterstrategies. Korea Rural Economic Institute, Seoul, p 282

Kliewer MW, Lider LA (1970) Effects of day temperature and light intensity on growth and composition of *Vitis vinifera* L. fruits. J Am Soc Hortic Sci 95:766–776

Kuden AB, Tuzcu O, Bayazit S, Yildirim B, Imrak B (2013) Studies on the chilling requirements of pecan nut (*Carya illionensis* Koch) cultivars. Afr J Agric Res 8(24):3159–3165

Kullaj E, Thomaj F, Kucera J (2017) Rapid screening of apple genotypes for drought tolerance by a simplified model of canopy conductance. J Agric Sci Technol 19:731–743

Kumar R (2015) Climate issues affecting sustainable litchi (*Litchi chinensis* Sonn) production in eastern India. In: Chaudhary ML, Patel VB, Siddiqui MW, Verma RB (eds) Climate dynamics in horticultural science, vol 2. Impact, adaptation and mitigation. Apple Academic Press, Oakville, p 351

Kumar R, Kumar KK (2007) Managing physiological disorders in litchi. Indian Hortic 52(1):22–24

Kumar R, Shukla O (2010) Effect of temperature on growth, development and reproduction of fruit fly Bractocera dorsalis. Hendel (Diptera Tephritidae) in mango. J Ecofriendly Agric 5 (2):150–153

Lal S, Singh DB, Sharma OC, Mir JI, Sharma A, Raja WH, Kumawat KL, Rather SA (2018) Impact of climate change on productivity and quality of temperate fruits and its management strategies. Int J Adv Res Sci Eng 7(4):1833–1844

Laucou V, Lacombe T, Dechesne F, Siret R, Bruno JP, Dessup M, Dessup T, Ortigosa P, Parra P, Roux C, Santoni S, Vares D, Peros JP, Boursiquot JM, This P (2011) High throughput analysis of grape genetic diversity as a tool for germplasm collection management. Theor Appl Genet 122:1233–1245

Legave JM, Richard JC, Fournier D (2006) Characterization and influence of floral abortion in French apricot crop area. Acta Hortic 701:63–68

Legave JM, Blanke M, Christen D, Giovannini D, Mathieu V, Oger R (2013) A comprehensive overview of the spatial and temporal variability of apple bud dormancy release and blooming phenology in Western Europe. Int J Biometeorol 57:317–331

Li Y, Cao K, Zhu G, Fang W, Chen C, Wang X, Zhao P, Guo J, Ding T, Guan L, Zhang Q, Guo W, Fei Z, Wang L (2019) Genomic analyses of an extensive collection of wild and cultivated accessions provide new insights into peach breeding history. Genome Biol 20:36. https://doi.org/10.1186/s13059-019-1648-9

Makhmale S, Bhutada P, Yadav L, Yadav BK (2016) Impact of climate change on phenology of mango: a case study. Ecol Environ Conserv Pap 22:119–124

Micheletti D, Dettori MT, Micali S, Aramini V, Pacheco I, Da Silva Linge C, Foschi S, Banchi E, Barreneche T, Quilot-Turion B, Lambert P, Pascal T, Iglesias I, Carbó J, Wang LR, Ma RJ, Li XW, Gao ZS, Nazzicari N, Troggio M, Bassi D, Rossini L, Verde I, Laurens F, Arús P,

Aranzana MJ (2015) Whole-genome analysis of diversity and SNP-major gene association in peach germplasm. PLoS One 10:1–19

Mitra SK (2018) Climate change: impact, and mitigation strategies for tropical and subtropical fruits. Acta Hortic 1216:1–9

Moretti CL, Mattose LM, Cal Bo AG, Sargent SA (2010) Climate changes and potential impacts on postharvest quality of fruit and vegetable crops: a review. Food Res Int 43:1824–1832

Morinaga K (2016) Impact of climate change on horticulture industry and technological countermeasures in Japan. Food and Fertilizer Technology Center, Taipei

Mullins MG, Bouquet A, Williams LE (1992) Biology of the grapevine. Cambridge University Press, Cambridge

Nakajima I, Ban Y, Azuma A, Onoue N, Moriguchi T, Yamamoto T, Toki S, Endo M (2017) CRISPR/Cas9-mediated targeted mutagenesis in grape. PLoS One 12:e0177966

Nath V, Kumar G, Pandey SD, Pandey S (2019) Impact of climate change on tropical fruit production systems and its mitigation strategies. In: Climate change and agriculture in India: impact and adaptation. Springer International, Cham

Osman MB, Milan AR (2006) Mangosteen (*Garcinia mangostana* L.). DFID, FRP, CUC, World Agroforestry Centre and IPGRI

Oukabli A, Mahhou A (2007) Dormancy in sweet cherry (*Prunus avium* L.) under Mediterranean climatic conditions. Biotechnol Agron Soc Environ 11(2):133–139

Patil SS, Kelkar TS, Bhalerao SA (2013) Mulching: a soil and water conservation practice. Res J Agric For Sci 1(3):26–29

Paull RE, Duarte O (2011) Introduction. In: Paull RE, Duarte O (eds) Tropical fruits, 2nd edn. CAB International, London, pp 1–10

Pereira A (2016) Plant abiotic stress challenges from the changing environment. Front Plant Sci 7:1123. https://www.frontiersin.org/article/10.3389/fpls.2016.01123

Prudencio AS, Werner O, Martínez-García PJ, Dicenta F, Ros RM, Martínez-Gómez P (2018) DNA methylation analysis of dormancy release in almond using epi-genotyping by sequencing. Int J Mol Sci 19:3542

Puchta H, Fauser F (2014) Synthetic nucleases for genome engineering in plants: prospects for a bright future. Plant J 78:727–741

Rajan S (2008) Impact assessment of climate change in mango and guava research. Malhotra Publishing, New Delhi, pp 54–60

Rajan S (2012) In: Sthapit BR, Ramanatha Rao V, Sthapit SR (eds) Phenological responses to temperature and rainfall: a case study of mango, tropical fruit tree species and climate change. Bioversity International, New Delhi

Ramirez J, Jarvis A, Van den Bergh I, Staver C, Turner D (2011) Changing climates: effects on growing conditions for banana and plantain (Musa spp.) and possible responses. In: Yadav SS, Redden RJ, Hatfield JL, Lotze-Campen H, Hall AE (eds) Crop adaptation to climate change, 1st edn. Wiley, Chichester, UK, pp. 426–438

Ranjitkar S, Sujakhu NM, Merz J, Kindt R, Xu J, Matin MA, Ali M, Zomer RJ (2016) Suitability analysis and projected climate change impact on banana and coffee production zones in Nepal. PLoS Onc 11:c0163916

Royo C, Torres-Pérez R, Mauri N, Diestro N, Cabezas JA, Marchal C, Lacombe T, Ibáñez J, Tornel M, Carreño J, Martínez-Zapater JM, Carbonell-Bejerano P (2018) The major origin of seedless grapes is associated with a missense mutation in the MADS-box gene VviAGL11. Plant Physiol 177:1234–1253

Satisha J, Prakash GS (2006) The influence of water and gas exchange parameters on grafted grapevines under conditions of moisture stress. S Afr J Enol Vitic 27(1):40–45

Sdoodee S, Lerslerwong L, Rugkong A (2010) Effects of climatic condition on off-season mangosteen production in Phatthalung Province. Department of Plant Science, Prince of Songkla University, Songkhla

Sharma RR (2002) Growing strawberries. International Book Distributing, Lucknow, pp 1–99

Shirasawa K, Isuzugawa K, Ikenaga M, Saito Y, Yamamoto T, Hirakawa H, Isobe S (2017) The genome sequence of sweet cherry (*Prunus avium*) for use in genomics-assisted breeding. DNA Res 24:499–508

Spayd SE, Tarara JM, Mee DL, Ferguson JC (2002) Separation of sunlight and temperature effects on the composition of *Vitis vinifera* cv. Merlot berries. Am J Enol Vitic 53:171–182

Stock M, Gerstengarbe F, Kartschall T, Werner P (2005) Reliability of climate change impact assessments for viticulture. In: Williams LE (ed) Proceedings of the VII international symposium on grapevine physiology and biotechnology, Acta horticulturae, vol 689. International Society for Horticultural Science, Leuven, pp 29–39

Sugiura T (2019) Three climate change adaptation strategies for fruit production, pp 277–292. https://www.naro.affrc.go.jp/english/laboratory/niaes/files/fftc-marco_book2019_277.pdf

Tandonnet JP, Marguerit E, Cookson SJ, Ollat N (2018) Genetic architecture of aerial and root traits in field-grown grafted grapevines is largely independent. Theor Appl Genet 131:903–915

Tarara JM, Lee J, Spayd SE, Scagel CF (2008) Berry temperature and solar radiation alter acylation, proportion, and concentration of anthocyanin in merlot grapes. Am J Enol Vitic 59:235–247

Tindall HD (1994) Rambutan cultivation. Food and Agriculture Organization of the United Nations, Rome. ISBN 9789251033258

Thomas MR, Cain P, Scott NS (1994) DNA typing of grapevines: a universal methodology and database for describing cultivars and evaluating genetic relatedness. Plant Mol Biol 25:939–949

Turner DW, Fortescue JA, Tomas DS (2007) Environmental physiology of the bananas (*Musa* spp.). Braz J Plant Physiol 19:463–484

Tyagi S, Ahmad M, Sahay S, Nanher AH, Nandan B (2015) Strawberry: a potential cash crop in India. Rashtriya Krishi 10(2):57–59

UDPSM (2002) Plant propagation and planting in uplands. Training manual for municipal extension staff. Upland Development Program in Southern Mindanaao, October 2002, Davao, Philippines

Uga Y, Sugimoto K, Ogawa S, Rane J, Ishitani M, Hara N, Kitomi Y, Inukai Y, Ono K, Kanno N, Inoue H, Takehisa H, Motoyama R, Nagamura Y, Wu J, Matsumoto T, Takai T, Okuno K, Yano M (2013) Control of root system architecture by DEEPER ROOTING 1 increases rice yield under drought conditions. Nat Genet 45:1097–1102

Van Asten PJA, Fermont AM, Taulya G (2011) Drought is a major yield loss factor for rainfed East African highland banana. Agric Water Manag 98:541–552

Varma V, Bebber DP (2019) Climate change impacts on banana yields around the world. Nat Clim Chang 9(10):752–757. https://doi.org/10.1038/s41558-019-0559-9

Verde I, Abbott AG, Scalabrin S, Jung S, Shu S, Marroni F (2013) The high-quality draft of peach (*Prunus persica*) identifies unique patterns of genetic diversity, domestication and genome evolution. Nat Genet 45:487–494

Verheij EWM, Coronel RE (1992) Edible fruits and nuts: plant resources of South-East Asia, vol 2. ETI, Amsterdam, pp 128–131

Verma MK (2014a) Walnut production technology. In: Training manual on teaching of post-graduate courses in horticulture (fruit science). Post Graduate School, Indian Agricultural Research Institute, New Delhi

Verma MK (2014b) Apple production technology. In: Training manual on teaching of post-graduate courses in horticulture (fruit science). Post Graduate School, Indian Agricultural Research Institute, New Delhi

Verma MK (2014c) Pear production technology. In: Training manual on teaching of post-graduate courses in horticulture (fruit science). Post Graduate School, Indian Agricultural Research Institute, New Delhi

Wang X, Tu M, Wang D, Liu J, Li Z, Wang Y, Wang X (2018) CRISPR/Cas9-mediated efficient targeted mutagenesis in grape in the first generation. Plant Biotechnol J 16:844–855

Chapter 5
Evaluating Adaptation Strategies to Coastal Multihazards in Sundarban Biosphere Reserve, India, Using Composite Adaptation Index: A Household-Level Analysis

Mehebub Sahana, Sufia Rehman, Shyamal Dutta, Samsad Parween, Raihan Ahmed, and Haroon Sajjad

Abstract Coastal areas are experiencing greater risk because of the increased frequency of multiple climate change-induced hazards. Thus, identification of suitable adaptation strategies for local communities has become imperative for lessening the impact of climate change. Although a large number of studies have touched upon the different components of vulnerability to coastal hazards, adaptation assessment has remained underemphasized in the existing literature. This paper assesses adaptation strategies to multiple coastal hazards in Sundarban Biosphere Reserve (SBR), India. We collected socioeconomic data and responses on adaptation strategies opted for extreme climate events, economic activities, and coping mechanism from 570 households in the Reserve. Factor analysis was utilized to identify the significant adaptation strategies opted by the sampled households. This analysis provided 14 significant adaptation strategies, explaining around 66% of the total variation among 45 adaptation strategies pursued by the households. Correlation was performed to establish the relationship between socioeconomic characteristics of the households and adaptation strategies. A composite adaptation index was then constructed using principal component analysis (PCA) to ascertain the level of adaptation. Findings revealed very high adaptation in Kakdwip, Kultali, Sagar, Patharpratima, Namkhana, and Gosaba blocks and a high degree of adaptation was

M. Sahana
School of Environment, Education & Development, University of Manchester, Manchester, UK

S. Rehman · R. Ahmed · H. Sajjad (✉)
Department of Geography, Faculty of Natural Sciences, Jamia Millia Islamia, New Delhi, India

S. Dutta
Department of Geography, University of Burdwan, Bardhaman, West Bengal, India

S. Parween
Department of Geography, Aligarh Muslim University, Aligarh, Uttar Pradesh, India

M. N. Islam, A. van Amstel (eds.), *India: Climate Change Impacts, Mitigation and Adaptation in Developing Countries*, Springer Climate,
https://doi.org/10.1007/978-3-030-67865-4_5

observed in Hingalganj and Basanti blocks of the SBR. These blocks are located along the coast and are frequently impacted by coastal hazards. The blocks connected to the mainland are less affected by the disasters and have moderate or low levels of adaptation. The adaptive measures should target group-oriented prioritizing of the vulnerable areas and communities. Livelihood diversification and provision of basic amenities and facilities may lessen vulnerability in the SBR. The principal component-based adaptation index analysis helped in understanding the relationship between climatic adaptation strategies and the socioeconomic condition of the coastal community.

Keywords Adaptation strategies · Composite adaptation index · Coastal multihazards · Sundarban Biosphere Reserve

Introduction

Climate change-induced coastal hazards have been found to be increasing globally, leading to vulnerability of coastal communities. The vulnerability of these communities is modulated by climate perturbations, variability in rainfall and temperature, rising sea level, and anthropogenic pressure on coastal ecosystems. Frequent striking climate change-induced cyclones, storm surge, and flood are pushing the coastal areas toward a bleak future. The coastal areas are fragile and susceptible to salinity intrusion, erosion, and fluctuations in rainfall and temperature (Priya Rajan et al. 2019; Rehman et al. 2020). Further, the community and its assets are becoming susceptible because of the increasing coastal hazards and increasing population and urbanization (IPCC; Wong et al. 2014). Rising sea level is expected to increase subsidence, coastal floods, damage to protecting structures, and, in extreme cases, coastal erosion (Koroglu et al. 2019). The World Economic Forum (2019) identified two major threats to humanity: climate change-induced disasters and inability to adapt to those disastrous events. In 2019, economic losses from weather-related disasters globally were accounted to be about USD 229 billion. Among these disasters, flood ranked first, followed by cyclones. Cyclones have caused huge devastation in Japan, the United States, and China, leading to combining economic losses of USD 10 billion. Monsoon floods in India also caused huge economic losses of about USD 10 billion, and cyclone Fani caused losses of about USD 8 billion in coastal areas of India and Bangladesh (Aon 2020). Sea level rise is expected to increase by 2100, depending on the magnitude of greenhouse gases (GHGs) emissions. Thus, decreasing sea level may allow more adaptation to the communities living in coastal areas and deltas (IPCC 2018a).

Adaptation refers to foreseeing the implications of climate change and taking appropriate actions to lessen the damages; adaptive capacity is defined as the capability of an individual or system to adapt to the implications of climate change. IPCC (2018b) defined adaptation as "*the process of adjustment to actual or expected climate and its effects, in order to moderate harm or exploit beneficial opportunities.*

In natural systems, the process of adjustment to actual climate and its effects; human intervention may facilitate adjustment to expected climate and its effects." Adaptation constitutes an essential part of policy formulation for reducing vulnerability to climate change. Although various activities and programs on adaptation have been carried out, turning these processes into action is slow (Swart et al. 2014). The Paris Agreement on climate change stressed its causes and consequences. The coastal areas globally are under severe risks from many coastal developmental processes, pollution, overfishing, loss of habitat, and other anthropogenic activities (USAID 2009). The Paris Agreement calls for reducing the emission of GHGs and investing in adaptation (UNFCCC 2015). Adaptation may guide humans toward resilience pathways, especially among the vulnerable coastal communities. Increase in GHG emissions and alteration in rainfall and temperature patterns may have implications for the coastal areas, causing ocean acidification, erosion, sea level rise, storm surges, and coastal flooding. Hence, effective adaptation is required to lessen the impacts of these changes and hazards among the coastal communities.

Various attempts have been made by scholars to assess adaptation to climate change in deltaic ecosystems (Maru et al. 2014; Houghton et al. 2017; de Campos et al. 2020). Ahammed and Pandey (2019) examined vulnerability to cyclones in the eastern coastal areas of Andhra Pradesh, India, using a coastal social vulnerability index. These authors suggested that improvement in early warning systems and coastal infrastructure may help reduce vulnerability in the coastal communities. Mallick et al. (2009) examined individual level adaptation strategies in coastal Bangladesh, using several research methods to analyze household level adaptation in surveyed villages. Ahamed (2013) examined the adaptation strategies in coastal Bangladesh using a community-based approach. His findings revealed that lack of financial resources and physical location are the major factors of vulnerability in this area. Onyeneke et al. (2020) analyzed the adaptation strategies for fish farmers in the Niger Delta using instrumental variable regression. Their findings indicated that increased adaptation among farmers has also increased the returns from fish farming. Baills et al. (2020) reviewed various adaptation measures in coastal areas using multicriteria analysis and suggested that indicators should be independent of area for easy comparison. It can be seen that studies have highlighted the importance of adaptation for lessening vulnerability and increasing livelihood opportunities for the coastal communities. Improvement of early warning systems, infrastructural development, strengthening local government institutions, flexible fisheries management, and an integrative approach have been identified as some efficacious adaptive measures for coastal communities (Gopalakrishnan et al. 2018; Ahammed and Pandey 2019; Greenan et al. 2019; Nichols et al. 2019). Early warning systems aim at reducing vulnerability to disasters at varied spatial scales. An effective early warning system should be timely, continuous, flexible, and transparent, integrating the community and constituting adequate human capacity (Kelman and Glantz 2014). Lack of an early warning system was identified as an important factor causing vulnerability in coastal areas (Oo et al. 2018). Losses caused by cyclones can be reduced through early warning systems in deltas (DECCMA 2017; Kuenzer and Renaud 2012). Ogundele and Ubaekwe (2019) advocated that ecosystem-based adaptation and early warning systems may help reduce flood vulnerability in Nigeria.

The Indian coastline, stretching over 7516 km, is the home of nearly 500 million people. This coastal stretch is dynamic and varies in physiographical characteristics, which increases its vulnerability to disasters (Sudha Rani et al. 2015; Sahana et al. 2019b; Rehman et al. 2020). Tropical cyclones are destructive in Indian coastal areas. The cyclones Idai, Fani, and Bulbul collectively led to economic losses of more than USD 14 billion (Aon 2020). The Indian Sundarban Biosphere Reserve (SBR), being the world's largest mangrove ecosystem, is fragile to varied impacts by climatic stressors. This region has evolved by three major river systems, the Ganga–Brahmaputra–Meghna (GBM) Rivers. The vulnerability of this deltaic ecosystem is high because of large population concentrations, low gradient, the complex river network, and supporting many species of fauna and flora (Sahana and Sajjad 2019). Most of the people are engaged in agriculture and fishing activities, which are adversely affected by cyclone, flood, storm surge, and sea level rise (Sahana et al. 2020a). Climate variability has also magnified the frequency of extreme weather events in the SBR, and thus effective adaptation strategies are urgently required for lessening the vulnerability of the SBR coastal communities (Sahana et al. 2019a, 2020b). Although a large volume of research has focused on vulnerability to climate change-induced disasters in the Reserve, no comprehensive disaster adaptation framework has appeared. This chapter aims to seek better insight for adaptation strategies at the household level in the Reserve.

Methods and Materials

Study Area

The Sundarban Biosphere Reserve (SBR) is located in the lower part of the Ganga–Brahmaputra–Meghna delta. It spreads over an area of 9630 km^2 and is bounded by the Dampier Hodges line in the north, the Bay of Bengal in the south, the Hooghly estuary in the west, and the Ichamati–Raimangal River in the east. It covers the South and North 24 Parganas districts of West Bengal extending over 21°31′N and 22°30′N latitude and 88°10′E and 89°51′E longitude (Fig. 5.1). Of the total islands in the Reserve, 48 islands are uninhabited, covered by mangrove forest, and 54 islands are inhabited (Fig. 5.1). Most of the Reserve consists of low-lying alluvial mudflats, tidal creeks, and multiple river channels (Sahana et al. 2016). The Gangetic deltaic region was formed in the Late Quaternary Period (2500–5000 years ago) by deposition of sediments brought by the Ganga, Brahmaputra, and Meghna Rivers (Allison et al. 2003). The SBR can be divided into a core zone (1692 km^2), a buffer zone (2233 km^2), and a transition zone (5705 km^2). The Reserve was badly hit by cyclone Aila during 2009, a cyclone identified as one of the most dramatic climatic disasters that occurred in Indian Sundarban (Danda 2010). The islands located in the extreme southern part of Sundarban are affected by sea level rise and storm surge

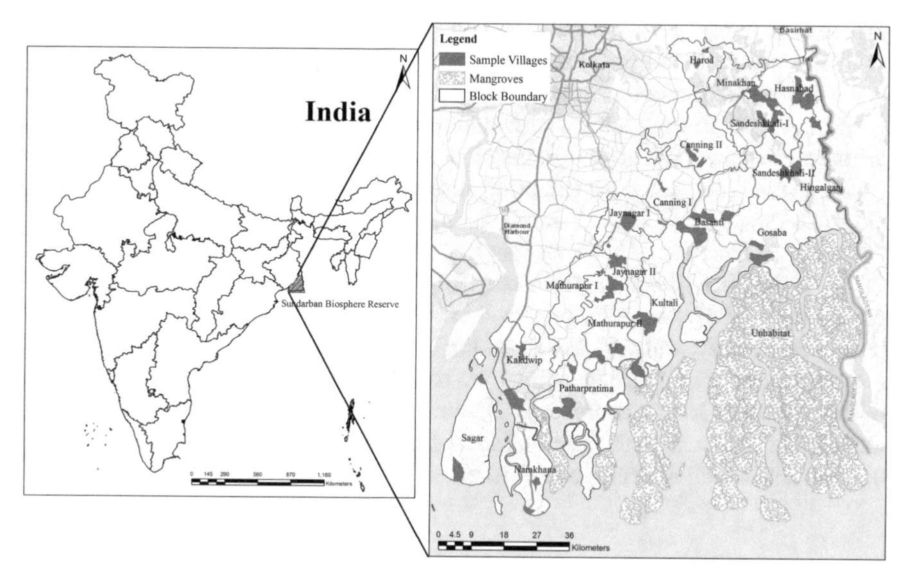

Fig. 5.1 Location of the study area: (**a**) location of Bengal delta in the India subcontinent; (**b**) location of Sundarban in West Bengal; (**c**) Sundarban Biosphere Reserve (SBR) and the sampled villages

(Hazra et al. 2002). An area of nearly 97 km^2 of these islands has been lost to flood, coastal erosion, and submergence (Danda 2010).

The Bengal delta has witnessed extreme tropical cyclones quite often, damaging livelihoods and mangrove plants (Sahana et al. 2015). The cyclones have a tendency to return after about 2 years or less. This delta is one of the most storm surge-prone areas. Nearly 26 of the world's 35 total deadliest cyclones have occurred in the Bay of Bengal. In the past two centuries, about 27% of world's tropical cyclone-associated deaths in India were recorded here. Cyclonic storms have caused a large number of deaths in this densely populated delta and rendered several thousand people homeless. The deadliest storm in world history, the Bhola Cyclone of 1970, killed around 300,000–500,000 people, caused by landfall in Bangladesh and high storm surge (estimated at up to 10.4 m or 34 ft) in the coastal areas. The 1991 Bangladesh cyclone was one of the most powerful tropical cyclones ever recorded in the basin. This tropical cyclone caused a 6.1 m (20 ft) storm surge, which inundated the coastline, causing at least 138,866 deaths. The Odisha super cyclone of 1999 that made landfall near Paradip had a wind speed of 260–270 km/h in the core area, which produced a huge storm surge and took away nearly 10,000 lives. Tropical cyclone Sidr (2007), which made landfall in Bangladesh, caused a storm surge as great as 5 m across Sundarban and killed nearly 4235 persons. Tropical Storm Rashmi (2008) swept through Bangladesh leaving behind many downed trees, damaged crops and houses, and power outages in several districts (Sahana et al. 2019a). The storm surge height of cyclone Aila (2009) was 2–3 m above tide levels, experienced along the coasts of West Bengal and the adjoining areas of Bangladesh. In 2015 cyclone Komen was accompanied by a storm surge of 1–2 m (3.3–6.6 ft) that affected the Sundarban in Bangladesh. In 2019 two important tropical cyclones occurred in the Bengal delta. The Bulbul cyclone (2019) was about 1–2 m above astronomical tide levels and inundated low-lying areas of South and North 24 Parganas. The maximum extent of inundation was about 2 km over South and North 24 Parganas. Cyclone Fani, which hit Odisha in May 2019, killed more than 70 people in Odisha and West Bengal (Sahana et al. 2021).

The Sundarban Biosphere Reserve (SBR) has a population of 4.37 million, with a population density of 975 persons/km^2 (Census of India 2011). Inadequate resources, poverty, and remoteness are major challenges for the Sundarban community. Nearly 43.5% of the population lives below the poverty line. Most of the islands of the SBR have low infrastructural facilities and inadequate amenities. The people are highly dependent on the forests. Most of the population resides in rural areas and depends on coastal agriculture and fisheries.

Preparation of Database

The data for assessing adaptation capacity and strategies for coping with multiple hazards were drawn from a comprehensive survey of 570 households in the study area during 2018–2019. The sample design adopted for the study had two stages. The first stage consisted of selection of the villages from SBR. From every block (the study area spreads over 19 blocks), 2 villages (1 from near to the coast/riverside and 1 away from the coast/riverside) were selected, so that 38 villages were selected. The second stage consisted of selecting households from these 38 villages. From every village, 15 households were selected randomly so that 570 households were selected from the study area. The respondents for the household survey were the heads of the households. Discussion was also held with the heads of the *Panchayat* (the local parliament). A questionnaire was designed to collect the relevant information regarding adaptation strategies. Sufficient care was taken to make the questionnaire communicable to the respondents. Responses on adaptation strategies opted for extreme climate events and economic activities were gathered from the sampled households. Responses on coping mechanisms adopted during disasters were also collected from these households. All the collected responses on adaptation strategies were authenticated by the *Gram Pradhan* (the Head of the village).

Analytical Framework for Adaptation Strategies

Understanding responses in anticipation of disasters is essential to lessen the degree of vulnerability. It is imperative to judge the responses of the individual households regarding their strategies to cope with the situation in any hazard-prone region to fully understand the situation of the whole region and identify lacunas for a proper management plan in general. The adaptation strategies vary for different natural hazards as well as with diversified economic activities. The effectiveness of these strategies also varies from place to place and time to time. Three important aspects were kept in mind while analyzing the adaptation strategies in SBR: type of hazard, level of social structure, and emergency response in case of disasters. The present study recorded the various adaptation strategies opted by sampled households for extreme weather events (cyclone, flood hazard, salinity intrusion, bank erosion), changes in economic activity (agriculture, fishing/shrimp cultivation), and emergency response during disasters. The detailed methodological framework is presented in Fig. 5.2.

Adaptation Strategies for Climate Extreme Events

The household responses to extreme weather events such as cyclones, floods, salinity intrusions, and river bank erosion were measured using binary-type (yes/no) responses. For each hazard event several questions were asked to respondents for analyzing the adaptation strategies to these hazardous events. Responses on no adaptation option, shifted to cyclone center, stayed at home, migrated to other places, prepared cyclone-protected homes, and make community action team adaptation strategies were examined for cyclones. To measure the responses against floods at household level, seven types of questions such as stored food, migrated to safe place, migrated to other village, link with other communities, maintain and improve the river bank, make community action team, and make home with higher foundation were examined. The responses on salinity were measured through asking five questions to the household, including do nothing, rainwater harvesting, conservation of pond water, use of pond sand filter, and digging of ponds. Five questions including no adaptation opted, sea wall/earthen embankment, soil bunds, waterways, stone bunds, grass strips, and mangroves replanting were asked to obtain responses on the adaptations against riverbank erosion in the study area.

Adaptation Measures for Changes in Economic Activities

Changes in two important economic activities, namely, agriculture and fish/shrimp farming practices, were analyzed among the sampled households. Adaptation in agricultural activities were addressed through varied responses, such as no adaptation strategy, crop diversification, change in cropping calendar, adaptation of new farming techniques, and transformation of land for fishery adaptation strategies, among the sampled households. Fish/shrimp farming is one of the important economic activities after agriculture. Adaption strategies against the changes occurring in this sector were assessed by the binary-type responses as no responses, increased embankment height, digging pond inside fish farms, liming, use of medicine, and placing net around shrimp field as adaptation strategies at the household level.

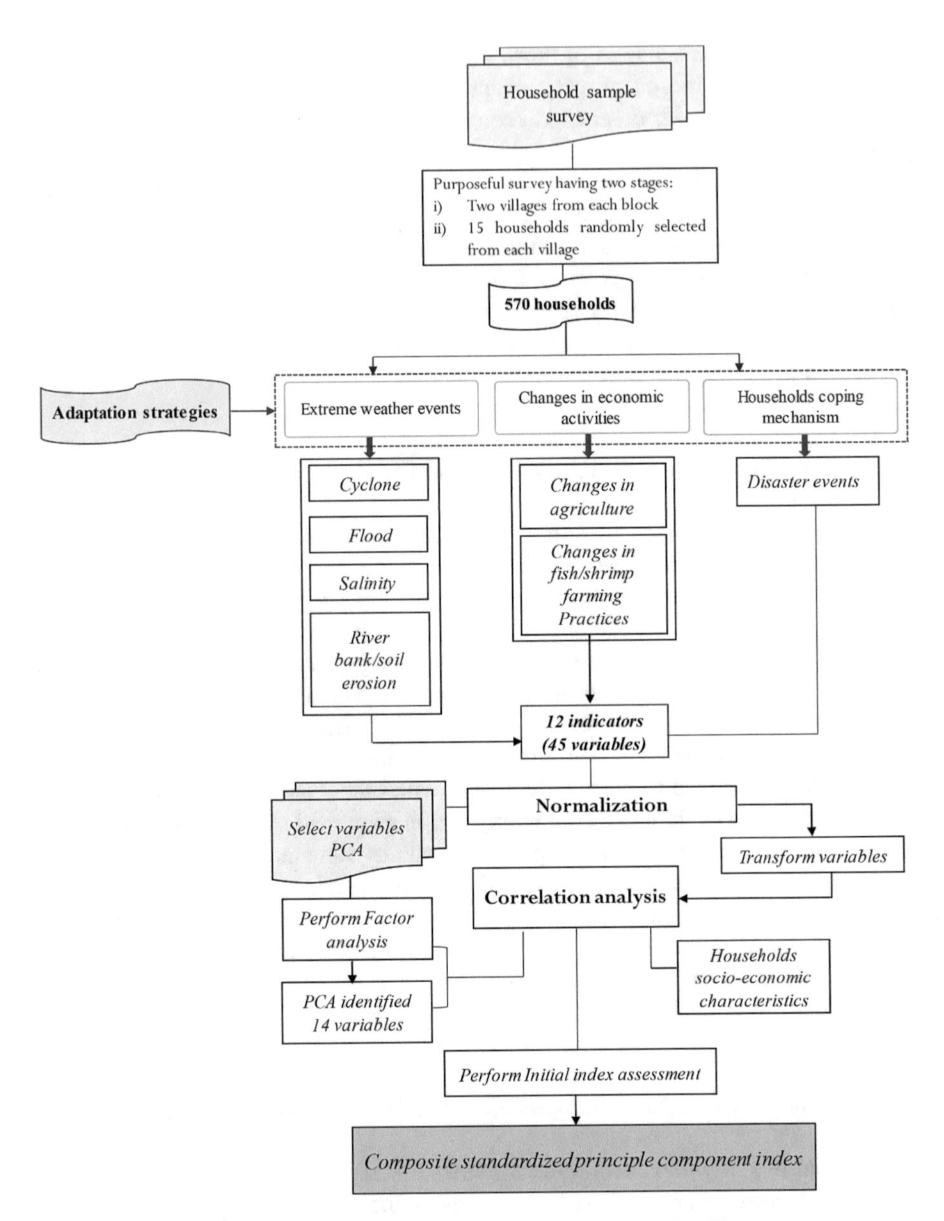

Fig. 5.2 Analytical framework for composite adaptation strategies in the SBR

Household Coping Mechanisms

Household responses on their coping mechanisms during disasters were recorded on the basis of 12 variables with their binary-type responses: these included personal savings, borrowing money from relatives, financial assistance from governmental sector, borrowing money from (NGOs), relief from NGOs, engaging children to work, borrowing money from bank, mortgage assets, stopping schooling of children, reducing expenditures on food and other consumption, selling household assets, and migrating to another place.

Composite Adaptation Strategies of SBR Villages

Principal component analysis (PCA) was used to develop a composite adaptation index. The PCA transforms numerous variables into a concise and coherent set of uncorrelated factors called principal components (Hotelling 1933). The principal components of each variable account for much of the variance among the series of actual variables. Each component is a linear weighted combination of the initial variables. Each component is structured in a way that the first component signifies the largest possible amount of variation in the actual variables. This method was first used by Boelhouwer and Stoop (1999) for integrating various indicators into a single index. It is noteworthy that the UNDP Human Development Index was also modified by PCA, replacing the simple aggregation procedures (Lai 2003). It is a multivariate statistical technique used as a data reductionist method that transforms the actual set of variables into a lesser number of linear varieties by illuminating the underlying factors (Kaźmierczak and Cavan 2011; Solangaarachchi et al. 2012). PCA has been effectively used by scholars, statisticians, and planners to prepare socioeconomic indices during the past few decades (Tata and Schultz 1988; Fotso and Kuate-defo 2005; Rygel et al. 2006; Antony and Rao 2007; Havard et al. 2008; Messer et al. 2008; Krishnan 2010).

Before applying the PCA framework, all the variables were tested and authenticated by qualitative interviews with village *Panchayat* (representatives elected by the village assembly) and the *Gram Sabha* (general assembly of village people) because variation was seen in the data collected from the households. PCA as a reductionist technique permitted a strong as well as reliable set of variables for assessing adaptation strategies in the SBR. A composite adaptation index was established by using 45 different adaptation strategies opted by the Sundarban community for extreme climate events (cyclone, flood, etc.), changes in economic activities (agriculture and fish collecting), and household coping mechanisms. The PCA was performed to resolve the number of factors to bc rctaincd for further analysis. Correlation analysis and screen plot were performed using these 45 parameters. Finally, factor analysis identified the strategies most often adapted from the 45 factors by the sampled households. After applying the factor analysis, 14 important factors were identified. The rank of these responses was then normalized and correlated with household socioeconomic characteristics. PCA constructed a composite adaptation strategies index for the sampled villages. Prinscore was calculated by using following formula:

$$\text{Pinscore} = \{\left(V1 \times v1 + V2 \times v2 + V3 \times v3 + V4 \times v5 + \ldots v35\sqrt{\text{Eigenvalue}}\right.$$

where $V1$, $V2 \ldots$ $V35$ = actual values of variables up to '35' number; $v1$, $v2 \ldots v35$ = loadings from the first principal component of each variable (Eigen vectors).

The level of adaptation strategy was determined using average standardized values of the composite factor of each household. The composite adaptation index was divided into low, moderate, high, and very high using a natural break classification scheme.

Results and Discussion

Household Socioeconomic Characteristics

Most of the sampled households in SBR were headed by males (80%) with an average age of 45 years. Analysis of social characteristics revealed that most households were practicing Hinduism (92%), followed by Islam (8%). The average family size of the sampled households was five members. Wide income inequalities were found among the respondents (0.3); the average monthly income of the households was calculated to be about 5466.8 rupees (US $72.29). Analysis of socioeconomic characteristics revealed that the majority of the respondents (39%) was living below the poverty line and only 31% of the sampled households had their own agricultural land (Table 5.1). Most of the sampled households had electricity, but only 37% of these households had sanitation facilities on their premises.

Table 5.1 Descriptive statistics of the sampled households in the SBR

Respondent (household) characteristic	Subcategory	Mean, range, or percent	SD
Average age of respondent (min–max)		44.72 (21–84)	14.81
Male respondent (%)		80.4	
Respondent religion (%)	Hindu	92.1	
	Muslim	7.9	
Average family size (min–max)		4.75 (2–13)	1.71
Average household income per month (rupees)		5466.84	3789.8
Households under below poverty line (BPL) (%)		39.2%	
Income inequality (Gini coefficient)		0.297	
Households owning agricultural land (%)		30.9	
Households with sanitation facility (%)		36.7	
Households with electricity connection (%)		68.4	

Source: Authors' own calculations based on field survey 2018–2019

Table 5.2 Socioeconomic characteristics of the sampled households in the SBR

Characteristics	Subcategory	Percentage
Social group	General	46.8
	OBC	13.2
	SC and ST	40.0
Occupation	Daily worker	36.2
	Fishing	15.3
	Cultivator	16.1
	Job and business	4.9
	Nonworker	27.5
Education	Illiterate	37.9
	Primary (1–5 standard)	25.0
	Secondary (6–10 standard)	30.4
	Higher secondary (11–12 standard)	6.7
Property	Own	70.5
	Khas	14.2
	Government land	7.2
	Road/riverside	8.1
Household size	0–3	21.1
	4–5	51.9
	6–8	21.6
	>8	5.4
Land	Landless	62.8
	Cultivator	31.8
	Land on lease	4.9
Types of house	Muddy	63.7
	Semi-cemented	6.5
	Fully cemented	11.2
	Grass/bamboo	13.2
	Other	5.4
Source of drinking water	Tube well/borehole	64.7
	Hand pump	0.5
	Pond/river	12.5
	Near house	22.3
Distance to source	100–500 m	57.9
	500–1000 m	24.6
	More than 1000 m	17.5

Source: Compiled from the household survey conducted in 2018–2019 for 570 total surveyed households with average income Indian rupees 1122

Data regarding socioeconomic characteristics were collected to relate with the adaptation strategies practiced by the sampled households. Table 5.2 shows socio-economic statistics of the sampled households.

Most of the households belonged to the backward and scheduled castes. Of the total workers in the sampled households, more than one third listed their occupation as daily wage work. A high proportion of the population (27%) was dependent, including children and elderly people. Agriculture and fishing were found to be the major occupations of the people. Most of the respondents in the sampled households were illiterate. An overwhelming majority of the respondents had their own house (70%). The remaining were found to live on *khas* (land provided by the government), vacant government land, and along a river/roadside. Most of the respondents had a large family size, consisting of five members. Most of the respondents were landless. Only 31% of respondents engaged in practicing agriculture. The condition of the structure of the houses of the sampled respondents was not very encouraging, as most were living in mud houses that are vulnerable to disasters. Nearly one third of the respondents had difficulty accessing drinking water. Most of the respondents used a tube well or borehole well for obtaining water.

Adaptation Strategies for Extreme Climate Events

The effectiveness of adaptation strategies varies at spatial and temporal scales. Adaptation strategies were found to vary with the type of disaster in the SBR. Adaptation strategies for extreme weather events, namely, cyclones, floods, salinity intrusion, and bank erosion, are analyzed in this section. The analyses revealed that most of the respondents (42%) prefer to stay at home during cyclone events, followed by migrating to a safer place (18%) and shifting to cyclone centers (15%). Nearly 15% of respondents did not have a preferred strategy during cyclones. However, nearly 8% of respondents formed a community action team whereas only 2% of the sampled households in the study area opted to prepare cyclone protections for their home (Fig. 5.3a). Storing food (28%) was identified as the most preferred strategy during a flood, followed by migrating to a safer place (22%), maintaining

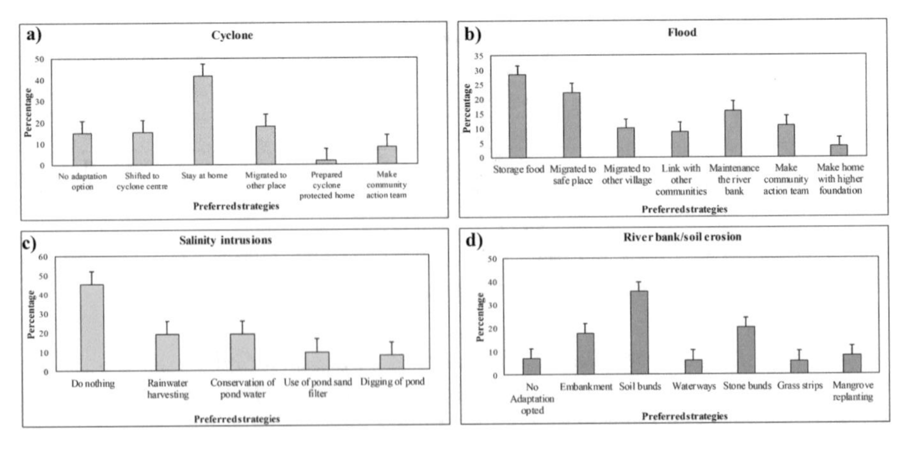

Fig. 5.3 Adaptation strategies for climate extreme events: (**a**) cyclone, (**b**) flood, (**c**) salinity intrusion, and (**d**) river bank/soil erosion

the river bank (16%), making a community action team (11%), migrating to other villages (9%), and constructing a home with a higher foundation (4%). It is important to notice that most of the sampled households preferred to store food and move to safer places during the flood period (Fig. 5.3b). In salinity intrusions, most of the sampled households (45%) did not have any adaptation strategy. Nearly 19% of the sampled households opted for rainwater harvesting, whereas conservation of pond water was carried out by 19% of the respondents. Around 10% of the respondents use a pond sand filter to reduce salinity intrusion and nearly 8% of the sampled households dig ponds (Fig. 5.3c).

Various adaptation strategies were reported by respondents for bank erosion. From Fig. 5.3d, it is observed that most of the respondents constructed soil bunds (35%) to reduce bank erosion, followed by construction of stone bunds (20%) or embankments (18%), or chose mangrove replantation (8%), no adaptation (7%), waterways (6%), and alignment of grass strips (6%).

The analysis of adaptation strategies for extreme weather events revealed that most of the sampled households did not choose any strategy during cyclones, salinity intrusions, and river bank erosion. It is also noteworthy that few respondents opted for disaster-resilient structures and embankments. Priority was given to staying at home, storing food, harvesting rainwater, and constructing soil bunds along the rivers and channels. Apart from these strategies, moving to safer places and migration to other villages were preferred by the sampled households. These strategies reflect the varying degree of vulnerability of the people to different weather events in the SBR and their responses to lessen the impacts on themselves.

Adaptation Measures of the Sampled Households for Changes in Economic Activities

Figure 5.4 illustrates the different adaptive measures taken by the sampled households in response to changes in agricultural and fish/shrimp activities. Nearly 41% of the sampled household did not use any strategy in case of changing agricultural

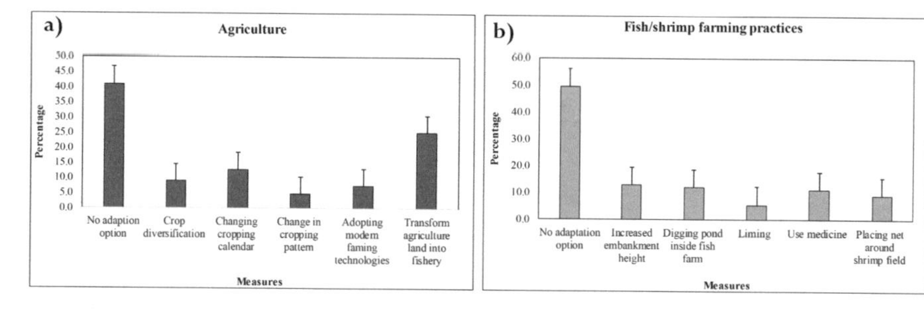

Fig. 5.4 Adaptation strategies for changes in economic activities: (**a**) agriculture and (**b**) fish/shrimp farming practices

activities. Most of the respondents (25%) transformed their agricultural fields into fish ponds for income generation. The other strategies adopted by respondents were changing the crop calendar (13%), crop diversification (9%), and adapting modern farm practices (7%). Nearly half the sampled respondents (50%) did not have any adaptation strategy for fish/shrimp practices, although nearly 13% of the sampled households increased the embankment height, followed by digging ponds inside fish farms (12%), using medicine (11%), placing nets around shrimp fields (9%), and used liming in the fish ponds (5%).

Moreover, analysis of adaptive measures showed that most of the respondents did not undertake any adaptation measures against the changes in these two economic activities, although a segregated pattern of adaptive measures was discerned among the other respondents. It is also important to observe that fishing was the second most preferred economic activity after agriculture. However, in agriculture very few respondents adapted modern farm practices to increase agricultural productivity.

Coping Mechanism of the Sampled Households

Analysis of household coping mechanisms revealed that nearly 21% of respondents borrowed money from their relatives to cope with losses that occurred from natural hazards (Fig. 5.5). Nearly 18% of respondents disclosed that they asked their children to work to supplement the family income. Apart from these measures, around 15% of respondents mortgaged assets to supplement their expenses, followed by selling their assets (14%), temporal migration (8%), receiving financial assistance from government (6%), stopping schooling of their children (6%), borrowing money from an NGO (5%), limiting their basic expenses (3%), and accepting relief from an NGO (2%). Only 2% of respondents were able to use their own saving to supplement

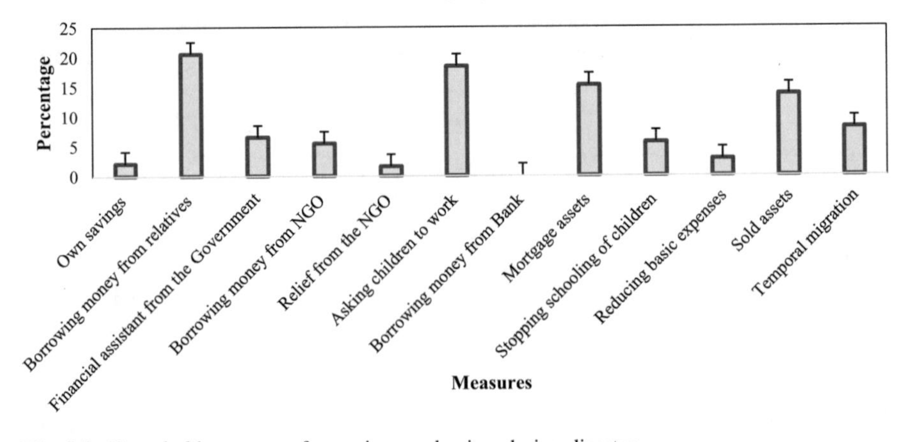

Fig. 5.5 Household responses for coping mechanism during disaster

Fig. 5.6 (**a**) Pond of sweet water in the Sundarban Reserve forest for wild animals. (**b**) Protection of shrimp field with net. (**c**) Agricultural land converted to pisciculture in Basanti block. (**d**) Solar panel on roof of a house in Kultali block. (Source: Field photographs by the author during January 2018)

their needs although none of the respondents borrowed money from a bank. Figure 5.6 illustrates the various adaptative measures taken by respondents in various blocks of the SBR. It is found that these measures were effective to some extent to reduce the vulnerability to disasters.

Screen Plot and PCA Varimax Rotation

Factor analysis was performed on 45 variables to identify the most influencing factors. The correlation matrix was used as an input to PCA to extract the factors. One of the most commonly used techniques is Kaiser's criterion, or the eigen value rule. Under this rule, only those factors with an eigen value (the variances extracted by the factors) of 1.0 or more are retained. After applying the PCA on the database, we obtained 14 important factors for adaptation strategies. We utilized a graphical method also known as the Catell (1966) scree test (Fig. 5.7); these are plots of each of the eigen values of the factors. One can inspect the plot to find the place where the smooth decrease of eigen values appears to level off. To the right of this point, only 'factorial scree' (meaning debris that collects on the lower part of a rocky slope) is found. After examining the scree plot, only 14 factors were extracted for analysis.

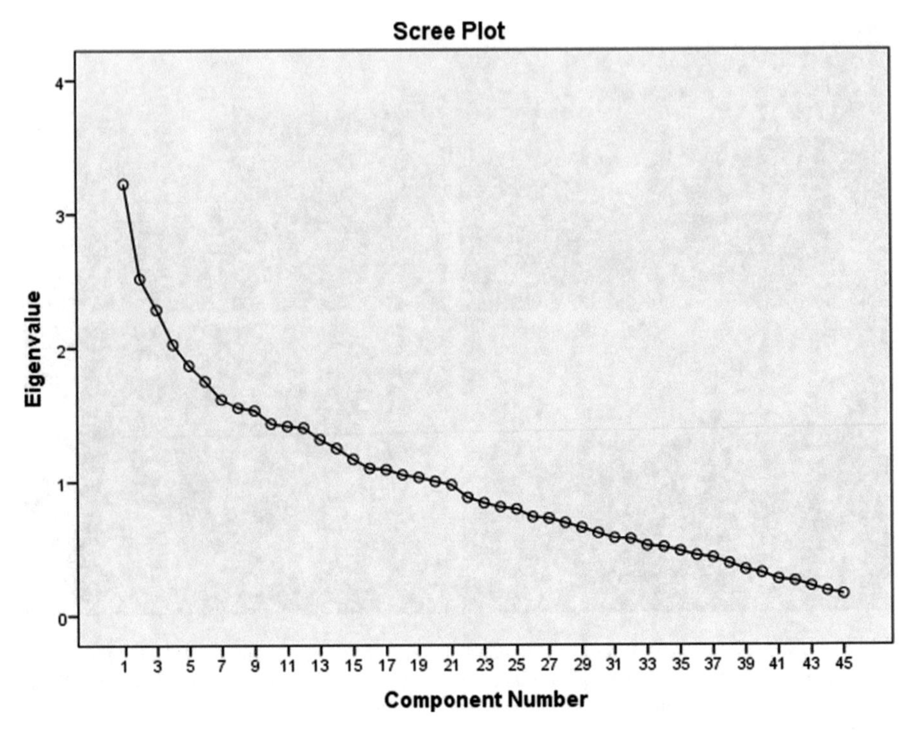

Fig. 5.7 Scree plot of the 45 variables selected for factor analysis

The screen plot showed the contribution of the variables used for the principal component analysis.

The results of PCA using varimax rotation are presented in Table 5.3. Of the total 45 variables, 14 factors accounted for 66.7% of the total variance. For the first factor, seawall/earthen embankment, increased embankment height, grass strips, mangrove replanting, and maintenance and improvement of the river bank showed markedly higher positive loadings and contributed more than 5% while asking their own children to work in a field or in any sector. Community action teams, higher home foundations, and stone bunds have the lowest contribution to the loading. These 14 factors were used for the final composite index of block level adaptation strategies.

Correlation Analysis Between Socioeconomic Conditions and the Adaptation Strategies

The socioeconomic characteristics of the sampled households, namely, social group, occupation, education, property, household size, land tenure, type of house, source of drinking water, and distance to source were correlated with the different

Table 5.3 PCA-identified factors and their variance

Sample no.	Factors	Initial eigen values	
		Total	Percent of variance
1	Seawall/earthen embankment	3.003	9.384
2	Increase embankment height	2.249	7.027
3	Grass strips and mangroves replanting	1.852	5.787
4	Maintenance and improving the river bank	1.735	5.423
5	Migration to safe place	1.579	4.935
6	Transformation of land for fishery	1.398	4.369
7	Adapting of new farming techniques	1.353	4.227
8	Digging pond inside fish farm	1.325	4.139
9	Placing net around shrimp	1.303	4.071
10	Asking one's own children to work in a field or in any sector	1.231	3.846
11	Make community action team	1.133	3.54
12	Make higher foundation for home	1.112	3.476
13	Stone bunds	1.058	3.306
14	Stop schooling of children	1.025	3.203
Total variance explained			66.732

Source: Authors' own calculations based on PCA analysis

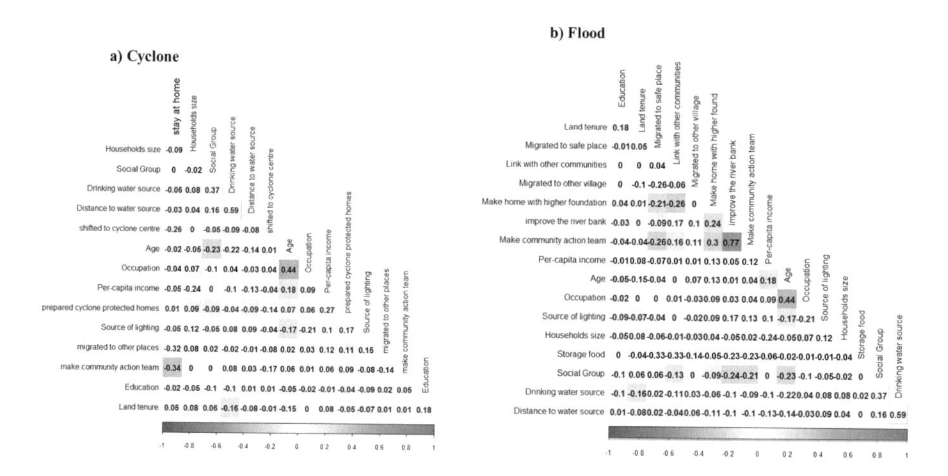

Fig. 5.8 Relationship between socioeconomic characteristics with extreme weather events: (**a**) cyclone and (**b**) flood

adaptation preferred by respondents for extreme weather events and changes in economic activities. The correlograms show the relationship between adaptation strategies and the socioeconomic characteristics of the household (Figs. 5.8, 5.9, and 5.10). Negative correlation is depicted by red and blue color shows the positive correlations. The legend color below the correlograms shows the correlation

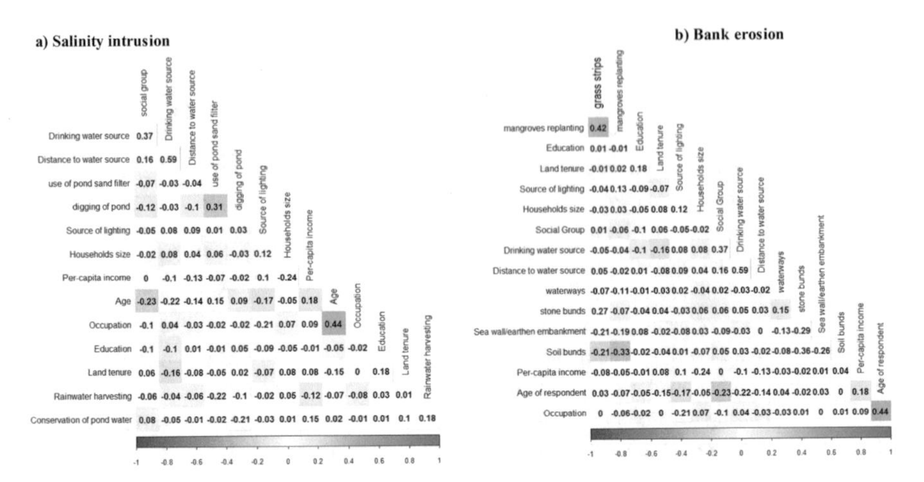

Fig. 5.9 Relationship between socioeconomic characteristics and extreme weather events: (**a**) salinity intrusion and (**b**) bank erosion

coefficient with the respective color scheme. In case of cyclone, the household size and average age are positively correlated with the decision of shifting to a cyclone center and adaptation strategies of preparing cyclone-protected homes. Social group, drinking water source, and distance to the drinking water source are negatively correlated with the decisions for a cyclonic storm (Fig. 5.8a).

Correlation analysis between the adaptation strategies opted for flood hazards and the socioeconomic conditions of the sampled households revealed that making a higher foundation for the home and improving the river bank are positively correlated with per capita income. The social group is also positively correlated with migration to a safer place during hazardous floods (Fig. 5.8b). In case of salinity intrusion, age, household size and per capita income were found positively correlated with the adaptation decision for salinization of agriculture land (Fig. 5.9a). Mangrove replantation is positively correlated with grass strips in the river bank (Fig. 5.9b). Adapting stone bunds were positively correlated with the household size and social group. Thus, the correlation analysis indicated that shifting to a cyclone center, preparing a cyclone-protected home, improving riverbanks, constructing a home with higher foundations, and mangrove replantation were preferred adaptation measures by the respondents during extreme weather events. Adaptation practices for decreased agricultural lands are very much correlated with socioeconomic conditions in the SBR households. Change in cropping patterns and crop diversification were positively associated with social group, distance from water source, and occupation (Fig. 5.10a). Adapting new farming techniques was positively correlated with household size. Increased embankment height, digging pond, and liming are positively correlated with the different selected socioeconomic conditions (Fig. 5.10b).

Fig. 5.10 Relationship between socioeconomic characteristics with changes in economic activities: (**a**) agriculture and (**b**) fish/shrimp farming

Composite Adaptation Index Using PCA

A composite adaptation index was constructed using principal component analysis (PCA) for analyzing adaptation measures for salinity intrusion, fish/shrimp farming, agriculture, bank soil erosion, floods, and cyclones (Fig. 5.11). The composite adaptation index revealed very high adaptation in the Kakdwip, Kultali, Sagar, Patharpratima, Namkhana, and Gosaba blocks of the SBR. High adaptation was found in Hingalganj and Basanti blocks. The very high and high adaptations, respectively, in these blocks are attributed to very high vulnerability to coastal disasters.

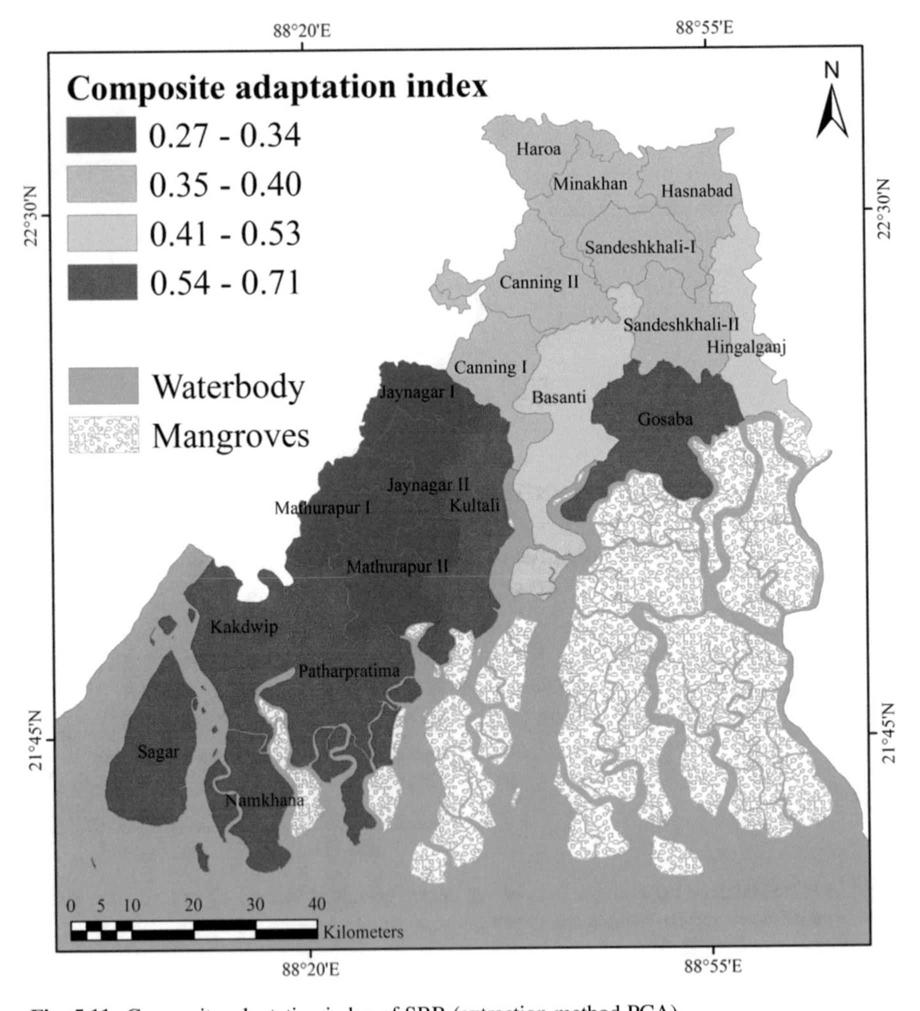

Fig. 5.11 Composite adaptation index of SBR (extraction method PCA)

Moderate adaptation was observed in Canning I, Canning II, Haroa, Sandeshkhali-I, Sandeshkhali-II, Hasnabad, and Minakhan blocks. These blocks are located in the northern part and away from the coast. A low level of adaptation was found in the Mathurapur I, Mathurapur II, Jaynagar I, and Jaynagar II blocks. These blocks are well connected with mainland and near to Kolkata city (Fig. 5.11). As the coastal blocks are very frequently highly susceptible to different natural hazard events, the adaptation strategies of these households are much higher than in the other blocks. The households of these blocks are practicing different adaptation strategies than the households situated in the upper part of the Reserve.

The analysis revealed that of 14 adaptive measures, nearly 11 adaptation strategies, namely, shifting to cyclone center, preparing cyclone-resilient structures, improving riverbanks, construction of home with higher foundation, mangrove replantation, change in cropping patterns, crop diversification, adapting new farming techniques, increased embankment height, digging pond, and liming were most often adopted by the sampled households. These measures are somewhat consistent and to some extent helped in lessening the impacts of climate change-induced extreme weather events. Thus, the adaptation strategies should be prioritized based on vulnerable people and places. The policies should be formulated to uplift the socioeconomic status of the affected communities in the SBR. Coastal management practices should be consistent to address the impacts of climate change, salinity, inundation, and changes in economic activities. Livelihood diversification and government employment schemes may be effective in supplementing the income of the farming communities. Provision of basic amenities such as healthcare centers, improved sanitation, connectivity among villages, education, awareness, and an early warning system may provide a way forward for efficacious coastal management and help in lessening the vulnerability of the coastal communities.

Conclusion

Frequent storms, surge floods, cyclones, salinity intrusion, sea level rise, land subsidence, waterlogging, and coastal erosion have caused significant socioeconomic vulnerability among the rural households of the SBR. Poor economy and lack of infrastructural facilities have made SBR people more vulnerable to disasters. Cyclone and flood have severely affected the SBR community. Agricultural and fishing activities have declined because of salinity intrusion. The findings revealed that the people of the SBR have relatively limited capacity to cope with climate change-induced extreme weather events, which are likely to have persistent effects on their livelihoods, economic activities, health, education, and social well-being. A holistic composite adaptation index was used to understand the capacity to handle multihazard events in each community development block of the SBR using demographic, social, economic, and infrastructural facilities. The multidimensional composite index developed here within the framework of PCA provided a better picture of economic, social, cultural, and related multihazard adaptation strategies. The blocks located near the coast are practicing a large number of adaptation strategies to cope with the disasters because these blocks are more vulnerable to coastal disasters. However, these adaptation strategies are not effective because of the low socioeconomic status of the communities. Government intervention in construction, protection of river embankments, and planting of riparian forests may help in reducing river bank erosion. New policy interventions oriented toward improving accessibility to food, healthcare, water, and sanitation may be effective in lessening vulnerability among these communities. The socioeconomic profiles and social networks of these communities need to be strengthened to enhance their adaptive capacity. Diversification in livelihood, and development of improved communication, transportation, and access to markets and services, would support existing and alternative livelihoods for the Sundarban community.

References

Ahamed M (2013) Community based approach for reducing vulnerability to natural hazards (cyclone, storm surges) in coastal belt of Bangladesh. Procedia Environ Sci 17:361–371

Ahammed KB, Pandey AC (2019) Coastal social vulnerability and risk analysis for cyclone hazard along the Andhra Pradesh, East Coast of India. KN J Cartogr Geogr Inf 69(4):285–303

Allison MA, Khan SR, Goodbred SL Jr, Kuehl SA (2003) Stratigraphic evolution of the late Holocene Ganges-Brahmaputra lower delta plain. Sediment Geol 155(3–4):317–342. https://doi.org/10.1016/S0037-0738(02)00185-9

Antony GM, Rao KV (2007) A composite index to explain variations in poverty, health, nutritional status and standard of living: use of multivariate statistical methods. Public Health 121 (8):578–587

Aon (2020) Weather, climate & catastrophe insight: 2019 annual report. http://thoughtleadership.aon.com/Documents/20200122-if-natcat2020.pdf?utm_source=ceros&utm_medium=storypage&utm_campaign=natcat20. Accessed 10th June 2020

Baills A, Garcin M, Bulteau T (2020) Assessment of selected climate change adaptation measures for coastal areas. Ocean Coast Manag 185:105059

Boelhouwer J, Stoop I (1999) Measuring well-being in the Netherlands: the SCP index from 1974 to 1997. Soc Indic Res 48(1):51–75

Census of India (2011) Primary census abstracts. Office of the Registrar General and Census Commissioner, Ministry of Home Affairs, Government of India. http://www.censusindia.gov.in/pca/Searchdata.aspx. Accessed 25th June 2020

Danda A (2010) Sundarbans: future imperfect climate adaptation report. World Wide Fund for Nature, New Delhi

de Campos RS, Codjoe SNA, Adger WN, Mortreux C, Hazra S, Siddiqui T et al (2020) Where people live and move in deltas. In: Deltas in the Anthropocene. Palgrave Macmillan, Cham, pp 153–177

DECCMA (2017) The Mahanadi delta: understanding the present state of climate change, adaptation and migration. https://generic.wordpress.soton.ac.uk/deccma/wp-content/uploads/sites/181/2017/10/68439-A4-DECCMA-MD_final_web.pdf. Accessed 10th June 2020

Fotso JC, Kuate-Defo B (2005) Measuring socioeconomic status in health research in developing countries: should we be focusing on households, communities or both? Soc Indic Res 72 (2):189–237

Gopalakrishnan S, Landry CE, Smith MD (2018) Climate change adaptation in coastal environments: modeling challenges for resource and environmental economists. Rev Environ Econ Policy 12(1):48–68

Greenan BJW, Shackell N, Ferguson K, Greyson P, Cogswell A, Brickman D et al (2019) Climate change vulnerability of American lobster fishing communities in Atlantic Canada. Front Mar Sci 6:579

Hazra S, Ghosh T, Das Gupta R, Sen G (2002) Sea level and associated changes in the Sundarbans. Sci Cult 68(9/12):309–321

Havard S, Deguen S, Bodin J, Louis K, Laurent O, Bard D (2008) A small-area index of socioeconomic deprivation to capture health inequalities in France. Soc Sci Med 67 (12):2007–2016

Hotelling H (1933) Analysis of a complex of statistical variables into principal components. J Educ Psychol 24(417–441):498–520

Houghton A, Austin J, Beerman A, Horton C (2017) An approach to developing local climate change environmental public health indicators in a rural district. J Environ Public Health 2017. https://doi.org/10.1155/2017/3407325

IPCC (2018a) Summary for policymakers. In: Masson-Delmotte V, Zhai P, Pörtner HO, Roberts D, Skea J, Shukla PR, Pirani A, Moufouma-Okia W, Péan C, Pidcock R, Connors S, Matthews JBR, Chen Y, Zhou X, Gomis MI, Lonnoy E, Maycock T, Tignor M, Waterfield T (eds) Global warming of 1.5°C. An IPCC special report on the impacts of global warming of 1.5°C above pre-industrial levels and related global greenhouse gas emission pathways, in the context of strengthening the global response to the threat of climate change, sustainable development, and efforts to eradicate poverty. World Meteorological Organization, Geneva, 32 pp

IPCC (2018b) Annex I: glossary [Matthews, J.B.R. (ed.)]. In: Masson-Delmotte V, Zhai P, Portner H-O, Roberts D, Skea J, Shukla PR, Pirani A, Moufouma-Okia W, Pean C, Pidcock R, Connors S, Matthews JBR, Chen Y, Zhou X, Gomis MI, Lonnoy E, Maycock T, Tignor M, Waterfield T (eds) Global Warming of 1.5°C. An IPCC special report on the impacts of global warming of 1.5°C above pre-industrial levels and related global greenhouse gas emission pathways, in the context of strengthening the global response to the threat of climate change, sustainable development, and efforts to eradicate poverty. World Meteorological Organization, Geneva

Kaźmierczak A, Cavan G (2011) Surface water flooding risk to urban communities: analysis of vulnerability, hazard and exposure. Landsc Urban Plan 103(2):185–197

Kelman I, Glantz MH (2014) Early warning systems defined. In: Reducing disaster: early warning systems for climate change. Springer, Dordrecht, pp 89–108

Koroglu A, Ranasinghe R, Jiménez JA, Dastgheib A (2019) Comparison of coastal vulnerability index applications for Barcelona Province. Ocean Coast Manag 178:104799. https://doi.org/10.1016/j.ocecoaman.2019.05.001

Krishnan V (2010) Constructing an area-based socioeconomic index: a principal components analysis approach. Early Child Development Mapping Project, Edmonton

Kuenzer C, Renaud FG (2012) Climate and environmental change in river deltas globally: expected impacts, resilience, and adaptation. In: The Mekong delta system. Springer, Dordrecht, pp 7–46

Lai D (2003) Principal component analysis on human development indicators of China. Soc Indic Res 61(3):319–330

Mallick BJ, Witte SM, Sarkar R, Mahboob AS, Vogt J (2009) Local adaptation strategies of a coastal community during cyclone Sidr and their vulnerability analysis for sustainable disaster mitigation planning in Bangladesh. J Bangladesh Inst Planners 2:158–168

Maru YT, Smith MS, Sparrow A, Pinho PF, Dube OP (2014) A linked vulnerability and resilience framework for adaptation pathways in remote disadvantaged communities. Glob Environ Chang 28:337–350

Messer LC, Vinikoor LC, Laraia BA, Kaufman JS, Eyster J, Holzman C et al (2008) Socioeconomic domains and associations with preterm birth. Soc Sci Med 67(8):1247–1257

Nichols CR, Wright LD, Bainbridge SJ, Cosby AG, Hénaff A, Loftis JD et al (2019) Collaborative science to enhance coastal resilience and adaptation. Front Mar Sci 6:404

Ogundele OM, Ubaekwe RE (2019) Early warning system and ecosystem-based adaptation to prevent flooding in Ibadan Metropolis, Nigeria. In: Handbook of climate change resilience. Springer, Cham

Onyeneke RU, Igberi CO, Aligbe JO, Iruo FA, Amadi MU, Iheanacho SC et al (2020) Climate change adaptation actions by fish farmers: evidence from the Niger Delta Region of Nigeria. Aust J Agric Resour Econ 64(2):347–375

Oo AT, Van Huylenbroeck G, Speelman S (2018) Assessment of climate change vulnerability of farm households in Pyapon District, a delta region in Myanmar. Int J Disaster Risk Reduct 28:10–21

Priya Rajan SM, Nellayaputhenpeedika M, Tiwari SP, Vengadasalam R (2019) Mapping and analysis of the physical vulnerability of coastal Tamil Nadu. Hum Ecol Risk Assess Int J 2019:1–17

Rehman S, Sahana M, Kumar P, Ahmed R, Sajjad H (2020) Assessing hazards induced vulnerability in coastal districts of India using site-specific indicators: an integrated approach. GeoJournal:1–22. https://doi.org/10.1007/s10708-020-10187-3

Rygel L, O'sullivan D, Yarnal B (2006) A method for constructing a social vulnerability index: an application to hurricane storm surges in a developed country. Mitig Adapt Strateg Glob Chang 11(3):741–764

Sahana M, Sajjad H, Ahmed R (2015) Assessing spatio-temporal health of forest cover using forest canopy density model and forest fragmentation approach in Sundarban reserve forest, India. Modeling Earth Systems and Environment 1(4):1–10

Sahana M, Ahmed R, Sajjad H (2016) Analyzing land surface temperature distribution in response to land use/land cover change using split window algorithm and spectral radiance model in Sundarban Biosphere Reserve, India. Modeling Earth Systems and Environment 2(2):81

Sahana M, Sajjad H (2019) Vulnerability to storm surge flood using remote sensing and GIS techniques: a study on Sundarban Biosphere Reserve, India. Remote Sens Appl Soc Environ 13:106–120

Sahana M, Rehman S, Paul AK, Sajjad H (2019a) Assessing socio-economic vulnerability to climate change-induced disasters: evidence from Sundarban Biosphere Reserve, India. Geol Ecol Lands 2019:1–13

Sahana M, Hong H, Ahmed R, Patel PP, Bhakat P, Sajjad H (2019b) Assessing coastal island vulnerability in the Sundarban Biosphere Reserve, India, using geospatial technology. Environ Earth Sci 78(10):1–22

Sahana M, Rehman S, Sajjad H, Hong H (2020a) Exploring effectiveness of frequency ratio and support vector machine models in storm surge flood susceptibility assessment: a study of Sundarban Biosphere Reserve, India. Catena 189:104450

Sahana M, Rehman S, Ahmed R, Sajjad H (2020b) Analyzing climate variability and its effects in Sundarban Biosphere Reserve, India: reaffirmation from local communities. Environ Dev Sustain 2020:1–28

Sahana M, Rehman S, Ahmed R, Sajjad H (2021) Assessing losses from multi-hazard coastal events using Poisson regression: empirical evidence from Sundarban Biosphere Reserve (SBR), India. J Coast Conserv 25(1):1–10

Solangaarachchi D, Griffin AL, Doherty MD (2012) Social vulnerability in the context of bushfire risk at the urban-bush interface in Sydney: a case study of the Blue Mountains and Ku-ring-gai local council areas. Nat Hazards 64(2):1873–1898

Sudha Rani NNV, Satyanarayana ANV, Bhaskaran PK (2015) Coastal vulnerability assessment studies over India: a review. Nat Hazards 77(1):405–428. https://doi.org/10.1007/s11069-015-1597-x

Swart R, Biesbroek R, Lourenço TC (2014) Science of adaptation to climate change and science for adaptation. Front Environ Sci 2:29

Tata RJ, Schultz RR (1988) World variation in human welfare: a new index of development status. Ann Assoc Am Geogr 78(4):580–593

UNFCCC (2015) Report of the conference of the parties on its twenty-first session, held in Paris from 30 November to 13 December 2015. Addendum. Part two: action taken by the conference of the parties at its twenty-first session. United Nations Framework Convention on Climate Change, Bonn

USAID (2009) Adapting to coastal climate change: a guidebook for development planners. US Agency for International Development, Washington, DC

Wong PP, Losada IJ, Gattuso J-P, Hinkel J, Khattabi A, McInnes KL, Saito Y, Sallenger A (2014) Coastal systems and low-lying areas. In: Field CB, Barros VR, Dokken DJ, Mach KJ, Mastrandrea MD, Bilir TE, Chatterjee M, Ebi KL, Estrada YO, Genova RC, Girma B, Kissel ES, Levy AN, MacCracken S, Mastrandrea PR, White LL (eds) Climate change 2014: impacts, adaptation, and vulnerability. Part A: global and sectoral aspects. Contribution of Working Group II to the Fifth Assessment Report of the Intergovernmental Panel on Climate Change. Cambridge University Press, Cambridge/New York, pp 361–409

World Economic Forum (ed) (2019) The global risks report 2019, 14th edn. World Economic Forum, Geneva

Chapter 6
Impact of Climate Change on Groundwater Resource of India: A Geographical Appraisal

Sutapa Mukhopadhyay and Amit Kumar Mandal

Abstract The inherent advantages of consistent temperature, widespread availability, decentralized access, limited vulnerability, outstanding natural quality, low extraction cost, and drought protection capacity as well as fluctuation in seasonal availability, result in monsoonal rainfall uses of groundwater in the domestic and agricultural sectors of India to be ever increasing. The rapid growth of groundwater-based irrigation systems in the past 50 years has multiplied groundwater extraction in India, resulting in its overexploitation. Recent study has revealed that climate change leads to the altered precipitation and evapotranspiration rate of the country, so it may be considered as tan additional major threat to this resource. Hence, understanding the behavior and dimensions of the groundwater regime under future climatic and other changes is significant in adopting an accurate water management strategy. This review presents an outline of the groundwater resource base of India and its susceptibility to ongoing and future trends of changes in climatic variables.

Keywords Groundwater resource · Climate change · Groundwater fluctuation · Projected groundwater level

Introduction

Water is the utmost important element of our environment. Without water, life cannot be sustained on Earth. It is fundamental for our existence, crucial for alleviating poverty, hunger, and disease, and critical for economic use (Chinnaiah 2015). Of the total water reserve of 386×10^9 Km^3 on the Earth, the total freshwater reserve accounts for 35.03×10^6 Km^3 (2.8%), and 30% of the world's freshwater (10.5×10^6 Km^3) is stored in the subsurface aquifers in the form of groundwater (UNESCO 1994; Das 2011).

S. Mukhopadhyay (✉) · A. K. Mandal
Department of Geography, Visva-Bharati, Santiniketan, West Bengal, India

M. N. Islam, A. van Amstel (eds.), *India: Climate Change Impacts, Mitigation and Adaptation in Developing Countries*, Springer Climate,
https://doi.org/10.1007/978-3-030-67865-4_6

Groundwater is among the most valuable water resources in the entire world, especially in arid and semiarid regions (Todd and Mays 2005; Mukherjee et al. 2012). It is commonly considered as the best quality water for drinking and irrigation everywhere (Hoque et al. 2009; Adiat et al. 2012). Besides these domestic, agricultural, and industrial uses, the in situ functions of groundwater are also significant: feeding the flow of springs and streams, as well as maintaining the stability of the land surface and preventing the migration of saline water or other poor-quality water into fresh aquifer domains (Gun and Lipponen 2010). Groundwater has several advantages such as consistent temperature, wide availability, decentralized access, limited vulnerability, outstanding natural quality, low extraction cost, and drought protection capacity (Jha et al. 2007; Arkoprovo et al. 2012). Another quality of groundwater in comparison to surface water is that it is the least affected by disastrous events and can be tapped when required (Manap et al. 2013). Consequently, exploitation of groundwater as the alternative to the insufficiency and irregularity of surface water is ever increasing (Manap et al. 2013).

Groundwater has a significant function in fulfilling the water requirements of the agriculture, industrial, and domestic sectors in India. About 85% of India's rural domestic water requirements, 50% of its urban water needs, and more than 60% of its irrigation requirements are being met from groundwater resources (Mall et al. 2006; Chinnaiah 2015; CGWB 2019a). Hence, India has become the world's largest consumer of groundwater by extracting 248.69 BCM (billion cubic meters) groundwater annually, one fourth of the total global annual groundwater extraction, about 90% (221.46 BCM) of which is used in irrigation (CGWB 2019a). In 1995, net groundwater draft from irrigation uses was about 115 BCM, which indicates almost doubling of groundwater extraction within a time span of 15 years (CGWB 2011). Although groundwater-based irrigation accounts for about 60% of the total irrigated area of India, 80% of the country's total agricultural production is still dependent on groundwater in one form or another (Dains and Pawar 1987; Wani et al. 2009). Also, crop productivity in groundwater-irrigated areas is higher (often double) in comparison to that in surface water-based irrigated areas (Shah 1993; Meinzen-Dick 1996; Wani et al. 2009). So, groundwater extraction in the irrigation sector has been accelerated significantly in India, mainly controlled through decentralized private activity rather than under strict government policy. Hence, the groundwater resources of the country are now facing crisis, urgently requiring attention and understanding of the situation (Gandhi and Bhamoriya 2011).

According to the United Nations Convention Framework on Climate Change, "a change of climate which is attributed directly or indirectly to human activity that alters the composition of the global atmosphere and which is in addition to natural climate variability observed over comparable time periods" (UNFCC 1992; Karl et al. 2009; Anupam and Shinjiro 2013). Climate change leads to changes in precipitation and evapotranspiration rates, which have a deep impact on the hydrological cycle, resulting in large-scale modifications in the amount and status of water present in glaciers, rivers, lakes, and oceans (Panwar and Chakrapani 2013). Such changes will influence subsurface hydrological dynamics and cause changes in groundwater quality, recharge, discharges, and storage in many aquifers (Panwar and Chakrapani 2013; Aizebeokhai et al. 2017).

Climate change directly influences the surface water resource; hence, its impacts on surface water have been analyzed and discussed (Singh and Kumar 2010; Aizebeokhai et al. 2017). As relationships between climate change and groundwater dynamics are much more complicated, the vulnerability of groundwater resources to climate change has not been sufficiently studied (IPCC 2007; Green 2016; Singh and Kumar 2010; Aizebeokhai et al. 2017). Climate changes may affect the quantity and quality of groundwater in several ways, but it is difficult to quantify the extent and nature of these impacts because of uncertainty at all stages of the assessment process (Dettinger and Earman 2007; Kundzewicz et al. 2007; Aizebeokhai et al. 2017).

In India, the rapid growth of groundwater-based irrigation systems has multiplied groundwater extraction, resulting in its overexploitation. Thus, incidences of climate change are imposing additional major threats to this resource (Wang et al. 2013). Hence, understanding the behavior and dimensions of the groundwater regime under future climatic and other changes is significant for adopting an appropriate water management strategy (Singh and Kumar 2010; Aizebeokhai et al. 2017).

This study presents an outline of the groundwater resource base of India and its susceptibility to ongoing and future trends of changes in climatic variables. The potential impact of climate change on different subsurface layers and groundwater recharge and discharge processes is discussed, and the nature of changing trends of groundwater resources in India in connection with changes in climatic variables have been analyzed by highlighting important findings from earlier research and assessing available recorded data.

Climate Change Scenario in India

In spite of huge regional variations in terms of patterns of wind, temperature, and rainfall, rhythm of seasons, and degree of wetness or dryness, in the broad sense the Indian climate is considered as the 'tropical monsoon type,' characterized by seasonal reversal of wind direction associated with alternating periods of wetness and dryness (Khullar 2005; Rajaram 2006). The average annual rainfall for the country as a whole is about 118 cm, the highest for land of comparable size in the world (Lal 2001; Khullar 2005). India receives three fourths of its total rainfall during the 'rainy season' or 'monsoon season' from June to mid-September, consequent to the onset and flows of southwest monsoon wind from the Indian Ocean toward the landmass. During the rest of the seasons, scanty rainfall occurs, so that in this time an almost dry condition prevails in most parts of the country. The origin and mechanism of the monsoon wind is a complex phenomenon determined by many local, regional, and global factors. Hence, the climate of India is also greatly diversified and erratic in nature. The issues of climate change will bring more complications to the spatiotemporal behavior of climatic variables in India, which will threaten the sustenance of the huge populations of the country.

Under this global climate change scenario, a number of studies have investigated the changing trends of climatic parameters in India. Some scholars have analyzed the past weather data recorded by meteorological stations to detect trends of climatic variables on a local, regional, or national scale. On the other hand, in a few studies projection of future climate has been based on present trends of climatic variables. So, the climate change scenario of India can be explored from two perspectives: observed trend of climate change and projected trend of climate change.

Observed Trend of Climate Change in India

Kothawale and Rupa Kumar (2005) have examined monthly maximum and minimum temperature data of 121 stations of the Indian Meteorological Department (IMD) for the time periods of 1901–2003 to identify seasonal and annual trends in temperature change in the country as a whole and in seven homogeneous regions, namely, Western Himalaya, Northwest, North Central, Northeast, West Coast, and Interior Peninsula. A parametric statistical technique, namely, linear regression analysis, was applied to detect trends of temperature change from the period 1901–2003 and make comparisons of the trends of two other time periods, 1901–2003 and 1971–2003. Following are some key findings of this study.

1. During 1901–2003, the annual mean temperature of India has increased at the rate of 0.05 °C/10 years, but in the relatively short 1971–2003 time period this trend has accelerated to the rate of 0.22 °C/10 years. The rising trend of annual maximum temperature (0.07 °C/10 years) is significantly higher than that of the annual minimum temperature (0.02 °C/10 years) during 1901–2003, although during 1971–2003 the annual minimum temperature (0.21 °C/10 years) increased at a slightly greater rate than annual maximum temperature (0.20 °C/10 years).
2. Annual maximum temperature significantly increased over India in all the four seasons during 1901–2003, but in this period a significant warming trend of mean temperature is observed only in winter and in post monsoon seasons. During 1971–2003 a statistically significant increasing trend of annual maximum temperature is observed in winter and post monsoon seasons, whereas in winter and monsoon seasons a rising trend of annual minimum temperature becomes significant.
3. During 1901–2003 all regions of India experienced rising trends in annual mean and annual maximum temperature, but in the northwestern region a decreasing trend of annual minimum temperature is noticed. However, during 1971–2003 no decreasing trend of annual mean or annual maximum or annual minimum temperature is observed anywhere in India.

Attri and Tyagi (2010) have explored the climate change scenario of India by analyzing climatic data of more than 100 years (1901–2009), recorded in the well-distributed network of meteorological stations of the India Meteorological Department (IMD). In this study, a parametric statistical technique, namely, linear regression analysis, has been employed to compute the magnitude of trends of climatic variables. Findings of this study can be summarized through the following points.

1. Annual mean temperature of the country as a whole is rising at the rate of 0.56 °C per hundred years. Increasing trend of annual mean maximum temperature (1.02 °C/100 years) is significantly higher than that of annual mean minimum temperature (0.12 °C/100 years), although since 1990 the annual mean minimum temperature is steadily rising at a higher rate than the annual mean maximum temperature. The maximum rising trend of annual mean temperature is observed in the post monsoon season (0.77 °C/100 years) followed by the winter season (0.70 °C/100 years), pre monsoon season (0.64 °C/100 years), and monsoon season (0.33 °C/100 years), respectively.
2. The average annual and monsoon rainfall of the entire country does not show any significant trend during the time period of 1901–2009, although over this time period alternating sequences of multi-decadal periods of 30 years with frequent droughts and floods are observed in the monsoon rainfall pattern of India. In the post monsoon season a trend of slightly increasing of rainfall is observed over most of India whereas in the winter season a decreasing trend is noticed.
3. During the time period of 1961–2000 a slight delay in onset and withdrawal of monsoon activity is observed, and its duration has increased by about a week in comparison to the normal duration in the past.

Kumar and Jain (2010) have conducted a river basin trend analysis of annual rainfall and annual rainy days, employing Sen's method and the Mann–Kendall test to estimate the magnitude of the trend and to assess its statistical significance. Findings of this study can be summarized through the following points.

1. Of 22 major river basins in India, 15 river basins have experienced a decreasing trend of annual rainfall ranging from 0.27 mm/year to 10.16 mm/year during 1951–2004. An increasing trend of annual rainfall ranging between 0.45 mm/year and 4.93 mm/year is observed in 6 river basins of India, and in the Ganga River basin no trend is found.
2. Annual rainy days have increased (0.024 days/year to 0.111 days/year) over 4 river basins during 1951–2004 whereas in 15 river basins a decreasing trend (0.044 days/year to 0.375 days/year) of annual rainy days is observed. In 3 river basins, no changes in terms of annual rainy days are indicated.

Rathore et al. (2013) outlined state level climate change trends over India during 1951–2010 with respect to temperature and rainfall under the Meteorological Monograph Series of the Indian Meteorological Department (IMD). They examined long-term continuous temperature and rainfall datasets as recorded in 282 and 1451 meteorological stations, respectively. In this study, two nonparametric statistical techniques, Sen's method and the Mann–Kendall test were employed to detect increasing or decreasing trends and to judge the presence of a statistically significant

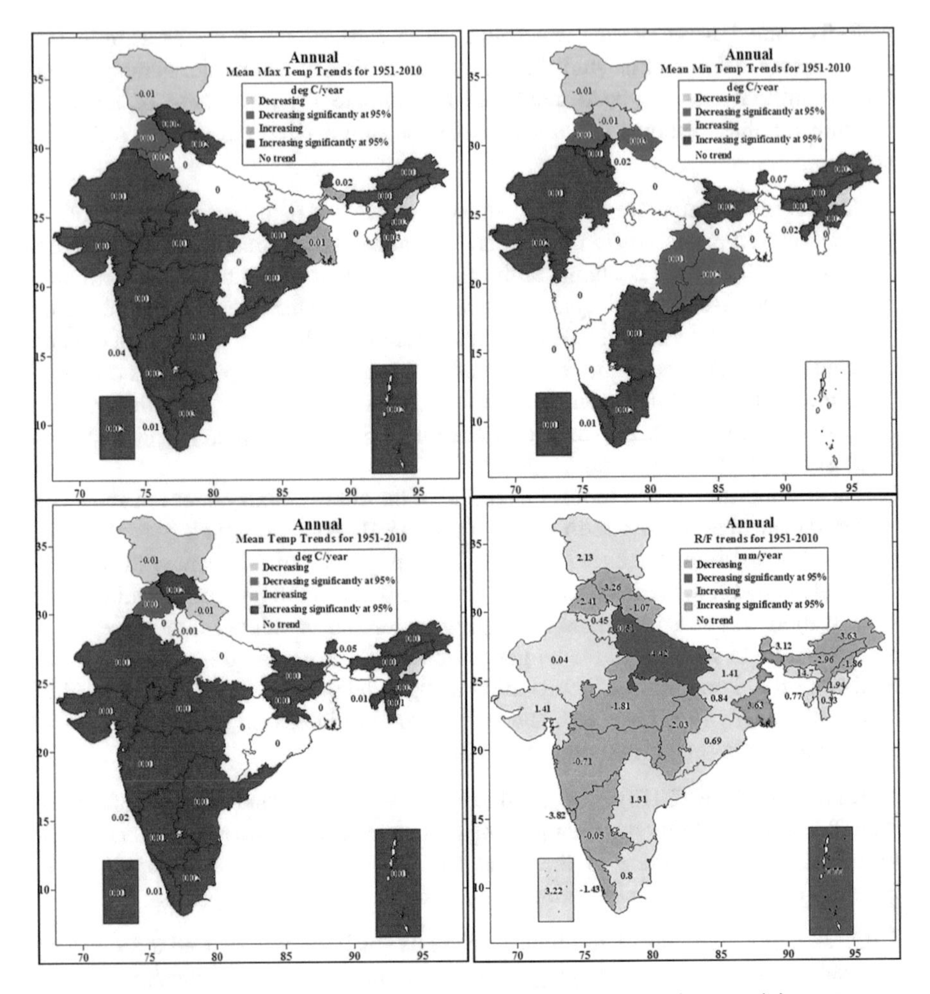

Fig. 6.1 State level trend of annual mean maximum temperature, annual mean minimum temperature, annual mean temperature, and annual rainfall (1951–2010). (Source: Rathore et al. 2013)

trend in rainfall and temperature on annual, seasonal, and monthly bases, with the following findings (Fig. 6.1).

1. Increasing trend of annual mean maximum temperature ranging from 0.01 °C per year to 0.06 °C per year is observed in most of the states of India. In some states of northern India, a decreasing trend (0.01 °C–0.02 °C per year) is noticed.
2. Annual mean minimum temperature has increased significantly (0.01 °C–0.07 °C per year) in most of the states of the eastern coastal part, northeastern provinces, and semiarid desert regions. A significant decreasing trend is noticed especially in

northern India (0.01 °C–0.07 °C per year). All other states have no noticeable trend.

3. A significant increasing trend (0.01 °C–0.07 °C per year) of annual mean temperature is observed in most of the states of India, especially over the western, central, southern, and eastern coastal parts of India. Moreover, increasing trends have also been observed in most of the northeastern provinces. A significant decreasing trend (0.01 °C per year) of annual mean temperature is noticed in northern India. No trend has been noted in the other six states of the country.
4. An increasing trend (0.04 mm/year–14.68 mm/year) of annual average rainfall is observed mostly in the southeastern, eastern, northwestern, and the riverine plain region of the northeastern states of India. A decreasing trend (0.05 mm/year–7.77 mm/year) of annual average rainfall is observed in the central as well as along the western coastal part of the country. A remarkable decreasing trend is also noticed along the Himalayan foothill region of northeast India (14.68 mm/year) and in the Andaman and Nicobar Islands (7.77 mm/year), respectively.

IPCC (2014), in its fifth assessment report, identified an increase in the number of monsoon break days and a decrease in the number of monsoon depressions over India, which has resulted in an overall decline in mean rainfall of the monsoon season. Increase in extreme rainfall events over central India and in many other parts of the country has also been observed. According to this report, the frequency of heavy precipitation events is increasing in South Asia whereas light rain events are decreasing.

Projected Trend of Climate Change

IPCC (2013) opines that "future climate will depend on committed warming caused by past anthropogenic emissions and natural climate variability" (Christensen et al. 2013). Continuous emission of greenhouse gas will lead to increased temperatures, resulting in changes in interrelated components of the climatic system. Based on different possible time series scenarios of emissions and concentrations of greenhouse gases, aerosols, and chemically active gases, known as representative concentration pathways (RCPs), IPCC has projected changes in mean surface temperature and other climatic parameters for different periods of the twenty-first century relative to the 1986–2005 period. According to this report, by the end of the twenty-first century (2081–2100), global mean surface temperature will rise in the range 0.3 °C–1.7 °C under RCP 2.6, 1.1 °C–2.6 °C under RCP 4.5, 1.4 °C–3.1 °C under RCP 6.0, and 2.6 °C–4.8 °C under RCP 8.5.

In a similar global trend, IPCC has projected a clear increase in temperature over India, especially in winter (Christensen et al. 2013). As per the RCP 4.5 scenario based on the CMIP5 model, annual mean surface temperature of South Asia will increase in the range of 0.2 °C–1.3 °C by 2035, relative to the 1986–2005 period. By 2065, the rise of annual mean surface temperature of South Asia will be in the range

Table 6.1 Projected temperature and precipitation by CMIP5 global models in South Asia relative to 1986–2005

		Temperature (°C)			Precipitation (%)		
Month	Yea	Minimum	Maximum	Median	Minimum	Maximum	Median
DJF	2035	0.1	1.4	1	−18	8	−1
	2065	0.6	2.6	1.8	−17	4	13
	2100	1.4	3.7	2.3	−14	28	8
JJA	2035	0.3	1.3	0.7	−3	9	3
	2065	0.9	2.6	1.3	−3	33	7
	2100	0.7	3.3	1.7	−7	37	10
Annual	2035	0.2	1.3	0.8	−2	7	3
	2065	0.8	2.5	1.6	−2	26	7
	2100	1.3	3.5	2.1	−3	27	10

Source: IPCC (2013)

of 0.8 °C–2.5 °C, and at the end of the twenty-first century annual mean surface temperature will increase in the range of 1.3 °C–3.5 °C. This report also projects the increase of hot days and nights in the summer season. This report predicts average 3%, 7%, and 10% increases of annual precipitation by 2035, 2065, and 2100, respectively. Increase in precipitation will be more prominent in summer season than winter season, as increased atmospheric moisture content will lead to more precipitation through southwest monsoon wind (Table 6.1).

Kumar et al. (2006) have analyzed climate change scenarios over India applying the state-of-art regional climate modeling system, known as PRECIS, developed by the Hadley Centre of Climate Prediction and Research. This study predicts a widespread increase of temperature throughout the country as well as an increase in extremes of maximum and minimum temperature, although the findings of this study reveal a nonuniform spatial pattern of precipitation changes in India. West central India is projected to experience the maximum increase in rainfall whereas the west coast of India and west central India will have substantial increase in the occurrences of extreme precipitation.

Salvi et al. (2013) have applied a statistical downscaling technique for projections of monsoon rainfall over India at a resolution of 0.5° latitude/longitude. This study predicts spatially nonuniform changes of rainfall over India. Rainfall surplus areas of India, especially the western coastline and northeastern India, are expected to receive more rainfall, whereas in rainfall deficit areas of India, particularly Western and Northern India and in the Southeastern Coastline, rainfall will likely be decreased.

Scenario of Groundwater Resources in India

India is a vast country with highly diversified hydrogeological settings resulting from its varied geological, topographic, and climatological conditions (CGWB 2017). The diversified geological formations with considerable lithological and chronological variations, complex tectonic framework, climatological dissimilarities, and various hydrochemical conditions lead to the complicated behavior of the groundwater system in India (Jha 2007).

The various rock formations with distinctive hydrogeological characteristics act as different aquifer systems of various dimensions (CGWB 2017). Based on the hydrogeological characteristics, the whole country has been categorized into 14 Principal Aquifer Systems and 42 Major Aquifers by the Central Groundwater Board of India (CGWB 2011). Alluvium is the most important aquifer system, covering a maximum area of about 30% of the entire country in the Indus–Ganga–Brahmaputra basin areas. Around 20% of the area of the country is covered by banded gneissic complex (BGC) and gneiss aquifers, which are accessible in almost all the peninsular states as well as the Himalayan states. The basalt aquifer, sandstone aquifer, shale aquifer, and limestone aquifer cover about 16%, 8%, 7%, and 2% of the area of the country, respectively. The rest (17%) of the entire area of India is covered by other aquifers: schist, granite, quartzite, charnockite, khondalite, laterites, and intrusive (Fig. 6.2).

Following GEC-2015 norms, the CGWB have estimated total annual groundwater recharge for the entire country as 432 billion cubic meters (BCM). After subtracting the total natural discharge of 39 BCM, the total annual extractable groundwater resources of the whole country have been estimated as 393 BCM (CGWB 2019a). Monsoon rainfall has been assessed as the main source of groundwater recharge, contributing about 58% (251.9 BCM) to total annual groundwater recharge. Overall, 67% (288.2 BCM) of annual groundwater recharge occurs from annual rainfall, both in monsoon and non-monsoon periods, and 33% (143.6 BCM) of annual recharge is caused from other sources: canal seepage, return flow from irrigation, recharge from tanks, ponds, and water conservation structures. Unit recharge of groundwater (meters) is higher (>0.25 m) in the Indus–Ganga–Brahmaputra alluvial belt in north, east, and northeast India covering the Punjab, Haryana, Uttar Pradesh, Bihar, and West Bengal states and the valley areas of the northeastern states from the occurrences of sufficient rainfall and the presence of recharge supportive unconsolidated alluvial formations. In the coastal alluvial belt, particularly on the east coast, recharge intensity is also higher. However, in western India, central India, and major parts of peninsular India, annual groundwater recharge is lower (<0.25 m) because less rainfall occurs and the hard rock terrain is unfavorable for infiltration (Tables 6.2 and 6.3).

Currently, groundwater depletion has become a major threat to the country, caused by its overextraction for irrigation and domestic purposes. As per estimation of CGWB, 248.69 BCM groundwater was extracted in India in 2017, about 89% (221.5 BCM) of which is utilized for irrigation (CGWB 2019a). Currently (2014–2015), groundwater contributes to irrigation of 42.9 million hectares (ha) (63%) net irrigated area of the country whereas in 1950–1951 only 5.9 million ha (29%) agricultural land was irrigated from this source (Fig. 6.3, Table 6.4). So, within the span of 65 years, the groundwater-dependent net irrigated area of the country has increased more than sevenfold and the proportion of groundwater irrigation to total irrigated area has essentially doubled. This rapid acceleration of groundwater use for irrigation purposes is facilitated by the huge improvement in

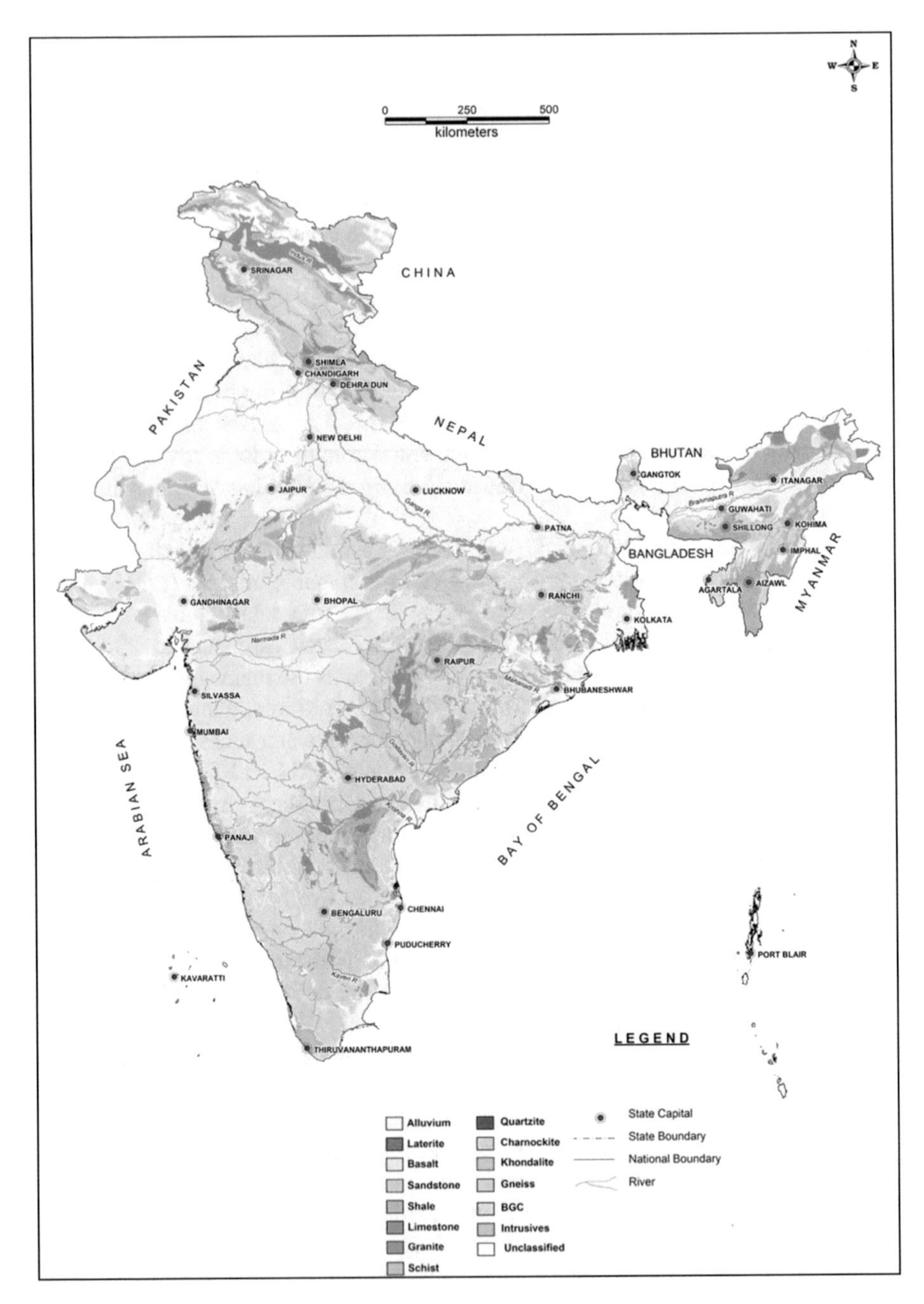

Fig. 6.2 Distribution pattern of principal aquifer system of India. (Source: CGWB 2019a)

Table 6.2 State-wise groundwater resources of India, 2017

States/ Union Territories	Groundwater recharge (bcm)					Unit recharge of groundwater (m)	Total natural discharges (bcm)	Annual extractable ground water resource (bcm)	Current annual ground water extraction (bcm)				Annual GW allocation for domestic use as on 2025 (bcm)	Net groundwater availability for future use (bcm)	Stage of ground water extraction (%)
	Monsoon season		Non-monsoon season		Total annual ground water recharge										
	Recharge from rainfall	Recharge from other sources	Recharge from rainfall	Recharge from other sources					Irrigation	Industrial	Domestic	Total			
Andhra Pradesh	9.96	5.62	1.21	4.42	21.22	0.13	1.07	20.15	7.85	0.14	0.9	8.9	1.48	12.31	44.15
Arunachal Pradesh	1.89	0.18	0.95	0.01	3.02	0.04	0.36	2.67	0	0	0.01	0.01	0.03	2.64	0.28
Assam	20.22	0.43	7.28	0.74	28.67	0.37	4.42	24.26	1.97	0.06	0.69	2.73	0.79	21.43	11.25
Bihar	19.83	3.95	3.14	4.5	31.41	0.33	2.43	28.99	10.78	0.66	1.83	13.26	1.83	15.78	45.76
Chhattisgarh	7.82	1.36	0.76	1.64	11.57	0.09	1	10.57	3.98	0.05	0.67	4.7	0.79	5.76	44.43
Delhi	0.13	0.06	0.03	0.11	0.32	0.22	0.02	0.3	0.09	0.02	0.24	0.36	0.29	0.02	119.61
Goa	0.19	0.03	0.01	0.05	0.27	0.07	0.11	0.16	0.02	[a]	0.03	0.05	0.04	0.07	33.5
Gujarat	15.95	3.4	0	3.02	22.37	0.11	1.12	21.25	12.84	0.11	0.63	13.58	0.9	7.98	63.89
Haryana	3.56	2.55	1.03	3	10.15	0.23	1.01	9.13	11.53	0.34	0.63	12.5	0.72	0.87	136.91
Himachal Pradesh	0.34	0.02	0.11	0.04	0.51	0.01	0.05	0.46	0.2	0	0.19	0.39	0.34	0.16	86.37
Jammu & Kashmir	1	0.5	0.88	0.51	2.89	0.01	0.29	2.6	0.2	0.07	0.5	0.76	0.5	1.84	29.47
Jharkhand	5.25	0.13	0.41	0.42	6.21	0.08	0.52	5.69	0.8	0.22	0.56	1.58	0.56	4.13	27.73
Karnataka	6.59	4.36	2.67	3.22	16.84	0.09	2.05	14.79	9.39	[a]	0.95	10.34	1.14	5.41	69.87
Kerala	3.91	0.04	0.68	1.13	5.77	0.15	0.56	5.21	1.22	0.01	1.44	2.67	1.57	2.41	51.27
Madhya Pradesh	27.1	1.51	0.82	6.99	36.42	0.12	1.95	34.47	17.43	0.22	1.24	18.88	1.72	15.84	54.76
Maharashtra	20.59	2.29	0.53	8.23	31.64	0.10	1.74	29.9	15.1	0.003	1.22	16.33	2.28	12.91	54.62
Manipur	0.23	0.01	0.17	0.02	0.43	0.02	0.04	0.39	0	0	0	0.01	0.04	0.34	1.44
Meghalaya	1.37	0.01	0.43	0.02	1.83	0.08	0.19	1.64	0.03	0	0.01	0.04	0.02	1.59	2.28
Mizoram	0.16	0	0.05	0	0.21	0.01	0.02	0.19	0	0	0.01	0.01	0.01	0.18	3.82
Nagaland	1.65	0.03	0.52	0	2.2	0.13	0.22	1.98	0	0	0.02	0.02	0.02	1.96	0.99
Odisha	10.53	2.34	1.5	2.37	16.74	0.11	1.17	15.57	5.28	0.14	1.15	6.57	1.3	8.85	42.18

(continued)

Table 6.2 (continued)

States/ Union Territories	Groundwater recharge (bcm)					Unit recharge of groundwater (m)	Total natural discharges (bcm)	Annual extractable ground water resource (bcm)	Current annual ground water extraction (bcm)				Annual GW allocation for domestic use as on 2025 (bcm)	Net groundwater availability for future use (bcm)	Stage of ground water extraction (%)
	Monsoon season		Non-monsoon season		Total annual ground water recharge				Irrigation	Industrial	Domestic	Total			
	Recharge from rainfall	Recharge from other sources	Recharge from rainfall	Recharge from other sources											
Punjab	5.54	11.83	1.31	5.25	23.93	0.48	2.35	21.58	34.56	0.2	1.01	35.78	1.41	1.09	165.77
Rajasthan	9.74	0.78	0.24	2.44	13.21	0.04	1.22	11.99	14.85	0	1.92	16.77	2.67	0.88	139.88
Sikkim	5.2	0	0.43	0	5.63	0.79	4.11	1.52	0	0	0	0	0.01	1.51	0.06
Tamil Nadu	6.67	9.41	1.89	2.26	20.22	0.16	2.02	18.2	13.06	0	1.67	14.73	1.85	5.66	80.94
Telangana	7.56	1.42	1.88	2.76	13.62	0.12	1.25	12.37	7.09	[a]	1	8.09	1.39	4.26	65.45
Tripura	0.8	0.06	0.4	0.26	1.53	0.15	0.29	1.24	0.02	0	0.08	0.1	0.11	1.11	7.88
Uttar Pradesh	37.73	11.67	1.59	18.93	69.92	0.29	4.6	65.32	40.89	[a]	4.95	45.84	5.96	20.36	70.18
Uttarakhand	1.15	0.93	0.09	0.87	3.04	0.06	0.15	2.89	1.3	0.13	0.22	1.64	0.22	1.25	56.83
West Bengal[b]	18.71	1.51	5.26	3.85	29.33	0.33	2.77	26.56	10.84	[a]	1	11.84	1.53	14.19	44.6
Andaman & Nicobar	0.35	0	0.02	0	0.37	0.04	0.04	0.33	0	0	0.01	0.01	0.01	0.32	2.74
Chandigarh	0.02	0.01	0	0.01	0.04	0.35	0	0.04	0	[a]	0.03	0.03	0.03	0	89
Dadra & Nagar Haveli	0.06	0	0	0.01	0.07	0.14	0	0.07	0.01	[a]	0.01	0.02	0.01	0.04	31.34
Daman & Diu	0.02	0	0	0	0.02	0.18	0	0.02	0.01	0	0	0.01	0	0	61.4
Lakshdweep	0.01	0	0	0	0.01	0.33	0.01	0.004	0	0	0.002	0.002	0	0	65.99
Puducherry	0.09	0.07	0.02	0.05	0.23	0.47	0.02	0.2	0.11	[a]	0.04	0.15	0.04	0.05	74.33
Grand Total	**251.9**	**66.49**	**36.34**	**77.13**	**431.86**	0.13	**39.16**	**392.7**	**221.46**	**2.38**	**24.87**	**248.69**	**31.62**	**173.25**	**63.33**

Source: CGWB (2019a, b)

[a]Industrial and domestic draft has not been estimated separately in Goa, Himachal Pradesh, Karnataka, Rajasthan, Tamil Nadu, Uttar Pradesh, Chandigarh, Dadra, Nagar Haveli, and Puducherry

[b]The Ground Water resources assessment as on 2013 has been considered for the state of West Bengal

Table 6.3 State-wise categorization of assessment units, 2017

States/Union Territories	Total no. of Assessed Unit	Safe		Semicritical		Critical		Overexploited		Saline	
		Nos.	%	Nos.	%	Nos.	%	Nos.	%	Nos.	%
Andhra Pradesh	670	501	74.8	60	9.0	24	3.6	45	6.7	40	6.0
Arunachal Pradesh	11	11	100.0	0	0.0	0	0.0	0	0.0	0	0.0
Assam	28	28	100.0	0	0.0	0	0.0	0	0.0	0	0.0
Bihar	534	432	80.9	72	13.5	18	3.4	12	2.2	0	0.0
Chhattisgarh	146	122	83.6	22	15.1	2	1.4	0	0.0	0	0.0
Delhi	34	3	8.8	7	20.6	2	5.9	22	64.7	0	0.0
Goa	12	12	100.0	0	0.0	0	0.0	0	0.0	0	0.0
Gujarat	248	194	78.2	11	4.4	5	2.0	25	10.1	13	5.2
Haryana	128	26	20.3	21	16.4	3	2.3	78	60.9	0	0.0
Himachal Pradesh	8	3	37.5	1	12.5	0	0.0	4	50.0	0	0.0
Jammu and Kashmir	22	22	100.0	0	0.0	0	0.0	0	0.0	0	0.0
Jharkhand	260	245	94.2	10	3.8	2	0.8	3	1.2	0	0.0
Karnataka	176	97	55.1	26	14.8	8	4.5	45	25.6	0	0.0
Kerala	152	119	78.3	30	19.7	2	1.3	1	0.7	0	0.0
Madhya Pradesh	313	240	76.7	44	14.1	7	2.2	22	7.0	0	0.0
Maharashtra	353	271	76.8	61	17.3	9	2.5	11	3.1	1	0.3
Manipur	9	9	100.0	0	0.0	0	0.0	0	0.0	0	0.0
Meghalaya	11	11	100.0	0	0.0	0	0.0	0	0.0	0	0.0
Mizoram	26	26	100.0	0	0.0	0	0.0	0	0.0	0	0.0
Nagaland	11	11	100.0	0	0.0	0	0.0	0	0.0	0	0.0
Odisha	314	303	96.5	5	1.6	0	0.0	0	0.0	6	1.9
Punjab	138	22	15.9	5	3.6	2	1.4	109	79.0	0	0.0
Rajasthan	295	45	15.3	29	9.8	33	11.2	185	62.7	3	1.0
Sikkim	4	4	100.0	0	0.0	0	0.0	0	0.0	0	0.0

(continued)

Table 6.3 (continued)

States/Union Territories	Total no. of Assessed Unit	Safe		Semicritical		Critical		Overexploited		Saline	
		Nos.	%	Nos.	%	Nos.	%	Nos.	%	Nos.	%
Tamil Nadu	1166	427	36.6	163	14.0	79	6.8	462	39.6	35	3.0
Telangana	584	278	47.6	169	28.9	67	11.5	70	12.0	0	0.0
Tripura	59	59	100.0	0	0.0	0	0.0	0	0.0	0	0.0
Uttar Pradesh	830	540	65.1	151	18.2	48	5.8	91	11.0	0	0.0
Uttarakhand	18	13	72.2	5	27.8	0	0.0	0	0.0	0	0.0
West Bengal[a]	268	191	71.3	76	28.4	1	0.4	0	0.0	0	0.0
Andaman and Nicobar	36	35	97.2	0	0.0	0	0.0	0	0.0	1	2.8
Chandigarh	1	0	0.0	1	100.0	0	0.0	0	0.0	0	0.0
Dadra and Nagar Haveli	1	1	100.0	0	0.0	0	0.0	0	0.0	0	0.0
Daman and Diu	2	1	50.0	0	0.0	1	50.0	0	0.0	0	0.0
Lakshdweep	9	6	66.7	3	33.3	0	0.0	0	0.0	0	0.0
Puducherry	4	2	50.0	0	0.0	0	0.0	1	25.0	1	25.0
Grand total	**6881**	**4310**	**62.6**	**972**	**14.1**	**313**	**4.5**	**1186**	**17.2**	**100**	**1.5**

Source: CGWB (2019a, b)

[a]The Ground Water resources assessment as on 2013 has been considered for the state of West Bengal

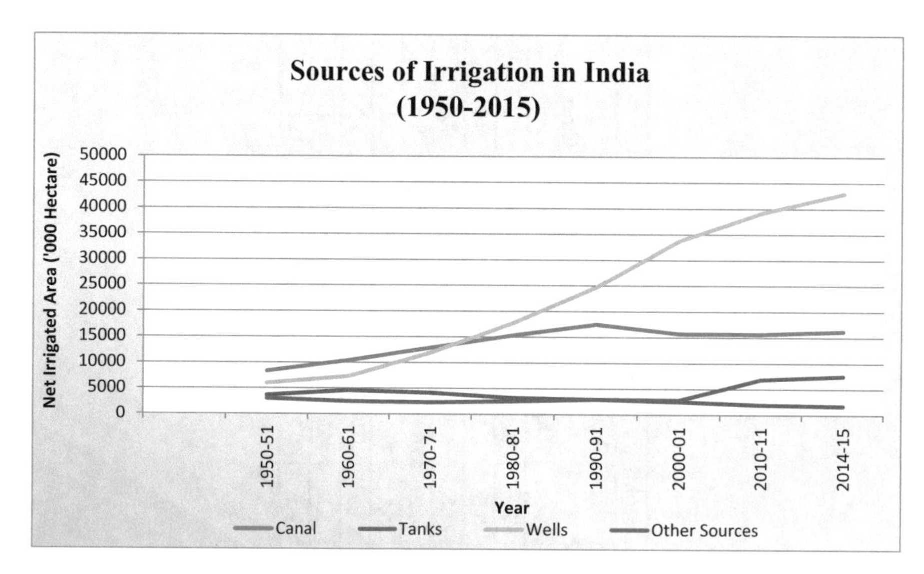

Fig. 6.3 Irrigated area in India according to sources

tubewell irrigation technology, which rose from zero in 1950–1951 to more than 46% in 2014–2015. In the early stage of groundwater extraction during the 1950s, groundwater was lifted mainly from traditional dug wells employing human labor or animal force-dependent devices such as the Persian wheel (Jeet 2005). From 1970 onward, shallow tubewells have emerged as the key technology of extracting groundwater for irrigation use, and during the 1980s, with the alteration of extraction technology, submersible pumps were introduced as deep tubewell irrigation technology, the depth of which increased to beyond 120 m in many areas (Fig. 6.4, Table 6.5). This successive advancement in water-lifting technology, with the government subsidies on electricity for operating electric pumps in the irrigation sector, has led to ever-increasing extraction of groundwater in India. Thus, in many areas groundwater extraction is exceeding annual recharge, lowering the groundwater table (Report of Expert Group 2007). As a consequence, the rate of groundwater extraction in relationship to net annual availability of groundwater in the country, termed stage of groundwater extraction, increased from 32% in 1995 to about 63% in 2017 (CGWB 2011, 2019a). The stage of groundwater extraction is becoming alarmingly higher (>100%) in the states of Delhi, Haryana, Punjab, and Rajasthan, which signifies that in these states annual groundwater withdrawal exceeds annual extractable groundwater resources (Table 6.2).

As per the guidelines of GEC-2015, on the basis of the stage of groundwater development and the long-term trends of pre and post monsoon groundwater levels, of 6881 assessment units (blocks/Mandals/Talkuas/Firkas) in the country, 1186, 313, and 972 units are categorized as overexploited, critical, and semicritical, respectively (Table 6.3; CGWB 2019a).

Table 6.4 Source-wise net irrigated area of India (1950–2015)

Year	Net area under different sources of irrigation in hectares							% to total net irrigated area					
			Groundwater irrigation							Groundwater irrigation			
	Canals	Tanks	Tubewells	Other wells	Total	Other sources	Total	Canals	Tanks	Tubewells	Other wells	Total	Other sources
1950–1951	8,295	3,613	0	5,978	59,78	2967	20,853	39.8	17.3	0.0	5,978.0	28.7	14.2
1960–1961	10,370	4,561	135	7,155	7,290	2440	24,661	42.1	18.5	0.5	7,155.0	29.6	9.9
1970–1971	12,838	4,112	4461	7,426	11,887	2266	31,103	41.3	13.2	14.3	7,426.0	38.2	7.3
1980–1981	15,292	3,182	9531	8,164	17,695	2551	38,720	39.5	8.2	24.6	8,164.0	45.7	6.6
1990–1991	17,453	2,944	14,257	10,437	24,694	2932	48,023	36.3	6.1	29.7	10,437.0	51.4	6.1
2000–2001	15,710	2,518	22,324	11,451	33,775	2831	54,833	28.7	4.6	40.7	11,451.0	61.6	5.2
2010–2011	15,646	1,979	28,543	10,629	39,172	6869	63,666	24.6	3.1	44.8	10,629.0	61.5	10.8
2014–2015	16,182	1,723	31,606	11,354	42,960	7519	68,384	23.7	2.5	46.2	11,354.0	62.8	11.0

Source: Agricultural Statistics, 2014–15

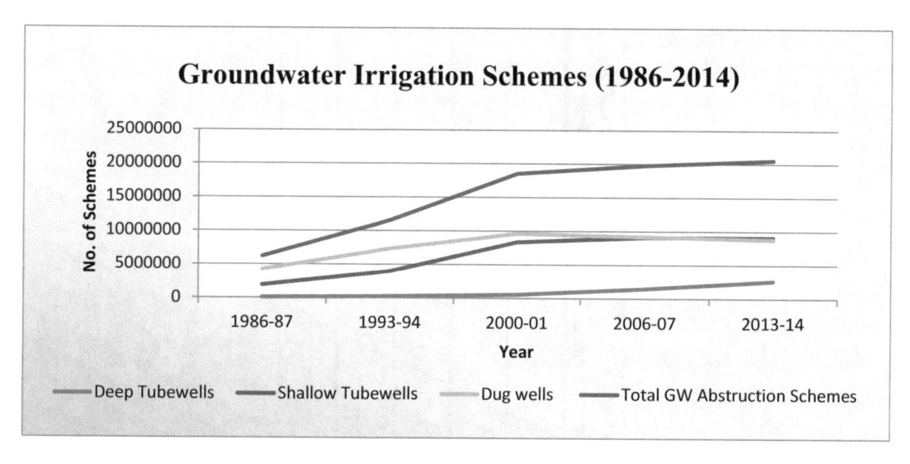

Fig. 6.4 Increasing trend of groundwater irrigation schemes (1986–2014)

Observed Changes in Groundwater Regimes in India

Analysis of spatial and temporal variability of groundwater level helps to understand the complex mechanism of the groundwater regime and its dynamic relationships with different natural and anthropogenic attributes (Chen and Feng 2013; Machiwal and Singh 2015). Natural factors that affect the groundwater regime include climatic parameters such as rainfall and evapotranspiration, and anthropogenic factors include pumping from the aquifer, recharge from irrigation systems, and other practices such as waste disposal (CGWB 2019a, b).

In pre monsoon, monsoon, post monsoon, and winter seasons of 2018–2019, the general depth of the groundwater level of the country ranged between 5 and 20 meters below ground level (MBGL), 0 and 5 MBGL, 2 and 10 MBGL, and 2 and 10 MBGL, respectively. The spatial pattern of groundwater level in four seasons indicates the shallow water level (>10 MBGL) in the Ganga–Brahmaputra alluvial belt and coastal plain areas of the country whereas deep aquifers (>10 MBGL) are observed in northwestern, western, and a few pockets of peninsular India. Groundwater level of more than 40 MBGL was noticed in some parts of Rajasthan, Delhi, and Tamil Nadu (CGWB 2019b).

Comparison of depth of groundwater level of pre monsoon, monsoon, post monsoon, and winter season of 2018–2019, with decadal mean water level (2008–2017) of these seasons indicates that more than 50% of the wells of the country have experienced decline in groundwater level, mostly in the range of 0–2 m. In a few pocket areas of northwestern, southeastern, northern, and central India, more than 4 m fall in groundwater level in comparison to decadal mean values is observed in most seasons (Fig. 6.5) (CGWB 2019b).

Temporal fluctuation in groundwater level has been calculated on the basis of CGWB and different State Agency monitored 12,144 numbers of well data across the country, applying the parametric statistical technique linear regression analysis to

Table 6.5 Types of groundwater irrigation schemes (1986–2014)

Year	Deep tubewells	% to Total GW Irrigation Schemes	Shallow tubewells	% to Total GW Irrigation Schemes	Dug wells	% to Total GW Irrigation Schemes	Total No. Groundwater irrigation Schemes
1st MI Census (1986–1987)	81,007	1.31	1,884,219	30.42	4,227,821	68.27	6,193,047
2nd MI Census (1993–1994)	227,070	1.97	3,944,724	34.22	7,354,905	63.81	11,526,699
3rd MI Census (2000–2001)	530,194	2.87	8,355,692	45.16	9,617,381	51.98	18,503,267
4th MI Census (2006–2007)	1,441,782	7.30	9,114,884	46.14	9,199,551	46.57	19,756,217
5th MI Census (2013–2014)	2,618,792	12.76	9,117,493	44.43	8,785,599	42.81	20,521,884

Source: Report of 1st, 2nd, 3rd, 4th, and 5th M.I. Census, MOWR, GOI

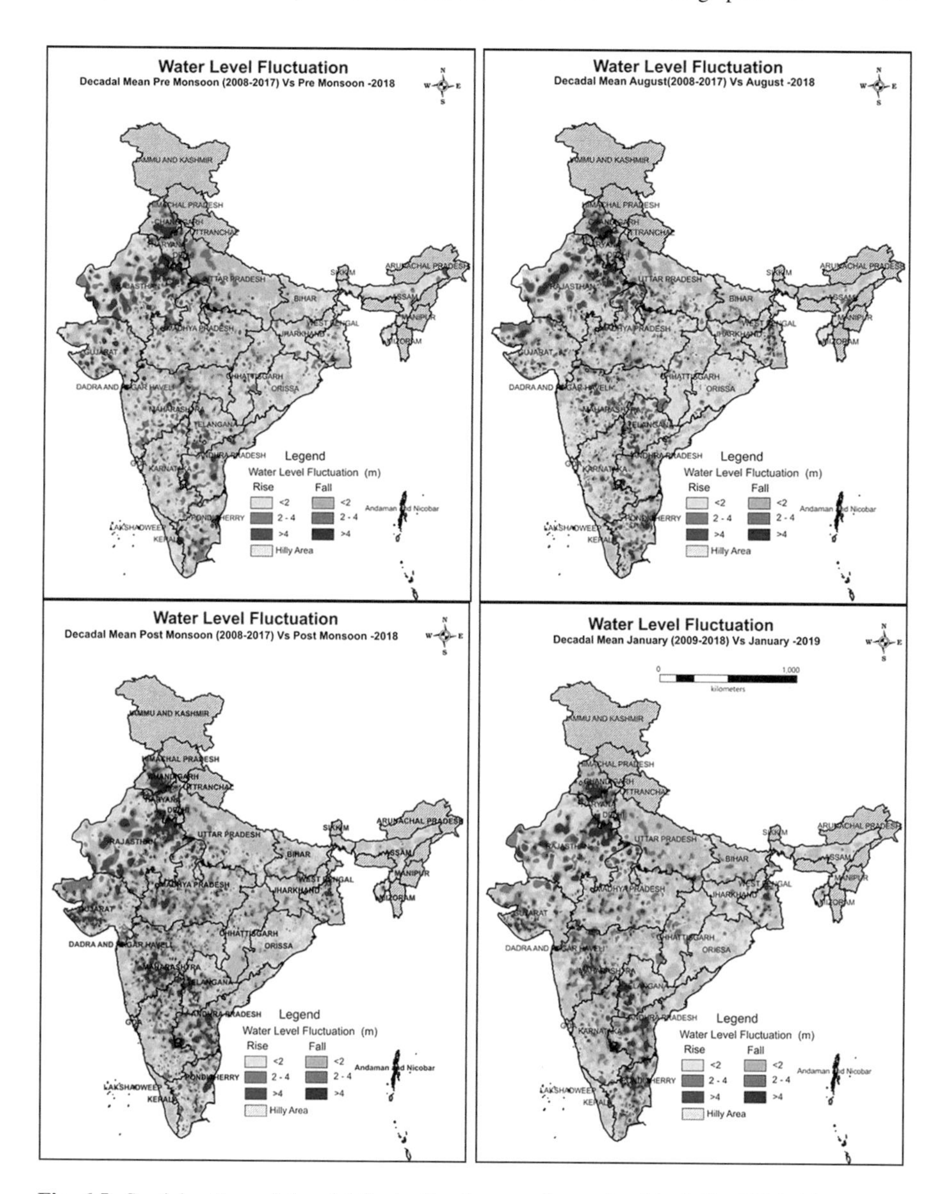

Fig. 6.5 Spatial pattern of decadal fluctuation in groundwater level in India (2009–2018 versus 2018–2019). (Source: CGWB 2019a, b)

detect trends of changes in groundwater level from 1995 to 2019 (Table 6.6). During this time span of 24 years, annual average groundwater level of the country as a whole declined at the rate of 4.7 cm/year (Fig. 6.6). A remarkable declining trend (>30 cm/year) in groundwater level is noticeable in the northwestern part of India in the states/UTs of Chandigarh, Delhi, Punjab, Haryana, Rajasthan, and in Andhra Pradesh and Pondicherry.

Table 6.6 State-wise pattern of temporal fluctuation in groundwater level (GWL)

States/Union Territories	Total no. of monitored wells	Rate of temporal fluctuation in depth to GWL (cm/year)	Nature of fluctuation
Andaman and Nicobar	90	11.4	Rise
Andhra Pradesh	1240	37.8	Fall
Arunachal Pradesh	8	11.8	Fall
Assam	118	2.9	Fall
Bihar	627	2.4	Fall
Chandigarh	12	59.5	Fall
Chhattisgarh	6	2.8	Rise
Dadra and Nagar Hav	17	0.4	Fall
Daman and Diu	8	7.1	Fall
Delhi	84	43	Fall
Goa	69	10.7	Rise
Gujarat	910	11.1	Fall
Haryana	319	36.7	Fall
Himachal Pradesh	105	18.8	Rise
Jammu and Kashmir	199	20.2	Rise
Jharkhand	298	1.7	Rise
Karnataka	5	0.2	Rise
Kerala	429	1.4	Fall
Lakshdweep	NA	NA	NA
Madhya Pradesh	1440	0.2	Rise
Maharashtra	1628	0.5	Fall
Manipur	NA	NA	NA
Meghalaya	13	3	Fall
Mizoram	NA	NA	NA
Nagaland	15	3.8	Rise
Odisha	1260	6.4	Fall
Pondicherry	11	42.6	Fall
Punjab	261	45.1	Fall
Rajasthan	641	32.2	Fall
Sikkim	NA	NA	NA
Tamilnadu	563	13.7	Fall
Telangana	915	0.2	Rise
Tripura	25	5.9	Fall
Uttar Pradesh	6	13	Fall
Uttarakhand	53	11.4	Rise
West Bengal	769	5.6	Fall
Total	12,144	4.7	Fall

Source: Calculated from WRIS-NRSC data

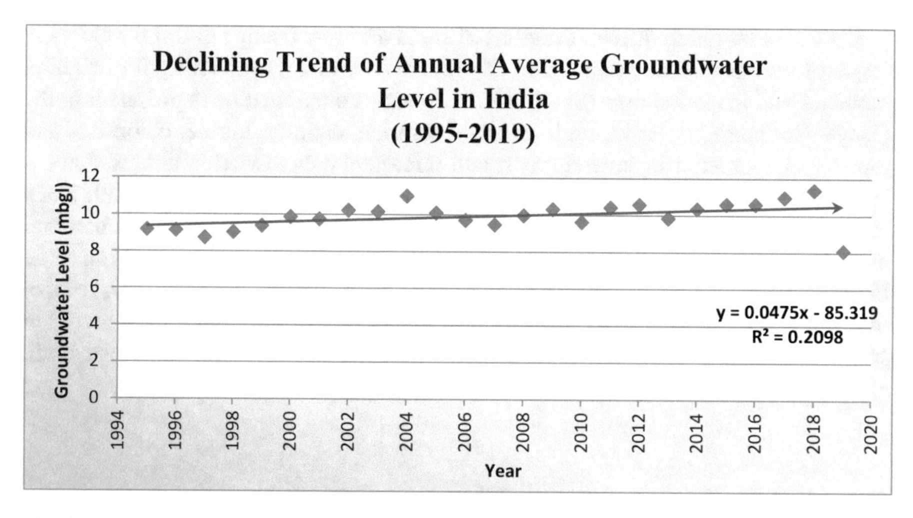

Fig. 6.6 Increasing trend of declining groundwater level in India

Analyzing terrestrial water storage change from the NASA Gravity Recovery and Climate Experiment Satellite (GRACE), Rodell et al. (2009) have identified a declining trend of groundwater level at the rate of 4.0 ± 1.0 cm/year and a depleting trend of groundwater storage at the rate of 17.7 ± 4.5 km^3/year over the northwestern states of Rajasthan, Punjab, Haryana, and Delhi during 2002–2008. They have estimated depletion of about 109 km^3 groundwater during this time span over these states.

Mukherjee et al. (2018) have mentioned that fall in water level in several lower reaches of the Ganges is related to reduction in base flow of groundwater, resulting from depletion of groundwater storage. They have estimated decline in groundwater level and depletion of groundwater storage in the Gangetic aquifer at the rate of −0.30 ± 0.07 cm/year and −2.39 ± 0.56 km^3/year.

Based on analysis of observed well water level data and satellite-based observation, Mukherjee and Bhanja (2019) have pointed out uneven changes of groundwater level (GWL) and groundwater storage (GWS) during 1996–2014 across India. They have observed a declining trend in groundwater level in maximum parts of the Indus–Ganga–Brahmaputra basin and slight rising trend of groundwater level in several parts of southern and western India. Satellite-based estimate indicates decline in groundwater level in northern and eastern parts of India at the rate of −1.40 ± 0.14 and −1.16 ± 0.35 cm/year; and depletion of groundwater storage in northern and eastern India at the rate of 14.02 ± 1.37 and 14.49 ± 4.36 km^3/year, respectively, during 2002–2016. Findings of this study reveal that in some states of western (Gujarat) and southern (Andhra Pradesh) India, groundwater storage has been reduced at the rate of −05.81 ± 0.38 km^3/year and −0.92 ± 0.12 km^3/year, but during 2002–2014 groundwater storage has started to increase at the rate of 2.04 ± 0.20 km^3/year and 0.76 ± 0.08 km^3/year.

On the other hand, different regions of India are now facing several problems in respect to the chemical composition of groundwater and its suitability for drinking purposes and irrigation uses (Singh 2015). Arsenic contamination is prevalent in the Ganga–Brahmaputra basin; high levels of fluoride, salinity, nitrate, chlorides, and iron are encountered in several areas; and microbial contamination affects shallow aquifers throughout the country (CGWB 2011; Singh 2015; Singh et al. 2017). In 2015, of 720 districts in India, EC (salinity), chloride, fluoride, iron, arsenic, and nitrate concentrations in groundwater of shallow aquifers beyond the permissible limits of the BIS standard is found in districts 166, 86, 191, 276, 27, and 337, respectively (CGWB 2018). The overall groundwater quality of this country is slowly but surely declining everywhere because of contamination, overexploitation, or both, associated with urban, industrial, and agricultural activities (CPCB 2008).

Potential Impact of Climate Change on Groundwater

The recent global trend of climate change, leading to changes in precipitation, temperature, and evapotranspiration rate, exhibits its deep impact on the quantity and quality of both surface and subsurface water (Panwar and Chakrapani 2013; Green 2016). The relationship between changes in climatic variables and groundwater is considered more complicated than its relationship with surface water as residence time of groundwater can range from days to tens of thousands of years, which postpone and diffuse the effects of climate change (Holman 2006; IPCC 2007; Chen et al. 2004; Green 2016). However, as part of the hydrological cycle it can be predicted that changes in precipitation pattern and evapotranspiration rate will affect the groundwater system through alteration in the process of recharge potentiality because of changes in the nature of interactions between groundwater and surface water systems and changes in groundwater use (Kumar 2014).

Todd divided the subsurface occurrence of groundwater into two zones: zone of aeration and zone of saturation (Todd 1980). The zone of aeration consists of interstices that are occupied partially by water and partially by air. In contrast, in the zone of saturation all interstices are filled with water and hydrostatic pressure (Todd 1980). The possible impacts of the changing patterns of climatic variables are dominant on these two vertical layers of groundwater in different geographical settings.

Vadose water occurs in the zone of aeration, which may be further subdivided into the soil water zone, intermediate vadose zone, and capillary zone (Todd 1980). Broadly, less infiltration, high evapotranspiration, and high runoff have a great negative impact in the water availability of this upper part of the vadose zone, which may be termed the soil water zone (Panwar and Chakrapani 2013).

Increase in surface temperature leads to a rise in groundwater temperature, which affects pore water chemistry, residence time, and volume of water in matrix and fractures, finally determining the composition of groundwater (Glassley et al. 2003; Panwar and Chakrapani 2013). The effects vary from aquifer to aquifer, and region to region, however, depending on both water composition and underground lithology. Moreover, the effect of climate change on this zone depends on its depth and thickness. The diurnal temperature fluctuation is effective to 1 m in this zone whereas the efficacy of seasonal fluctuation is noticeable to a depth of 10 m or more (Taylor and Stefan 2009; Panwar and Chakrapani 2013).

Because of the large storage capacity of water and its occurrence in greater depth, the zone of saturation in many aquifers is relatively less sensitive to changes in the climate than surface water bodies (Aizebeokhai et al. 2017). Depth of the water table is more strongly associated with the precipitation pattern, but the effect of temperature is important in shallow aquifers (Kundzewicz et al. 2007). The response of the saturated zone to climate change is thus dependent on the depth of the water table in the aquifer; shallow aquifers are more affected by climate change than deep aquifers (Panwar and Chakrapani 2013). Some scholars have opined that overall climate-induced changes have less effect on groundwater regimes in comparison to the nonclimatic factors such as contamination, reduction in stream flow, recharge and lowering of the water table, and the loss of storage because of the huge withdrawals for human use (Hanson et al. 2006; Gurdak et al. 2007; Aribisala et al. 2015; Aizebeokhai et al. 2017; Panwar and Chakrapani 2013). IPCC has stated that 'there is no evidence for ubiquitous climate related trends in groundwater' (IPCC 2007; Green 2016).

Changes in climatic variables have long-term effects on recharge rate and mechanisms (Kundzewicz et al. 2007; Aguilera and Murillo 2009; Green 2016). The response of the groundwater recharge process to climate change depends on the combination of precipitation and temperature. Rise in temperature and evapotranspiration rate may lead to longer persistence of soil moisture deficiency, equalizing the effect of increasing rainfall to replenish groundwater (Singh and Kumar 2010). Overall recharge is predicted to increase in most of the subtropical region (Eckhardt and Ulbrich 2003; Holman 2006; Aizebeokhai et al. 2017). However, in a dry climate, temperature and precipitation are inversely related as rise in temperature will cause failure of moisture supply to the atmosphere because little or no moisture content is available in the soil for evapotranspiration, resulting in decreased rainfall (Vita et al. 2012; Panwar and Chakrapani 2013; Singh and Kumar 2010). Hence, the effect of climate change in arid and semiarid climates is considered to be worse as the amount of groundwater recharge will be reduced (Dettinger and Earman 2007; Aguilera and Murillo 2009; Barthel et al. 2009; Novicky et al. 2010; Panwar and Chakrapani 2013; Green 2016).

Under the climate change scenario, spatiotemporal variations in precipitation, evapotranspiration, recharge, and runoff will directly influence groundwater discharge (Panwar and Chakrapani 2013). Increasing temperature as well as rate of evapotranspiration is responsible for decreasing both groundwater recharge and discharge in all seasons, causing decline in depth of the groundwater table (Woldeamlak et al. 2007). It will adversely influence aquatic life in wetlands and riverine ecosystems that rely on groundwater discharge to support base flow (Woldeamlak et al. 2007; Aizebeokhai et al. 2017).

Groundwater quality is expected to respond to changes in climatic parameters because of the consequences on the recharge and discharge of groundwater (Green 2016). Changes of intensity and duration of recharge and increase in underground temperature in the vadose zone affect pore water chemistry, residence time and volume of water in matrix and fractures, and downward transfer of contaminants, which together modify the chemical composition of groundwater (Glassley et al. 2003; Panwar and Chakrapani 2013). Increased infiltration mobilizes large porewater chloride and nitrate reservoirs in the vadose zone of arid and semiarid regions, resulting in the salinization of groundwater (Gurdak et al. 2007; Green 2016).

Under the climate change scenario, recharge during a dry period may generate a higher concentration of total dissolved solids (TDS), whereas recharge during wet periods may produce lower TDS concentrations in groundwater (Sukhija et al. 1998; Green 2016). In a wet climatic scenario, accelerated infiltration will cause the dissolution of more carbonate rock, resulting in higher calcium content, which increases groundwater hardness (Gurdak et al. 2007; Aizebeokhai et al. 2017).

Intrusion of salinity into coastal aquifers will be enhanced by climate change-induced factors of sea level rise and spatiotemporal changes in precipitation and evapotranspiration that determine recharge (Sharif and Singh 1999; Ranjan et al. 2006; Green 2016). Shallow aquifers of the low-lying coastal areas will be affected more because of the rise in sea level and resultant changes in groundwater flow pattern (Singh and Kumar 2010). Uncontrolled withdrawal and the resultant rapid drop in groundwater level leads to land subsidence in coastal areas, which accelerates the intrusion of salt seawater into the freshwater aquifer (Kundzewicz et al. 2007; Taniguchi et al. 2008; Konikow 2011; Panwar and Chakrapani 2013).

Future Groundwater Scenario in India with Respect to Projected Climatic Change Trends

From the previous analysis it is evident that the groundwater resource in India is now facing crisis from perspectives of both quantity and quality. The northwestern, western, southern, and some coastal areas are the most vulnerable areas of the country in respect to sustainability of this precious resource. In spite of adequate recharge, huge extractions for irrigation purposes is causing overexploitation of groundwater in the northwestern parts of India including the states of Punjab, Haryana, Delhi, and Western Uttar Pradesh. The occurrence of scanty rainfall has resulted in limited recharge of groundwater in the western part of the country, particularly in parts of Rajasthan and Gujarat, leading to stress on this resource. The crystalline aquifer system of southern peninsular India retards the recharge process, limiting availability of the groundwater resource. In several coastal areas of the country salinization of groundwater is an alarming threat to this precious freshwater resource.

The declining trend of groundwater level is very prominent in the states of Punjab, Rajasthan, Haryana, Delhi, Chandigarh, Andhra Pradesh, Pondicherry, Uttar Pradesh, Tamil Nadu, and Gujarat. The projected groundwater levels have been estimated for these critical states (Table 6.7). If the present trend continues, within the year 2050, groundwater level of these states/UTs will have fallen in the range of 4 to 26.5 m, resulting in immense depletion of shallow aquifer storage. The IPCC climatic projection predicts a slight increase in precipitation over India but its spatial pattern will not be uniform (IPCC 2014). In rainfall-deficient areas of India, particularly in western and northern India and in southeastern India, rainfall is likely to be decreased (Salvi et al. 2013). It is evident from previous analysis, in states or UTs of this region, that the groundwater reservoir is already shrinking because of excessive withdrawal; therefore, increase in temperature and decrease in precipitation will multiply the intensity of the crisis.

On the other hand, in the west coastal, west central, eastern, and northwestern parts of India, increased rainfall in the monsoon season may boost recharge to the aquifers (Salvi et al. 2013). However, rise in evapotranspiration resulting from elevated temperature and the persistence of prolonged dry weather in the non-monsoon seasons may necessitate huge extraction of groundwater through pumping for agricultural production, resulting in overexploitation of this resource. Rise in sea level is the most predicted outcome of climate change that will directly affect the coastal low-lying aquifers of India through salinity intrusion, causing reduction in available fresh groundwater storage.

As a consequence of increasing temperature, melting of the Himalayan glaciers will accelerate during the upcoming decades, increasing surface runoff, which may increase potential recharge to the aquifer. As the snowmelt-based runoff begins to decrease, decline in surface runoff may decrease groundwater recharge (Shah 2009). Under these climate change scenarios, the groundwater resource base of the country will be spatiotemporally more diversified and complicated.

Conclusion

To cater to the huge and growing population of the country, groundwater is a proven key resource for its multidimensional usage in agricultural, industrial, and domestic sectors. In different states of India, however, aquifers have already reached a critical condition as extraction has outstripped the replenishment of groundwater. Long-term trends of groundwater level and groundwater storage show that in the near future many areas of the country will face problems of groundwater shortage that will eventually disrupt regional food security, economic growth, and livelihood.

Although the incidence of global climate change is an established fact, relatively limited research has been conducted on the impacts of climate change on the groundwater system. It is high time to give more attention to the forthcoming complex relationships of these two dynamic variables. All that have been studied so far highlight that effects of climate change are significantly greater on the

Table 6.7 Projected groundwater level and climatic parameters in some Indian states

States	Current annual average groundwater level in mbgl (2019)	Projected groundwater level (mbgl) in 2020	Projected groundwater level (mbgl) in 2030	Projected groundwater level (mbgl) in 2040	Projected groundwater level (mbgl) in 2050	Observed trend of precipitation	Observed trend of temperature	Projected trend of precipitation	Projected trend of temperature
Punjab	17.8	18.4	23.0	27.5	32.0	Decreasing	Decreasing	Decreasing	Increasing
Rajasthan	24.9	26.8	27.6	28.4	29.2	Increasing	Increasing	Decreasing	Increasing
Haryana	15.6	17.8	21.5	25.2	28.9	Increasing	Increasing	Decreasing	Increasing
Delhi	16.3	19.9	24.2	28.5	32.8	Decreasing	Increasing	Decreasing	Increasing
Chandigarh	15.0	23.8	29.7	35.6	41.6	Increasing	Increasing	Decreasing	Increasing
Andhra Pradesh	13.8	15.0	18.8	22.5	26.3	Increasing	Increasing	Decreasing	Increasing
Pondicherry	12.7	12.4	16.7	20.9	25.2	Increasing	Increasing	Decreasing	Increasing
Uttar Pradesh	9.3	9.5	10.8	12.1	13.4	Decreasing	Increasing	Decreasing	Increasing
Tamil Nadu	6.8	9.8	11.2	12.6	14.0	Increasing	Increasing	Decreasing	Increasing
Gujarat	11.2	16.5	17.6	18.7	19.8	Increasing	Increasing	Decreasing	Increasing

Source: Estimated from WRIS-NRSC data and previous studies
mbgl meters below ground level

semiarid, arid, and coastal areas. As agriculture is the main source of livelihood for the people of these regions of India, impacts of future climate change will be severe there because of the scarcity of water. Proper understanding of the groundwater dynamics in relationship to the future climate change scenario and economic, demographic, and sociocultural aspects of human society is very much a challenging issue for scientific communities to establish effective guidelines for sustainable management of this precious resource, maintaining parity between economic performance and ecological balance.

In this context, the policy on participatory groundwater management is an urgent measure for conserving this age-old resource of the Earth by involving a collaborative approach among government departments, researchers, NGOs, and community members. Among all the programs, awareness campaigning for mobilizing the local community will be an effective path toward the conservation of groundwater. Otherwise, in future we may find ourselves faced with an environmental catastrophe of our own making. So, the viable solution of today is the practicing of restricted or limited use of this natural resource. Certainly, it will be acknowledged by future generations instead of only creating theoretical moral politics.

References

Adiat KAN, Nawawi MNM, Abdullah K (2012) Assessing the accuracy of GIS-based elementary multi criteria decision analysis as a spatial prediction tool—a case of predicting potential zones of sustainable groundwater resources. J Hydrol 440:75–89. https://doi.org/10.1016/j.jhydrol.2012.03.028

Aguilera H, Murillo J (2009) The effect of possible climate change on natural groundwater recharge based on a simple model: a study of four karstic aquifers in SE Spain. Environ Geol 57(5):963–974

Aizebeokhai AP, Oyeyemi KD, Adeniran A (2017) An overview of the potential impacts of climate change on groundwater resources. J Inform Math Sci 9(2):437–453

Anupam K, Shinjiro K (2013) Climate change and groundwater: vulnerability, adaptation and mitigation opportunities in India. Int J Environ Sci Dev 4(3):272–276. https://doi.org/10.7763/IJESD.2013.V4.352

Aribisala JO, Awopetu MS, Ademilua OI, Okunadel EA, Adebayo WO (2015) Effect of climate change on groundwater resources in Southwest Nigeria. J Environ Earth Sci 5(12):2224–3216

Arkoprovo B, Adarsa J, Prakash SS (2012) Delineation of groundwater potential zones using satellite remote sensing and geographic information system techniques: a case study from Ganjam district, Orissa, India. Res J Recent Sci 1(9):59–66

Attri SD, Tyagi A (2010) Climate profile of India. Met monograph no. Environment Meteorology-01/2010, India Meteorological Department

Barthel R, Sonneveld BGJS, Goetzinger J, Keyzer MA, Pande S, Printz A, Gaiser T (2009) Integrated assessment of groundwater resources in the Oueme basin, Benin, West Africa. Phys Chem Earth 34(4–5):236–250

CGWB (2011) Dynamic ground water resources of India. Central Ground Water Board, Ministry of Water Resources, Government of India, New Delhi

CGWB (2017) Report of the Ground Water Resource Estimation Committee (GEC-2015). Central Ground Water Board, Ministry of Water Resources, RD & GR, Government of India, New Delhi

CGWB (2018) Ground water quality in shallow aquifers of India. Central Ground Water Board, Ministry of Jal Shakti, Department of Water Resources, RD & GR, Government of India, Faridabad

CGWB (2019a) National compilation on dynamic ground water resources of India, 2017. Central Ground Water Board, Ministry of Jal Shakti, Department of Water Resources, RD & GR, Government of India, Faridabad

CGWB (2019b) Ground water year book–India 2018–19. Central Ground Water Board, Ministry of Jal Shakti, Department of Water Resources, RD & GR, Government of India, Faridabad

Chen L, Feng Q (2013) Geostatistical analysis of temporal and spatial variations in groundwater levels and quality in the Minqin oasis, Northwest China. Environ Earth Sci 70:1367. https://doi.org/10.1007/s12665-013-2220-7

Chen Z, Grasby SE, Osadetz KG (2004) Relation between climate variability and groundwater levels in the upper carbonate aquifer, southern Manitoba, Canada. J Hydrol 290(1–2):43–62

Chinnaiah Y (2015) Analysis of ground water levels and water quality in Musi River Basin, Telangana State. Unpublished doctoral thesis. Sri Krishnadevaraya University, Anantapuram, A.P., India

Christensen JH, Krishna Kumar K, Aldrian E, An S-I, Cavalcanti IFA, de Castro M, Dong W, Goswami P, Hall A, Kanyanga JK, Kitoh A, Kossin J, Lau N-C, Renwick J, Stephenson DB, Xie S-P, Zhou T (2013) Climate phenomena and their relevance for future regional climate change. In: Stocker TF, Qin D, Plattner G-K, Tignor M, Allen SK, Boschung J, Nauels A, Xia Y, Bex V, Midgley PM (eds) Climate change 2013: the physical science basis, Contribution of Working Group I to the fifth assessment report of the Intergovernmental Panel on Climate Change. Cambridge University Press, Cambridge, UK and New York

CPCB (2008) Status of groundwater quality in India, part II, Groundwater quality series: GWQS/10/2007–2008. CPCB, Ministry of Environment and Forests, New Delhi

Dains SR, Pawar JR (1987) Economic returns to irrigation in India. Report prepared by SDR Research Groups Inc. for the US Agency for International Development, New Delhi

Das S (2011) Groundwater resources of India. National Book Trust, New Delhi

Dettinger M, Earman S (2007) Western groundwater and climate change – pivotal to supply sustainability or vulnerable in its own right? Ground Water News 4:4–5

Eckhardt E, Ulbrich U (2003) Potential impacts of climate change on groundwater recharge and streamflow in a central European low mountain range. J Hydrol 284(1–4):244–252

Gandhi VP, Bhamoriya V (2011) Groundwater irrigation in India: growth, challenges, and risks. India Infrastruct Rep 90:91–117

Glassley WE, Nitao JJ, Grant CW, Johnson JW, Steefel CI, Kercher JR (2003) The impact of climate change on vadose zone pore waters and its implication for long-term monitoring. Comput Geosci 29:399–411

Green TR (2016) Linking climate change and groundwater. In: Jakeman AJ et al (eds) Integrated groundwater management. Springer, Cham, pp 97–141. https://doi.org/10.1007/978-3-319-23576-9_5

Gun JV, Lipponen A (2010) Reconciling groundwater storage depletion due to pumping with sustainability. Sustainability 2(11):3418–3435. https://doi.org/10.3390/su2113418

Gurdak JJ, Hanson RT, McMahon PB, Bruce BW, McCray JE, Thyne GD, Reedy RC (2007) Climate variability controls on unsaturated water and chemical movement, high plains aquifer, USA. Vadose Zone J 6:533–547

Hanson RT, Dettinger MD, Newhouse V (2006) Relations between climatic variability and hydrologic time series from four alluvial basins across the southwestern United States. Hydrogeol J 14(7):1122–1146

Holman IP (2006) Climate change impacts on groundwater recharge-uncertainty, shortcomings, and the way forward? Hydrogeol J 14(5):637–647

Hoque MA, Khan AA, Shamsudduha M, Hossain MS, Islam T, Chowdhury SH (2009) Near surface lithology and spatial variation of arsenic in the shallow groundwater: southeastern Bangladesh. Environ Geol 56:1687–1695. https://doi.org/10.1007/s00254-008-1267-3

IPCC (2007) In: Solomon S et al (eds) Climate change 2007: the physical science basis, Contribution of Working Group I to the fourth assessment report of the Intergovernmental Panel on Climate Change. Cambridge University Press, Cambridge/New York

IPCC (2013) In: Stocker TF, Qin D, Plattner G-K, Tignor M, Allen SK, Boschung J, Nauels A, Xia Y, Bex V, Midgley PM (eds) Climate change 2013: the physical science basis, Contribution of Working Group I to the fifth assessment report of the Intergovernmental Panel on Climate Change. Cambridge University Press, Cambridge, UK and New York. 1535 pp

IPCC (2014) In: Barros VR, Field CB, Dokken DJ, Mastrandrea MD, Mach KJ, Bilir TE, Chatterjee M, Ebi KL, Estrada YO, Genova RC, Girma B, Kissel ES, Levy AN, MacCracken S, Mastrandrea PR, White LL (eds) Climate change 2014: impacts, adaptation, and vulnerability. Part B: regional aspects, Contribution of Working Group II to the fifth assessment report of the Intergovernmental Panel on Climate Change. Cambridge University Press, Cambridge, UK and New York

Jeet I (2005) Groundwater resources of India: occurrence, utilization and management. Mittal Publications, New Delhi

Jha BM (2007) Groundwater development and management strategies in India. Bhu-Jal News 22:1–15

Jha MK, Chowdhury A, Chowdary VM, Peiffer S (2007) Groundwater management and development by integrated remote sensing and geographic information systems: prospects and constraints. Water Resour Manag 21(2):427–467

Karl TR, Melillo JM, Peterson TC (2009) Global climate change impacts in the United States. Cambridge University Press, New York

Khullar DR (2005) India: a comprehensive geography. Kalyani, New Delhi

Konikow LF (2011) Contribution of global groundwater depletion since 1900 to sea-level rise. Geophys Res Lett 38:L17401

Kothawale DR, Rupa Kumar K (2005) On the recent changes in surface temperature trends over India. Geophys Res Lett 32:L18714. https://doi.org/10.1029/2005GL023528

Kumar CP (2014) Impact of climate change on groundwater resources. GRIN Verlag, Munich. https://www.grin.com/document/281606

Kumar V, Jain SK (2010) Trends in rainfall amount and number of rainy days in river basins of India (1951–2004). Hydrol Res 42(4):290–306

Kumar KR, Sahai AK, Kumar KK, Patwardhan SK, Mishra PK, Revadekar JV, Kamala K, Pant GB (2006) High-resolution climate change scenarios for India for the 21st century. Curr Sci 90 (3):334–345

Kundzewicz ZW, Mata LJ, Arnell NW, Doll P, Kabat P, Jimenez B, Miller KA, Oki T, Sen Z, Shiklomanov AI (2007) Freshwater resources and their management. In: Parry ML, Canziani OF, Palutikof JP, van der Linden PJ, Hanson CE (eds) Climate change 2007: impacts, adaptation and vulnerability. Cambridge University Press, Cambridge

Lal M (2001) Climatic change–implications for India's water resources. J Soc Econ Dev III (1):57–87

Machiwal D, Singh PK (2015) Understanding factors influencing groundwater levels in hardrock aquifer systems by using multivariate statistical techniques. Environ Earth Sci 74 (7):5639–5652. https://doi.org/10.1007/s12665-015-4578-11

Mall RK, Gupta A, Singh R, Singh RS, Rathore LS (2006) Water resources and climate change: an Indian perspective. Curr Sci 90(12):1610–1624

Manap MA, Sulaiman WNA, Ramli MF, Pradhan B, Surip N (2013) A knowledge-driven GIS modeling technique for groundwater potential mapping at the Upper Langat Basin, Malaysia. Arab J Geosci 6(5):1621–1637

Meinzen-Dick R (1996) Groundwater markets in Pakistan: participation and productivity. International Food Policy Research Institute, Washington, DC

Mukherjee A, Bhanja SA (2019) An untold story of groundwater replenishment in India: impact of longterm policy interventions. In: Singh A et al (eds) Water governance: challenges and prospects. Springer, Singapore. Retrieved from https://doi.org/10.1007/978-981-13-2700-1_11

Mukherjee P, Singh CK, Mukherjee S (2012) Delineation of groundwater potential zones in arid region of India—a remote sensing and GIS approach. Water Resour Manag 26:2643–2672

Mukherjee A, Bhanja SN, Wada Y (2018) Groundwater depletion causing reduction of baseflow triggering Ganges river summer drying. Sci Rep 8:12049. https://doi.org/10.1038/s41598-018-30246-7

Novicky O, Kasparek L, Uhlik J (2010) Vulnerability of groundwater resources in different hydrogeology conditions to climate change. In: Taniguchi M, Holman IP (eds) Groundwater response to changing climate. International Association of Hydrogeologists. CRC Press/Taylor & Francis, London

Panwar S, Chakrapani GJ (2013) Climate change and its influence on groundwater resources. Curr Sci 105(1):37–46

Rajaram K (2006) Geography of India. Spectrum Books, New Delhi

Ranjan SP, Kazama S, Sawamoto M (2006) Effects of climate and land use changes on groundwater resources in coastal aquifers. J Environ Manag 80(1):25–35

Rathore LS, Attri SD, Jaswal AK (2013) State level climate change trends in India. Meteorological monograph no. ESSO/IMD/EMRC/02/2013, India Meteorological Department

Report of the Expert Group (2007) Ground water management and ownership. Planning Commission, Government of India, New Delhi

Rodell M, Velicogna I, Famiglietti JS (2009) Satellite-based estimates of groundwater depletion in India. Nature 460:999–1002. https://doi.org/10.1038/nature08238

Salvi K, Kannan S, Ghosh S (2013) High-resolution multisite daily rainfall projections in India with statistical downscaling for climate change impacts assessment. J Geophys Res Atmos 118:3557–3578. https://doi.org/10.1002/jgrd.50280

Shah T (1993) Groundwater markets and irrigation development: political economy and practical policy. Oxford University Press, Mumbai

Shah T (2009) Climate change and groundwater: India's opportunities for mitigation and adaptation. Environ Res Lett 4(13):175–195

Sharif MM, Singh VP (1999) Effect of climate change on sea water intrusion in coastal aquifers. Hydrol Process 13(8):1277–1287

Singh EJ (2015) Trace element analysis of groundwater in Manipur Valley with special reference to arsenic contamination (unpublished doctoral thesis). Assam University, Silchar, Assam, India

Singh RD, Kumar CP (2010) Impact of climate change on groundwater resources. Retrieved from https://www.researchgate.net/publication/215973855

Singh EJ, Singh NR, Gupta A (2017) Hydrochemistry of groundwater and quality assessment of Manipur Valley, Manipur, India. Int Res J Environ Sci 6(9):8–18

Sukhija BS, Reddy DV, Nagabhushanam P (1998) Isotopic fingerprints of paleoclimates during the last 30,000 years in deep confined groundwaters of southern India. Quat Res 50(3):252–260

Taniguchi M, Burnett WC, Ness GD (2008) Integrated research on subsurface environments in Asian urban areas. Sci Total Environ 404:377–392

Taylor CA, Stefan HG (2009) Shallow groundwater temperature response to climate change and urbanization. J Hydrol 375:601–612

Todd DK (1980) Groundwater hydrology, 2nd edn. Wiley India, New Delhi

Todd DK, Mays LW (2005) Groundwater hydrology, 3rd edn. Wiley, Hoboken

UNESCO (1994) Groundwater, IHP humid tropics programme serial no. 8. UNESCO, Paris, pp 29–30

UNFCC (1992) United Nations framework convention on climate change 1992, art 1

Vita DP, Allocca V, Manna F, Fabbrocino S (2012) Coupled decadal variability of the North Atlantic Oscillation, regional rainfall and karst spring discharges in the Campania region (southern Italy). Hydrol Earth Syst Sci 16:1389–1399

Wang X, Farooq SH, Keramat M, Lubis RF, Akhter MG (2013) Climate change vulnerability and adaptation in ground water dependent irrigation system in Asia-Pacific Region. Asia-Pacific Network for Global Change Research, Project Reference Number: ARCP2013-25NSY-Shahid

Wani SP, Sudi R, Pathak P (2009) Sustainable groundwater development through integrated watershed management for food security. International Crops Research institute for the Semi Arid Tropics (ICRISAT), Patancheru

Woldeamlak ST, Batelaan O, Smedt FD (2007) The effects of climate change on groundwater system in the Grote-Nete catchment, Belgium. Hydrogeol J 15(5):891–901

Chapter 7
Comparison of Classical Mann–Kendal Test and Graphical Innovative Trend Analysis for Analyzing Rainfall Changes in India

Tapash Mandal, Apurba Sarkar, Jayanta Das, A. T. M. Sakiur Rahman, and Pradip Chouhan

Abstract The analysis of rainfall trends is of the utmost importance for sustainable management and planning of agriculture and water resources under changing climate. Therefore, this study aims to detect the long-term rainfall trends of 36 meteorological subdivisions of India for the period 1901–2015. The graphical innovative trend analysis (ITA) technique and the classical statistical Mann–Kendall (MK) or modified Mann–Kendall (mMK) tests are used to detect historical rainfall changes. The nonparametric Sen's slope (Q) estimator is also applied to measure the magnitude of change. The ITA results show about 61.11% of the subdivisions exhibited decreasing trends in annual rainfall. On the seasonal scale, about 58.33%, 50%, 69.44%, and 88.88% of the subdivisions are experiencing decreasing trends in pre-monsoon (Mar–May), monsoon (Jun–Sep), post-monsoon (Oct–Nov), and winter (Dec–Feb) seasons, respectively. The results of the Z statistic show a good match (about 90%) with the graphical ITA method. However, the ITA method is able to detect subtrends while, MK/mMK shows only the monotonic trend. Such a trend analysis of rainfall can provide significant information that will be useful to build up adaptive capacity and community resilience against climate change. This study can be helpful for regional-scale planning about pre and post-disaster floods, drought mitigation, and agricultural development.

T. Mandal
Department of Geography and Applied Geography, University of North Bengal, Darjeeling, West Bengal, India

A. Sarkar (✉) · P. Chouhan
Department of Geography, University of Gour Banga, Malda, West Bengal, India

J. Das
Department of Geography, Rampurhat College, Birbhum, West Bengal, India

A. T. M. S. Rahman
Department of Earth and Environmental Science, Kumamoto University, Chuo-ku, Kumamoto, Japan

M. N. Islam, A. van Amstel (eds.), *India: Climate Change Impacts, Mitigation and Adaptation in Developing Countries*, Springer Climate,
https://doi.org/10.1007/978-3-030-67865-4_7

Keywords Hydrometeorological time series · Temporal changes · Graphical trend analysis · Classical trend analysis · Climate change · India

Introduction

Precipitation, an important component of the global hydrological circle, is significant in maintaining energy balance between terrestrial and atmospheric ecosystems (Mandal et al. 2020; Das et al. 2020b; Satishkumar and Rathnam 2020). Variability in rainfall has considerable influence on agriculture, forest, and water resources, and therefore the changes in rainfall, temperature, and humidity are important aspects that indicate climate change (Huang et al. 2020). According to Krishnan et al. (2016), changes in rainfall and temperature alter the hydrodynamic and thermal regime of the rivers, stream discharge, surface runoff, and the freshwater ecosystem. The amount of rainfall is an important part of the water resources of a country with multidimensional utility and considerable impact on socioeconomic conditions (Tabari and Talaee 2011). A significant association is observed between rainfall variability and the occurrence of extreme events such as floods, droughts, and landslides that make the community more vulnerable and cause huge economic loss (Roxy et al. 2017). Extreme or low rainfall events affect the availability of groundwater, soil moisture, runoff, and stream discharge, ultimately affecting the occurrence of floods and droughts. Furthermore, spatiotemporal changes in precipitation affect the agricultural system: crop type, cropping intensity, productivity, and the increased risk of crop failure (Farook and Kannan 2016; Guntukula and Goyari 2020; Rahman et al. 2020).

The Indian climate is controlled by monsoon winds: 75–90% of the rainfall in the Indian subcontinent is determined by the summer monsoon, which contributes about 60% of gross water demand for agriculture (Guhathakurta et al. 2011). Seasonal variation and geographical distribution of the rainfall in India is highly heterogeneous (Sreelekha and Babu 2019; Sinha et al. 2019; Varikoden et al. 2019). The Indian summer monsoon rainfall (ISMR) greatly influences the Indian economy because about 55% of the rural workforce is directly linked with agriculture and 60% of the net area is in the rainfed zone (Das et al. 2020a). The extension of agriculture and the rapid increase of population swell the demands for water resources (Islam and van Amstel 2018; Sarkar and Chouhan 2019, 2020). It is also observed that many hydrological phenomena are directly associated with extreme rainfall events or a low rainfall event and that such events are cyclic (Chang et al. 2009). Time series analysis can capture this periodicity of events and helped to forecast near-future phenomena (Islam and van Amstel 2018; Khan et al. 2018). For the researcher's evaluation of long-term trends in meteorological data are an important aspect of understanding climate change and its impact on human communities (Nisansala et al. 2020). Therefore, analyses of rainfall trends, patterns, and their spatial distribution are crucial not only for agriculture and food security, but also for hydrological aspects (Gedefaw et al. 2018; Nikumbh et al. 2019).

Extensive research has addressed the spatiotemporal dynamism of rainfall in the Indian subcontinent, emphasizing seasonality, periodicity, and global and local trends using a nonparametric statistical approach such as the Mann–Kendall (MK) or modified Mann–Kendall (mMK) test, rank correlation test, Sen's slope estimation, and the Spearman rank correlation test (Gedefaw et al. 2018; Das and Bhattacharya 2018; Das et al. 2019, 2020c, d). A few studies have also detected trends in annual and seasonal rainfall in India using both parametric methods (i.e., linear regression) and nonparametric methods (i.e., MK). Many researchers have documented that there is no evidence of a considerable trend in annual rainfall in India as a whole (Krishnan et al. 2016; Wang et al. 2020), although some studies show a decreasing trend in summer monsoon rainfall (Dash et al. 2009) with pre-monsoon, post-monsoon, and winter rainfall showing an increasing trend at the national scale. Many studies show that the frequency of extreme rainfall events has increased in the past 50 years, but the seasonal mean rainfall does not show any significant trend in central India (Sahoo et al. 2020). However, these trend analysis methods are purely statistical and do not detect trends at high, medium and low values at a single computation process. Hence, for effective and efficient water resource management, an advanced and flexible graphical technique of trend analysis at high, medium, and low values with such traditional methods is mandatory (Şen 2012). For that reason, a new approach, innovative trend analysis (ITA), was introduced by Şen for hydrometeorological analysis. ITA is a new and robust trend detection techniques that have already been used several studies in the field of hydrometeorological data such as rainfall (Ahmad et al. 2018; Das et al. 2020b; Ay and Kisi 2015; Öztopal and Şen 2017; Mandal et al. 2020), temperature (Cui et al. 2017; Mohorji et al. 2017), evapotranspiration (Kisi 2015; Pour et al. 2020; Tabari et al. 2011), streamflow (Changnon and Demissie 1996; Machiwal and Jha 2012; Şen 2012), groundwater level (Das et al. 2020a, b; Satishkumar and Rathnam 2020), and water quality parameters (Kisi and Ay 2014) in several parts of the world.

The main objectives of this study are to identify rainfall trends using the ITA for quantifying the subtrends considering 36 meteorological subdivisions in India (1901–2015) and make a comparison with the traditional nonparametric methods. The graphical trend detection method classify the given hydrometeorological data into "low," "medium," and "high" cluster for more convenient interpretation. In this study, the nonparametric Mann–Kendall (MK) test (Mann 1945; Kendall 1975) or the modified MK test (Hamed and Rao 1998) has been used to assess the reliability of the ITA (De Leo et al. 2020). Furthermore, the percent bias (P_{BIAS}) technique is also used for estimating the changes in rainfall following the first and second half of the ITA pre-process data, as well as the slope of changes estimated by Sen's estimator method (Şen 1968).

Study Area

India is the tropical monsoon country of Southeast Asia, particularly well known for agricultural production. The whole study area lies between latitudes 8°04′N and 37°06′ N and longitude 68°07′E and 97°25′E, with an area of 3,434,833 km^2. The country is characterized by four seasons: pre-monsoon (March to May) with high temperature and high evaporation, monsoon (June to September) with high intensity of rainfall, post-monsoon (October to November), the hot and humid period with less rainfall, and finally the winter season (December to February), the coldest and driest period. India can be separated into five physiographic divisions: (i) the Himalayan Mountain, (ii) the Great Plain of North India, (iii) the Peninsular Plateau, (iv) the Coastal Plains, and (v) the Islands. The Himalaya and the Thar Desert strongly affect the weather of the country, making it vulnerable to climate, economic, and cultural change. The largest share of the semiarid rainfed agriculture system and nearly 700 million of her 1 billion population living in rural areas directly depend on climate-sensitive sectors. This large section of the population is likely to be most affected by climate change because their livelihoods are critically correlated with agriculture, and natural resources are highly sensitive to regional climate. The map of the study area is seen in Fig. 7.1.

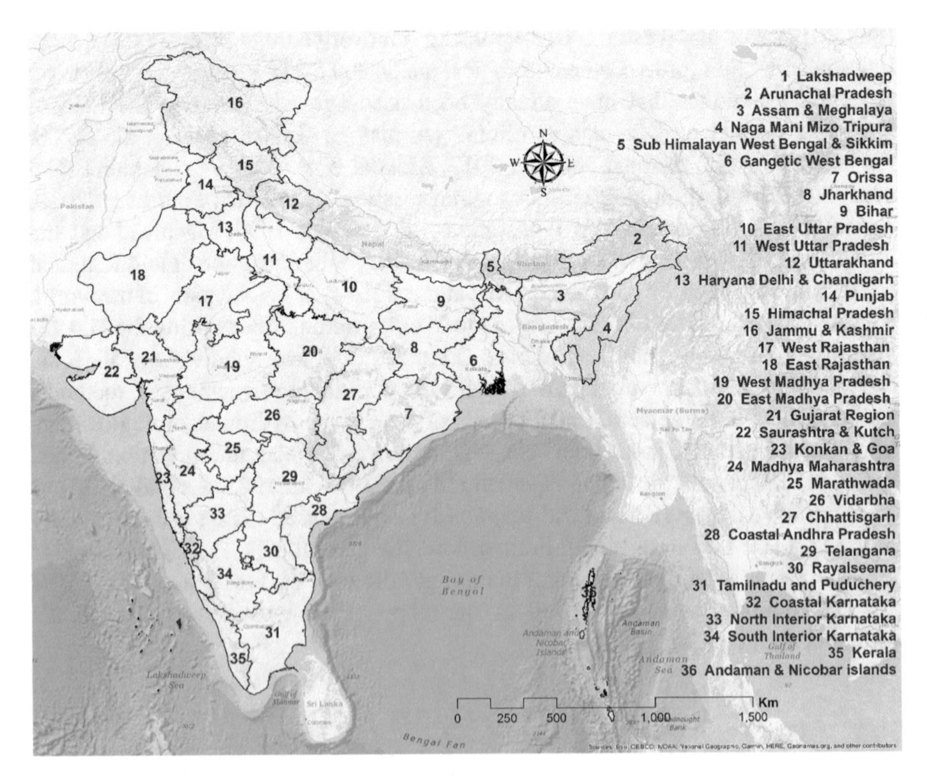

Fig. 7.1 Location map of the study area: numbers refer to the meteorological subdivisions of India

Database

The data sets used for the current investigation consist of the average monthly rainfall of 36 meteorological subdivisions from 1901 to 2015. The meteorological subdivision rainfall data were collected from the open-source platform-https://water.india.gov.in/content/water/en/waterresourcesdepartment/WaterManagement/IWRM.html. The hydrological year in India starts in April and ends in March. The missing data in the time series were observed less than 1%, which have been calculated using multiple imputation methods. R 3.5.1 and Arc 10.3 software have been used for statistical analysis and mapping purposes, respectively. The various methods used in this study are described in the following sections.

Methods

In this study, to detect the trends in historical rainfall data, ITA (Şen 2012) is used. Further, the results of ITA are equated with the MK test (Mann 1945; Kendall 1975) or mMK test (Hameed and Rao). The slope of changes is estimated by Sen's slope estimator (Şen 1968). However, a modified Mann–Kendell test (mMK) (Güçlü 2020; Phuong et al. 2020) has been performed for the serially correlated dataset as the performance of the ordinary MK test in such conditions is very poor and gives erroneous estimations (Sharma and Goyal 2020) because these statistics increase the risk of overestimates or underestimates of the Z statistic if the time series has significant autocorrelation (Ay and Kisi 2015; Rahman et al. 2017). Therefore, MK autocorrelation lag-1 should be executed first to confirm the reliability of the Z statistic. If significant autocorrelation exists in the data series, researchers suggests moving to the modified Mann–Kendall test (Güçlü 2020).

Innovative Trend Analysis

To perform ITA, first the monthly mean rainfall of a subdivision (j) is split into two subseries of equal numbers of observations. Then, each series is rearranged in ascending order and plotted against the other into a Cartesian coordinate system (Girma et al. 2020) wherein the first half is plotted on the x-axis and the second half on the y-axis. In the third step, a straight line is fitted with the scatter plot that represents 'monotonic trend or no trend.' If the scatter is concentrated above the 45° line (1:1), the time series exhibits an increasing trend; if the scatter points concentrate below the 1:1 line, a decreasing trend is indicated (Cui et al. 2017). If the scatter points are concentrated along the trend line, this indicates there is no trend in the data series (Şen 2012). The calculation of indicator of trend (TI) (Şen 2012) is derived from the following equation:

$$B = \frac{1}{n} \sum_{i=1}^{n} \frac{10(x_j - x_k)}{\overline{x}} \tag{7.1}$$

where B represents ITA slope, n denotes the extent of individual subseries, x_j and x_k represent the values of the consecutive subseries, and $\overline{x}$ represents the mean of the first subseries (x_k).

The positive slope of the B value indicates an increasing trend in the series, whereas the negative value of the slope signifies a decreasing tendency in the time series.

Mann–Kendall (MK)Test

The test statistic (S) of a time series m_1, m_2, m_3. . ., and m_n can be achieved by using the MK test:

$$S = \sum_{k=1}^{n-1} \sum_{j=k+1}^{n} \text{sign}(m_j - m_k) \tag{7.2}$$

where n denotes the length of the data sets, and m_j and m_k denote the observations at times j and k.

$$\begin{aligned} \text{Sign}(m_j - m_k) &= +1 \text{ if } m_j - m_k > 0 \\ &= \quad 0 \text{ if } m_j - m_k = 0 \\ &= -1 \text{ if } m_j - m_k < 0 \end{aligned} \tag{7.3}$$

Positive S values show an increasing or upward trend, and negative values of S indicate a decreasing or downward trend in the time series data.

The variance of test statistics VAR(S) can be achieved by the equation

$$\text{VAR}(S) = \frac{1}{18} \left\{ n(n-1)(2n+5) - \sum_{i=1}^{\rho} \tau_i(\tau_i - 1)(2\tau_i + 5) \right\}. \tag{7.4}$$

Here, ρ denotes the tied group number of observations to group I, which is a set of sample data with similar value, and τ_i indicates the extent of the ith ties number.

The estimated S and VAR(S) (Hamed and Rao 1998) are used to estimate the test statistic Z when n is >10 (Gilbert 1987):

$$Z = \begin{cases} \frac{S-1}{\sqrt{\mathrm{VAR}(S)}} & , \text{if } S > 0 \\ 0 & , \text{if } S = 0 \\ \frac{S+1}{\sqrt{\mathrm{VAR}(S)}} & , \text{if } S < 0 \end{cases} \tag{7.5}$$

A positive (+) and negative (−) value of the Z statistic signifies the directions of trends; the (+) values indicate an increasing trend and (−) values indicate a decreasing trend.

Modified Mann–Kendall (mMK) Test

The modified VAR (S) statistics can be estimated for the equation

$$\mathrm{VAR}(S) = \left(\frac{n\,(n-1)(2n+5)}{18}\right)\cdot\left(\frac{n}{n_e^*}\right) \tag{7.6}$$

Here, the correction factor $\left(\frac{n}{n_e^*}\right)$ is adjusted to the auto-correlated data as follows:

$$\left(\frac{n}{n_e^*}\right) = 1 + \left(\frac{2}{n^3 - 3n^2 + 2n}\right)\sum_{\mathrm{f}=1}^{n-1}(n-f)(n-f-1)(n-f-2)\,\rho_e(f) \tag{7.7}$$

$\rho_e(f)$ signifies the autocorrelation between ranks of observations and can be estimated as

$$\rho(f) = 2\sin\left(\frac{\pi}{6}\,\rho_e(f)\right) \tag{7.8}$$

Sen's Slope Estimator

Sen's slope (Şen 1968) estimator is also used to estimate the magnitude of change of (slope θ). The slope θ can be driven from N pairs of data as follows:

$$\theta_i = \frac{x_k - x_j}{k - j}, \qquad i = 1, 2, \cdots\cdots N, \quad k > j \tag{7.9}$$

where x_k and x_j characterize the values of data at k, j times, and θ_i is the median slope, respectively.

Percent Bias

In this study, in order to estimate the percentage of change of rainfall in the second half compared to first half of the time series, the percent bias method (Moriasi et al. 2007) was used following this formula:

$$\text{PBIAS} = 100 - \sum_{i-1}^{n} \frac{Yi}{Xi} \times 100 \tag{7.10}$$

where P_{BIAS} represents percent bias, n is the total extent of the subseries separately, *Xi* and *Yi* are the values of the observational data in the first and second subseries, respectively. The positive and negative values of P_{BIAS} indicate an increasing and decreasing trend in respect to first subseries.

Results and Discussion

Descriptive Analysis of Annual Rainfall

Table 7.1 depicts some statistical parameters of annual rainfall, such as mean minimum, mean maximum, mean, standard deviation (SD), coefficient of variation (CV), skewness (C_S), and kurtosis (C_K) of the annual rainfall of 36 meteorological subdivisions. The average annual rainfall varied from 292.64 mm $\pm$ 108.85 (West Rajasthan) to 3405.96 mm $\pm$ 481.3.97 (Coastal Karnataka), with CV of 27.20% and 14.13%, respectively, whereas the country mean rainfall is 1384.59 mm $\pm$ 106.97. The SD of annual rainfall varied from 108.85 (CV 37.20%) (West Rajasthan) to 485.10 (CV of 16.29%) (Kankan & Goa). The mean minimum and mean maximum rainfall varied from 62 mm (West Rajasthan) to 2511 mm (Coastal Karnataka) and from 769 mm (West Rajasthan) to 5554 mm (Coastal Karnataka), respectively. The CV of the datasets indicates the medium to high variability of rainfall in the study region as it varied from 12.02 (Assam & Meghalaya) to 40.78 (Saurashtra & Kutch). The western and northwestern parts of India show maximum rainfall variability; however, the annual rainfall is quite low in these parts. The other parts show minimum rainfall variability with medium to high annual rainfall. Similarly, Fig. 7.2 shows the annual rainfall variability and CV of the 36 subdivisions in India. This figure reveals low annual rainfall over the northwestern, extreme north, and southeastern portions (rain shadow zone), and high annual rainfall over the northeastern and southwestern portions of the study area. However, the CV is high in the northwestern and low in the northeastern parts of the country (Fig. 7.2).

Skewness is basically a measure of the degree of symmetry or asymmetry in any given dataset. As per results, skewness of the Indian rainfall varied from -0.74 (Assam and Meghalaya) to 2.16 (Arunachal Pradesh), with a country mean skewness of 0.09. Table 7.1 also shows that the data are skewed in nature because the value of

Table 7.1 Meteorological subdivisicn-wise statistical properties of annual rainfall (1901–2015)

Serial No.	Subdivisions	Area (km^2)	Minimum	Maximum	Mean	SD	CV (%)	C_S	C_K
1	Lakshadweep	32.69	993	2362	1599.8	264.05	16.51	0.38	0.25
2	Arunachal Pradesh	83,743	1427	3441	1937.53	258.21	13.33	2.16	10.12
3	Assam & Meghalaya	109,096	1744	3404	2580.75	310.21	12.02	−0.07	0.14
4	Naga Mani Mizo Tripura	70,495	1354	4316	2433.66	431.71	17.74	0.65	2.48
5	Sub-Himalayan West Bengal & Sikkim	21,625	1988	3655	2752.24	337.89	12.28	0.4	0.04
6	Gangetic West Bengal	66,228	1015	2100	1490.57	229.48	15.4	0.27	−0.48
7	Orissa	155,842	987	1945	1458.23	187.09	12.83	0.13	0.43
8	Jharkhand	79,638	697	1899	1309.37	198.23	15.14	−0.1	0.3
9	Bihar	94,235	629	1661	1197.7	194.66	16.25	−0.31	0.24
10	East Uttar Pradesh	146,509	494	1545	979.25	203.6	20.8	0.17	0.08
11	West Uttar Pradesh	96,782	372	1244	827.13	189.64	22.93	−0.13	−0.28
12	Uttarakhand	53,483	804	2103	1465.75	263.98	18.01	0.08	−0.31
13	Haryana Delhi & Chandigarh	45,698	235	987	530.52	142.16	26.8	0.51	0.12
14	Punjab	50,376	275	1223	593.59	163.38	27.52	0.94	1.72
15	Himachal Pradesh	55,673	776	1919	1260.4	247.36	19.63	0.49	−0.24
16	Jammu & Kashmir	78,114	657	1732	1138.17	234.49	20.6	0.67	−0.1
17	West Rajasthan	195,086	62	769	292.64	108.85	37.2	0.95	2.52
18	East Rajasthan	147,128	274	1351	655.25	166.43	25.4	0.45	1.75
19	West Madhya Pradesh	175,317	510	1434	941.59	184.82	19.63	0.4	−0.09
20	East Madhya Pradesh	135,156	654	1747	1205	219.5	18.22	0.01	−0.45
21	Gujarat Region	86,034	393	1620	918.3	278.4	30.32	0.25	−0.38
22	Saurashtra & Kutch	109,950	93	1120	495.25	201.96	40.78	0.51	0.15
23	Konkan & Goa	34,095	1683	4000	2977.74	485.1	16.29	−0.2	0.03
24	Madhya Maharashtra	115,306	438	1396	880.3	159.1	18.07	0.02	0.53
25	Marathwada	64,525	347	1198	790.75	189.21	23.93	0.25	−0.43

(continued)

Table 7.1 (continued)

Serial No.	Subdivisions	Area (km^2)	Minimum	Maximum	Mean	SD	CV (%)	C_S	C_K
26	Vidarbha	97,536	578	1606	1095.55	203.49	18.57	0.1	−0.31
27	Chhattisgarh	146,138	905	1974	1371.76	210.68	15.36	0.09	0.17
28	Coastal Andhra Pradesh	93,045	703	1713	1052.93	190.42	18.08	0.77	0.95
29	Telangana	114,726	437	1545	953.44	205.93	21.6	0.37	0.22
30	Rayalseema	69,043	433	1278	766.22	150.56	19.65	0.53	0.26
31	Tamilnadu and Puduchery	130,068	318	1365	943.79	165.96	17.58	−0.74	2.61
32	Coastal Karnataka	18,717	2511	5554	3405.96	481.3	14.13	0.87	2.41
33	North Interior Karnataka	79,895	470	1096	717.83	134.08	18.68	0.54	0.34
34	South Interior Karnataka	93,171	733	1410	1040.39	150.84	14.5	0.03	−0.56
35	Kerala	38,864	2069	4258	2925.58	422.12	14.43	0.5	0.68
36	Andaman & Nicobar islands	8250	1849	3938	2929.45	403.21	13.76	0.04	−0.18
India		2,942,090	1125	1666	1384.59	106.97	7.73	0.09	0.06

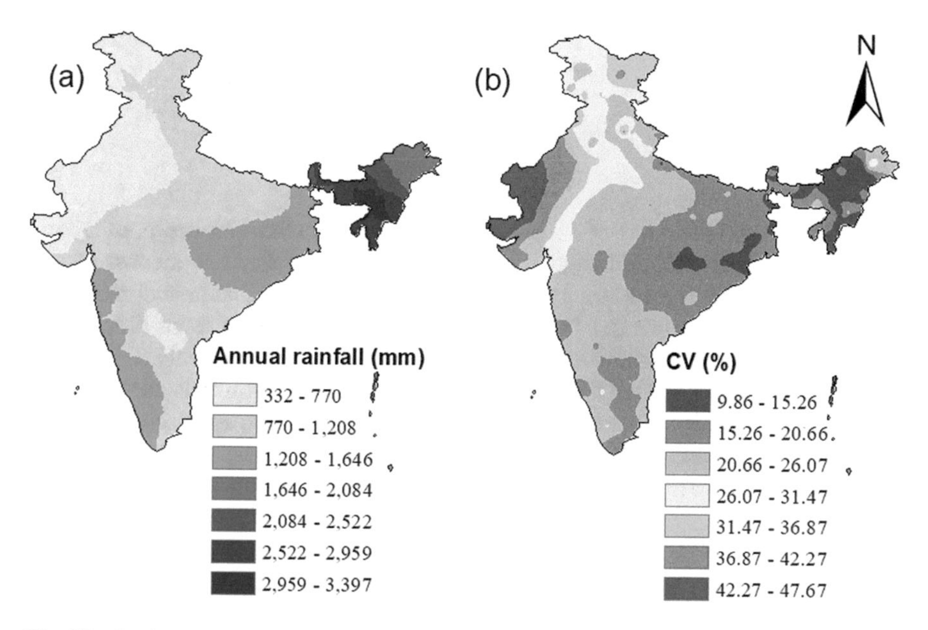

Fig. 7.2 Spatial distribution of (**a**) annual rainfall and (**b**) coefficient of variation

the skewness coefficient (Pearson's) is not smaller than zero and the majority are positively skewed; also, the mean, median, and mode of the datasets are different from each other. On the other hand Kurtosis of the time series dataset varied from −0.56 (in south interior Karnataka) to 10.12 (in Arunachal Pradesh). As the values are lesser and greater than zero, the kurtosis is not mesokurtic (normal distribution); rather, the kurtosis specifies platykurtic (thin tails) and leptokurtic (fat tails) i.e. non normal distribution.

Trends in Rainfall Pattern

In this study, a comparative assessment has been carried out between the advanced graphical method of trend detection, that is, the innovative trend analysis (ITA) and a traditional nonparametric statistical approach (MK and mMK tests), for investigation of long-term (1901–2015) trends in the annual and seasonal rainfall of India. As space is inadequate, only the ITA graph of some selected meteorological subdivisions from the annual and seasons scale is shown in this paper. The ITA graphs are presented in Fig. 7.4 (annual) and Fig. 7.7 (monsoon). The figures also show the significance of trends at the 0.05 level. If the data points are inside the significance band (along the 1:1 line), it considered as no trend in the data series. It is regarded as a significant trend if the data points are below or above the influential band. However, the data points above the line 1:1 indicate the increasing trend and vice-versa. On the other hand, trends obtained from MK/mMK are presented in Figs. 7.3d, 7.5d, 7.5h, 7.6d and 7.6h.

Annual Rainfall

The annual rainfall trends of the 36 meteorological subdivisions are presented in Table 7.2 and Fig. 7.3. Although the obtained results exhibited few inconsistencies between the methods, the results of all the methods showed that annual rainfall in most of the subdivisions (61.11%) is experiencing a decreasing trend (Table 7.2). About half (27.16%) is statistically significant at 95% confidence interval (Fig. 7.4). Spatial distribution of the ITA slope shows that the declining trends are circulated over all India except for some pockets of coastal Karnataka, Arunachal Pradesh, Gangetic West Bengal, and Konkan & Goa. However, Andaman & Nicobar Island, Naga Mani Mizo Tripura, and Gangetic West Bengal show an increasing trend in monotonic fashion during the study period. Also, the subdivisions in northeastern India (Arunachal Pradesh & Naga, Mani, Mizo, Tripura) show strong slopes (more than 6 mm/year), whereas the subdivisions in northwestern and southeastern region show a weak slope (less than 1 mm/year) (Fig. 7.3a). It is also observed from the results of MK/mMK that more than half (i.e., 52.78%) of the subdivisions had decreasing trends during the study period; of this, only 25% is statistically significant at a different significance level. Thus, the few significant decreasing trends ignored by the MK/mMK test can be identified by ITA, justifying the reliability of this method to detect unseen trends in the data series. However, the highest increasing (3.76 mm/year) and decreasing (4.00 mm/year) rates obtained from Sen's slope are found in Coastal Karnataka and Naga Mani Mizo Tripura, respectively (Table 7.2). The percent bias (P_{BIAS}) showed similar results with ITA with little discrepancy in spatial distribution, whereas Saurashtra & Kutch show maximum positive (+11.65%) change and Naga Mani Mizo Tripura show maximum negative (−13.25%) change in the second half (1959–2015) of the study period. This increase in extreme rainfall events has increased the potential occurrence of floods in this region. However, the trends in annual rainfall obtained by using the three methods are highly correlated; the slope of ITA and Sen's slope ($r = 0.728$) is quite highly correlated compared to the slope of ITA and Z of MK/mMK ($r = 0.579$) (Fig. 7.4).

Pre-Monsoon Rainfall

The ITA slope, P_{BIAS}, MK/mMK, and Sen's slope of rainfall in the pre-monsoon season are denoted in Table 7.3 and Fig. 7.5a–d. The estimated results of all three methods demonstrate most of the subdivisions had a decreasing tendency in pre-monsoon rainfall. Results of the ITA show that the majority of the subdivisions (55%) showed decreasing trends and about 50% of them are statistically significant at 95% confidence level whereas only 25% registered as significant (α=0.05) positive trend. The significant positive trend was mainly observed in the subdivisions

Table 7.2 Details of annual rainfall trend of India (1901–2015)

Sl. No	Subdivisions	Slope IT	P_{BIAS}	Z statistics of MK/mMK	Sen slope (Q)	ITD
1	Lakshadweep	−0.11	−0.40	−1.25	−0.92	∇
2	Arunachal Pradesh	2.52	3.88	0.44	0.19	Δ
3	Assam & Meghalaya	−3.17	−6.75	−2.70***	−2.56	∇
4	Naga Mani Mizo Tripura	−6.06	−13.25	−2.02**	−4.00	∇
5	Sub–Himalayan West Bengal & Sikkim	−0.14	−0.30	−0.51	−0.63	∇
6	Gangetic West Bengal	1.89	7.51	2.26**	1.69	Δ
7	Orissa	−1.42	−5.41	−1.24	−0.68	∇
8	Jharkhand	−2.03	−8.44	−2.84***	−1.58	∇
9	Bihar	−1.50	−6.89	−2.55**	−1.38	∇
10	East Uttar Pradesh	−2.01	−11.05	−2.81***	−1.99	∇
11	West Uttar Pradesh	−0.88	−5.91	−1.90*	−1.07	∇
12	Uttarakhand	−1.31	−4.97	−1.46	−1.11	∇
13	Haryana Delhi & Chandigarh	0.16	1.69	−0.13	−0.07	Δ
14	Punjab	−0.14	−1.32	0.02	0.00	∇
15	Himachal Pradesh	−1.88	−8.14	−0.98	−1.11	∇
16	Jammu & Kashmir	0.97	4.99	1.24	0.91	Δ
17	West Rajasthan	0.18	3.52	1.44	0.37	Δ
18	East Rajasthan	−0.62	−5.20	−0.48	−0.32	∇
19	West Madhya Pradesh	0.02	0.11	0.70	0.33	Δ
20	East Madhya Pradesh	−2.15	−9.67	−2.79***	−1.86	∇
21	Gujarat Region	−0.17	−1.05	0.18	0.13	∇
22	Saurashtra & Kutch	0.96	11.65	1.62	0.93	Δ
23	Konkan & Goa	1.73	3.36	2.05**	3.40	Δ
24	Madhya Maharashtra	0.82	5.43	1.88*	1.28	Δ
25	Marathwada	−0.36	−2.55	−0.49	−0.18	∇
26	Vidarbha	−0.79	−4.00	−0.90	−0.52	∇
27	Chhattisgarh	−2.49	−9.85	−3.29***	−1.85	∇
28	Coastal Andhra Pradesh	0.00	0.02	0.95	0.50	Δ
29	Telangana	1.04	6.38	1.75*	1.05	Δ
30	Rayalseema	0.60	4.57	1.35	0.57	Δ
31	Tamilnadu and Puduchery	−0.39	−2.34	0.28	0.09	∇
32	Coastal Karnataka	4.15	7.19	2.85***	3.76	Δ
33	North Interior Karnataka	0.39	3.12	1.03	0.48	Δ
34	South Interior Karnataka	1.25	7.11	2.89***	1.34	Δ
35	Kerala	−2.77	−5.26	−3.18***	−2.39	∇
36	Andaman & Nicobar islands	−1.34	−2.58	−0.95	−1.27	∇

(continued)

Table 7.2 (continued)

ITD innovative trend detection

*	The trend is significant at 10% significance or 90% confidence level
**	The trend is significant at 5% significance or 95% confidence level
***	The trend is significant at 1% significance or 99% confidence level
[shaded]	The trend was detected by modified Mann–Kendall test based on lag 1 autocorrelation

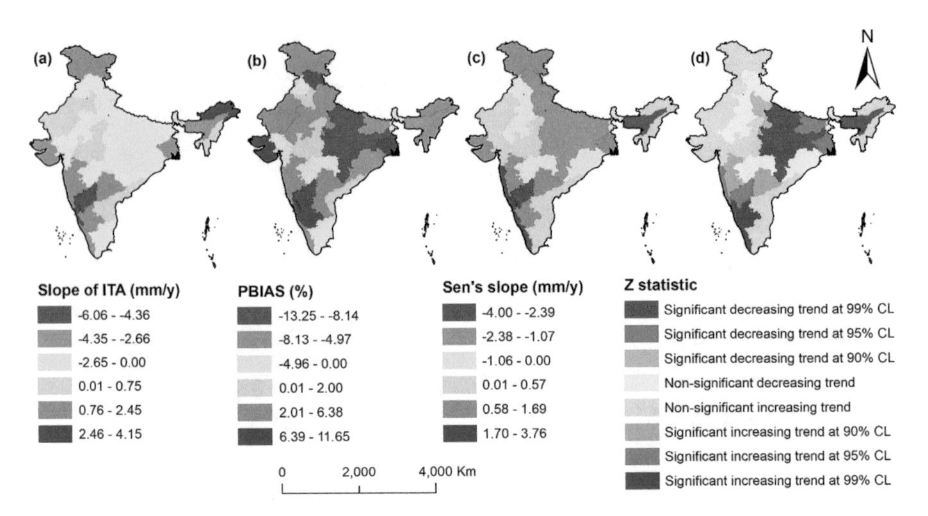

Fig. 7.3 Long-term (1901–2015) trends and magnitudes of the slope of annual rainfall: (**a**) slope of ITA; (**b**) slope of P_{BIAS}; (**c**) Sen slope; (**d**) Z statistics of Mk/mMK

located in the sub-Himalayan region, northwestern and southeastern India, and the Gangetic basin. The rest of the vast tract of central and Deccan India followed a decreasing trend except some isolated pockets (Fig. 7.5a). However, the subdivisions in northeastern India (Arunachal Pradesh and Assam & Meghalaya) show strong slopes (more than 1 mm/year), whereas the subdivisions in the whole study region, except for a few pockets of the northeastern and extreme northern portions, show a weak slope (less than 1 mm/year) (Fig. 7.5a). The orographic burden of monsoon rainfall by hills and mountains and the numerous types of land cover are the reasons for such regional variation in the distribution pattern of rainfall (Banerjee et al. 2020). The results of MK/mMK reveal half (50%) of the subdivisions have decreasing trends during the study period; of this, only 20% is statistically significant at different significant levels. Sen's slope analysis revealed that average rainfall decreases at a rate of −0.178 mm/year in the pre-monsoon season. However, the highest increasing rate (1.09 mm/year) and decreasing rate (−1.07 mm/year) obtained from Sen's slope are found in sub-Himalaya, West Bengal, and Assam &

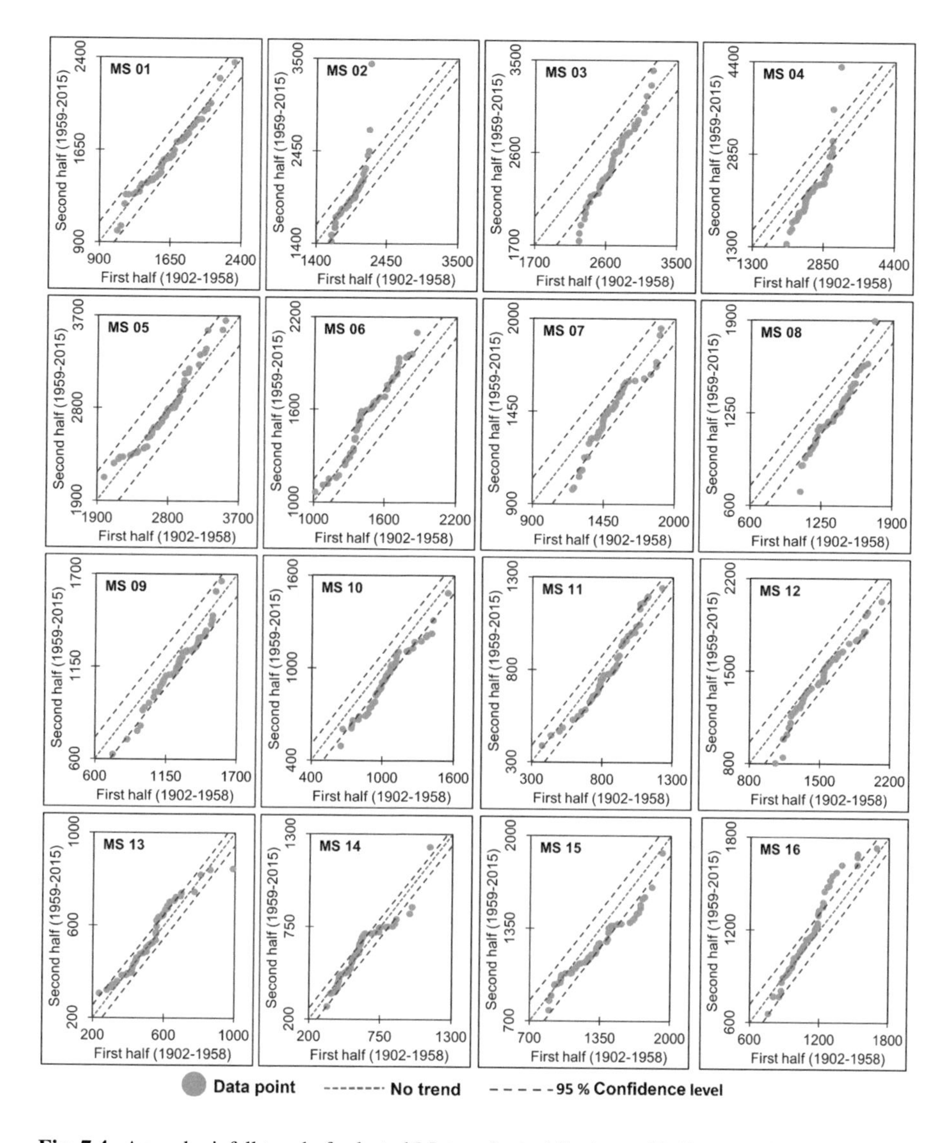

Fig. 7.4 Annual rainfall trend of selected Meteorological Stations of India using the ITA method

Meghalaya, respectively (Table 7.3). The percent bias (P_{BIAS}) showed similar results with ITA with a little discrepancy in spatial distribution, whereas West Rajasthan showed maximum positive (+48.96%) change and the Gujarat region showed a maximum negative (−42.39%) change in the second half (1959–2015) of the study period. The trends in pre-monsoon rainfall in this season obtained by the three methods are moderately correlated, and the slope of ITA and Sen's slope (r = 0.849) is quite highly correlated compared to the slope of ITA and Z of MK/mMK ($r = 0.624$).

Table 7.3 Details of pre-monsoon rainfall trend of India (1901–2015)

Sl. No	Subdivisions	Slope IT	P_{BIAS}	Z statistics of MK/mMK	Sen slope (Q)	ITD
1	Lakshadweep	−0.33	−8.06	−1.83*	−0.58	∇
2	Arunachal Pradesh	1.35	2.52	1.11	0.22	Δ
3	Assam & Meghalaya	−1.69	−14.30	−2.54**	−1.07	∇
4	Naga Mani Mizo Tripura	−0.66	−6.79	-0.69	−0.29	∇
5	Sub–Himalayan West Bengal & Sikkim	0.83	11.87	3.78***	1.09	Δ
6	Gangetic West Bengal	0.13	4.17	0.53	0.11	Δ
7	Orissa	−0.07	−3.46	−0.94	−0.12	∇
8	Jharkhand	−0.10	−6.12	−0.73	−0.09	∇
9	Bihar	0.09	6.48	0.87	0.08	Δ
10	East Uttar Pradesh	0.01	0.70	−0.23	−0.02	Δ
11	West Uttar Pradesh	0.05	10.71	0.65	0.02	Δ
12	Uttarakhand	0.60	26.14	1.85*	0.34	Δ
13	Haryana Delhi & Chandigarh	0.16	29.78	1.37	0.07	Δ
14	Punjab	0.09	10.91	−0.06	−0.01	Δ
15	Himachal Pradesh	0.38	10.22	1.23	0.21	Δ
16	Jammu & Kashmir	0.85	18.02	1.14	0.36	Δ
17	West Rajasthan	0.12	48.96	2.35**	0.06	Δ
18	East Rajasthan	0.01	−0.45	0.03	0.00	Δ
19	West Madhya Pradesh	−0.05	−16.53	−1.14	−0.03	∇
20	East Madhya Pradesh	−0.22	−34.65	−2.92***	−0.16	∇
21	Gujarat Region	−0.08	−42.39	−3.32***	−0.03	∇
22	Saurashtra & Kutch	−0.05	−34.60	−3.16***	−0.02	∇
23	Konkan & Goa	−0.05	−7.46	0.20	0.00	∇
24	Madhya Maharashtra	−0.08	−12.52	−1.65*	−0.09	∇
25	Marathwada	−0.04	−6.66	−0.95	−0.05	∇
26	Vidarbha	−0.16	−24.21	−2.26**	−0.12	∇
27	Chhattisgarh	−0.38	−34.16	−3.02***	−0.25	∇
28	Coastal Andhra Pradesh	−0.01	−0.44	0.19	0.03	∇
29	Telangana	−0.03	−3.32	0.73	0.06	∇
30	Rayalseema	−0.12	−8.27	1.21	0.10	∇
31	Tamilnadu and Puduchery	−0.35	−13.80	−1.17	−0.16	∇
32	Coastal Karnataka	0.22	8.10	1.70*	0.33	Δ
33	North Interior Karnataka	0.02	1.66	0.52	0.05	Δ
34	South Interior Karnataka	−0.18	−6.72	−0.45	−0.08	∇
35	Kerala	−0.35	−5.15	−0.02	−0.02	∇
36	Andaman & Nicobar islands	0.67	8.71	1.21	0.63	Δ

(continued)

Table 7.3 (continued)

ITD innovative trend detection

*	The trend is significant at 10% significance or 90% confidence level
**	The trend is significant at 5% significance or 95% confidence level
***	The trend is significant at 1% significance or 99% confidence level
	The trend was detected by modified Mann–Kendall test based on lag 1 autocorrelation

Monsoon Rainfall

The results of the ITA slope of monsoon rainfall have documented the dominance of both increasing and decreasing trends during the study period. Half the meteorological subdivisions (50%) registered a negative trend, and in the rest of the subdivisions a positive trend is observed in this season. Wherein, about 42 % are statistically significant for increasing and decreasing trends at 95% confidence interval (Fig. 7.7). The results of ITA also revealed that remarkable positive trends are observed in Coastal Karnataka (slope = +3.89), Konkan & Goa (slope = +1.98), and South Interior Karnataka (slope = +1.81), whereas notable decreasing trends are observed in Arunachal Pradesh (slope = −18.51), Naga Mani Mizo Tripura (slope = −4.39), and Himachal Pradesh (slope = −2.32), respectively (Table 7.4 and Fig. 7.5). The results obtained from the Z statistic test and Sen's slope estimator specify nearly similar results for increasing and decreasing trends with 0.01, 0.05, and 0.1 levels of significance. However, the maximum increasing (3.10 mm/year) and decreasing rate (−3.26 mm/year) obtained from Sen's slope are found in Konkan & Goa and Naga, Mani, Mizo, and Tripura, respectively (Table 7.4). Percentage bias of monsoon rainfall shows that in South Interior Karnataka (16%), the rainfed region of India has gained the highest increase in rainfall in the second half (1959–2015) in comparison to the first half (1901–1958). This increasing trend may amplify the vulnerability of landslides and flash flood in mountain regions and riverbank erosion in coastal areas of India (Mandal et al. 2020). Similarly, Himachal Pradesh (15.71%) experienced the greatest decrease in rainfall in the second half (1959–2015) in comparison to the first half (1901–1958) of the study period.

Post-Monsoon Rainfall

The results of ITA, P_{BIAS}, MK/mMK, and Sen's slope of post-monsoon rainfall are presented in Table 7.5 and Fig. 7.6. In this period, most of the subdivisions (69.44%,) showed decreasing trends (ITA) over the entire Gangetic basin, central India, and the western part of Deccan plateau (Fig. 7.6a), however, about 55 % of them are statistically significant at 0.05 significance level. On the other hand approximately one third (27%) of the subdivision showed significant positive trends in the study period. The degree of the increasing trend for the major part of the area varied

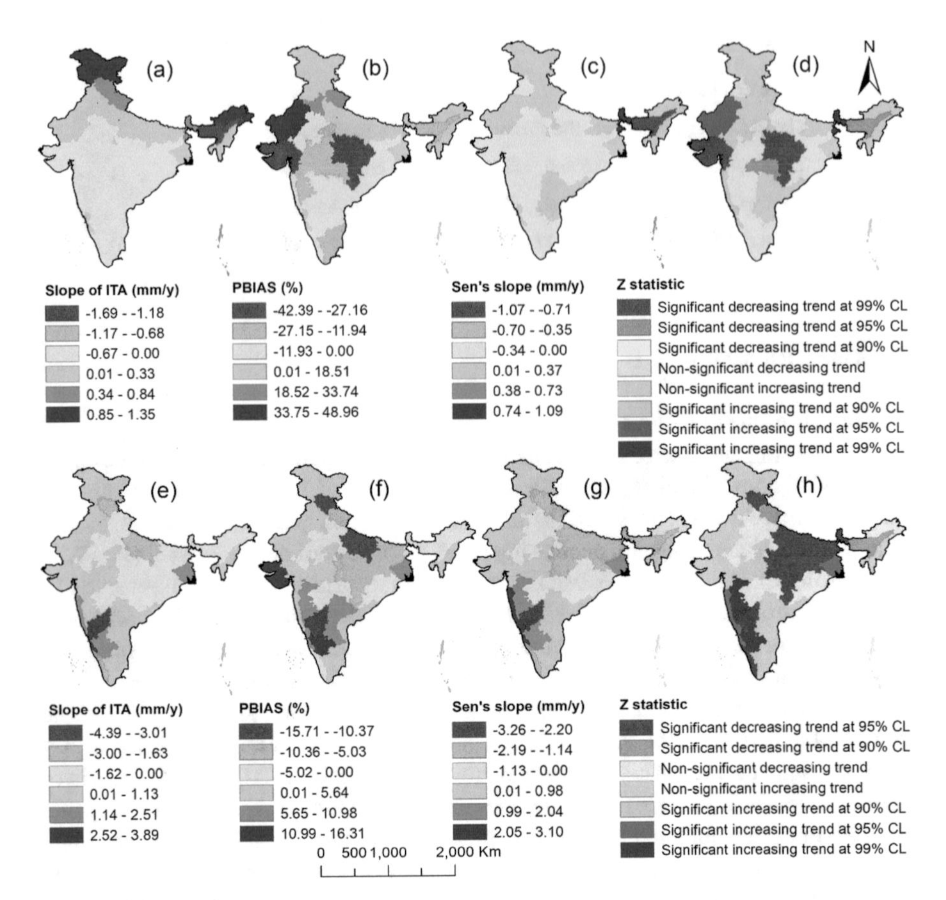

Fig. 7.5 Long-term (1901–2015) trends and magnitudes of the slope of seasonal rainfall in pre-monsoon (**a**)–(**d**) and monsoon (**e**)–(**h**) periods

between 0.01 and 0.29 mm/year, which is very marginal (Fig. 7.6a). However, the highest increasing trend (slope = +0.29 mm/year) obtained from ITA is observed at Rayalseema of Andhra Pradesh and the lowest is observed in Arunachal Pradesh (slope = −1.65). On the other hand, a few of the subdivisions (8.33%) for post-monsoon showed a significant trend (Z statistics) at different levels of significance levels; among these, nearly 6% exhibited a significant decreasing trend (Table 7.5). Sen's slope estimation revealed that the average rainfall decreases at a rate of 0.20 mm/year for the post-monsoon season. However, the highest decreasing and increasing trends are observed at Lakshadweep (0.66 mm/year) and Coastal Karnataka (0.29 mm/year), respectively. Moreover, the high moderate correlations between the slope of ITA and Sen's slope ($r = 0.542$) and high correlation between slope of ITA and Z of MK/mMK ($r = 0.842$) methods confirmed the reliability of the ITA method. Percentage bias showed that most of the region received less rainfall in the second half (1959–2015). On the other hand, Arunachal Pradesh, Uttarakhand,

Table 7.4 Details of monsoon rainfall trend of India (1901–2015)

Sl. No	Subdivisions	Slope IT	P_{BIAS}	Z statistics of MK/mMK	Sen slope (Q)	ITD
1	Lakshadweep	1.35	8.13	1.57	0.91	Δ
2	Arunachal Pradesh	−1.51	−3.52	−0.40	−0.18	∇
3	Assam & Meghalaya	−1.13	−3.69	−1.69*	−1.17	∇
4	Naga Mani Mizo Tripura	−4.39	−14.42	−2.51**	−3.26	∇
5	Sub–Himalayan West Bengal & Sikkim	−1.31	−3.45	−2.05**	−1.78	∇
6	Gangetic West Bengal	1.69	8.89	2.56**	1.35	Δ
7	Orissa	−0.84	−4.03	−1.37	−0.60	∇
8	Jharkhand	−1.48	−7.50	−2.56**	−1.31	∇
9	Bihar	−1.50	−8.05	−2.37**	−1.24	∇
10	East Uttar Pradesh	−1.77	−11.03	−2.51**	−1.44	∇
11	West Uttar Pradesh	−0.64	−4.90	−1.46	−0.75	∇
12	Uttarakhand	−1.57	−7.60	−1.84*	−1.18	∇
13	Haryana Delhi & Chandigarh	0.21	2.78	−0.08	−0.03	Δ
14	Punjab	0.15	1.84	0.32	0.16	Δ
15	Himachal Pradesh	−2.32	−15.71	−1.97**	−1.71	∇
16	Jammu & Kashmir	0.20	2.23	0.94	0.36	Δ
17	West Rajasthan	0.07	1.55	0.84	0.20	Δ
18	East Rajasthan	−0.47	−4.33	−0.38	−0.20	∇
19	West Madhya Pradesh	0.13	0.88	0.86	0.36	Δ
20	East Madhya Pradesh	−1.56	−7.93	−2.31**	−1.43	∇
21	Gujarat Region	0.00	0.01	0.36	0.27	Δ
22	Saurashtra & Kutch	0.92	11.96	1.59	0.92	Δ
23	Konkan & Goa	1.98	4.12	2.22**	3.10	Δ
24	Madhya Maharashtra	1.08	8.66	3.15***	1.33	Δ
25	Marathwada	−0.43	−3.61	−0.43	−0.14	∇
26	Vidarbha	−0.48	−2.78	−0.43	−0.22	∇
27	Chhattisgarh	−1.55	−7.09	−2.24**	−1.08	∇
28	Coastal Andhra Pradesh	0.37	3.28	1.39	0.44	Δ
29	Telangana	0.85	6.41	1.56	0.81	Δ
30	Rayalseema	0.45	6.68	1.15	0.34	Δ
31	Tamilnadu and Puduchery	0.19	3.40	0.65	0.16	Δ
32	Coastal Karnataka	3.89	7.71	2.42**	3.05	Δ
33	North Interior Karnataka	0.47	5.53	0.60	0.26	Δ
34	South Interior Karnataka	1.81	16.31	3.79***	1.39	Δ
35	Kerala	−1.67	−4.61	−2.19**	−2.13	∇
36	Andaman & Nicobar Islands	−1.84	−5.95	-1.35	−1.06	∇

(continued)

Table 7.4 (continued)

ITD innovative trend detection

*	The trend is significant at 10% significance or 90% confidence level
**	The trend is significant at 5% significance or 95% confidence level
***	The trend is significant at 1% significance or 99% confidence level
	The trend was detected by modified Mann–Kendall test based on lag 1 autocorrelation

Haryana Delhi & Chandigarh, East Madhya Pradesh, and West Uttar Pradesh received more than 20% less rainfall in comparison to the first half (1901–1958).

Winter Rainfall

The trends in winter –rainfall obtained from ITA reveal that most of the subdivisions (86.1%) experienced decreasing trends during the study period (1901–2015). About 69% of them are statistically significant at 95% confidence level. These decreasing trends are found in the whole study area except for a few pocket areas of Gangetic West Bengal, Himachal Pradesh, Konkan & Goa, and Coastal Andhra Pradesh. The highest decreasing and increasing trend (ITA) was observed in Arunachal Pradesh (slope = −0.53) and Sub-Himalaya West Bengal and Sikkim (slope + 0.30), respectively. On the other hand, the results of MK/mMK reveal that more than three fourths (89%) of the subdivisions showed decreasing trends during the study period; out of this only 39% are statistically significant at 10%, 5%, and 1% significance level. (Table 7.6 and Fig. 7.6h). Sen's slope analysis revealed that an average rate of decrease is −0.154 mm/year over India in the winter season. The spatiotemporal variation obtained from Sen's slope is shown in Fig. 7.6g, which depicts homogeneity in decreasing trends in all of central India. However, the highest decreasing trend is observed in Andaman & Nicobar Island (−0.84 mm/year) and the highest increasing trend in Himachal Pradesh (+0.30 mm/year). The coefficient of correlation results between the slope of ITA and Sen's slope ($r = 0.728$) and slope of ITA and Z of MK/mMK ($r = 0.503$) indicate the new robust ITA is reliable in trend analysis of the hydrometeorological series. Percentage bias of winter rainfall showed that sub-Himalaya West Bengal and Sikkim (49.09%) have gained the highest increase in rainfall in the second half (1959–2015) in comparison to the first half (1901–1958). Similarly, Saurashtra & Kutch (62.22%) have gained the highest decrease in rainfall in the second half (1959–2015) in comparison to the first half (1901–1958).

Table 7.5 Details of post-monsoon rainfall trend of India (1901–2015)

Sl. No	Subdivisions	Slope IT	P_{BIAS}	Z statistics of MK/mMK	Sen slope (Q)	ITD
1	Lakshadweep	−0.68	−12.40	−1.79*	−0.66	∇
2	Arunachal Pradesh	−1.65	−32.91	−0.11	0.00	∇
3	Assam & Meghalaya	−0.21	−6.41	−0.94	−0.17	∇
4	Naga Mani Mizo Tripura	−0.60	−14.34	−2.05**	−0.52	∇
5	Sub–Himalayan West Bengal & Sikkim	0.03	1.18	0.13	0.02	Δ
6	Gangetic West Bengal	0.12	5.01	0.48	0.14	Δ
7	Orissa	−0.39	−14.58	−1.14	−0.27	∇
8	Jharkhand	−0.09	−5.47	−0.17	−0.02	∇
9	Bihar	0.07	5.82	−0.06	0.00	Δ
10	East Uttar Pradesh	−0.12	−12.88	−0.15	0.00	∇
11	West Uttar Pradesh	−0.15	−22.64	−0.14	0.00	∇
12	Uttarakhand	−0.31	−31.49	−0.95	−0.07	∇
13	Haryana Delhi & Chandigarh	−0.10	−28.75	−0.31	0.00	∇
14	Punjab	−0.15	−38.25	0.10	0.00	∇
15	Himachal Pradesh	−0.10	−11.41	0.44	0.03	∇
16	Jammu & Kashmir	0.24	26.14	2.27**	0.21	Δ
17	West Rajasthan	0.02	13.98	1.26	0.00	Δ
18	East Rajasthan	−0.05	−12.75	−0.22	0.00	∇
19	West Madhya Pradesh	−0.05	−6.81	−0.07	0.00	∇
20	East Madhya Pradesh	−0.25	−23.81	−0.46	−0.06	∇
21	Gujarat Region	−0.04	−7.81	0.39	0.01	∇
22	Saurashtra & Kutch	0.15	54.05	1.19	0.02	Δ
23	Konkan & Goa	−0.21	−8.29	0.97	0.21	∇
24	Madhya Maharashtra	−0.13	−7.41	0.30	0.05	∇
25	Marathwada	0.19	14.24	1.12	0.18	Δ
26	Vidarbha	−0.04	−3.47	0.58	0.09	∇
27	Chhattisgarh	−0.27	−18.34	−0.54	−0.06	∇
28	Coastal Andhra Pradesh	−0.40	−8.24	−0.30	−0.05	∇
29	Telangana	0.23	14.87	1.23	0.23	Δ
30	Rayalseema	0.31	7.65	0.52	0.09	Δ
31	Tamilnadu and Puduchery	−0.02	−0.38	−0.04	−0.05	∇
32	Coastal Karnataka	0.02	0.57	1.23	0.29	Δ
33	North Interior Karnataka	−0.06	−2.67	0.48	0.10	∇
34	South Interior Karnataka	−0.32	−9.05	−0.52	−0.12	∇
35	Kerala	−0.51	−6.13	−1.03	−0.35	∇
36	Andaman & Nicobar islands	0.23	2.55	0.20	0.10	Δ

(continued)

Table 7.5 (continued)

ITD innovative trend detection

*	The trend is significance at 10% significance or 90% confidence level
**	The trend is significance at 5% significance or 95% confidence level
***	The trend is significance at 1% significance or 99% confidence level
[shaded]	The trend was detected by modified Mann–Kendall test based on lag 1 autocorrelation

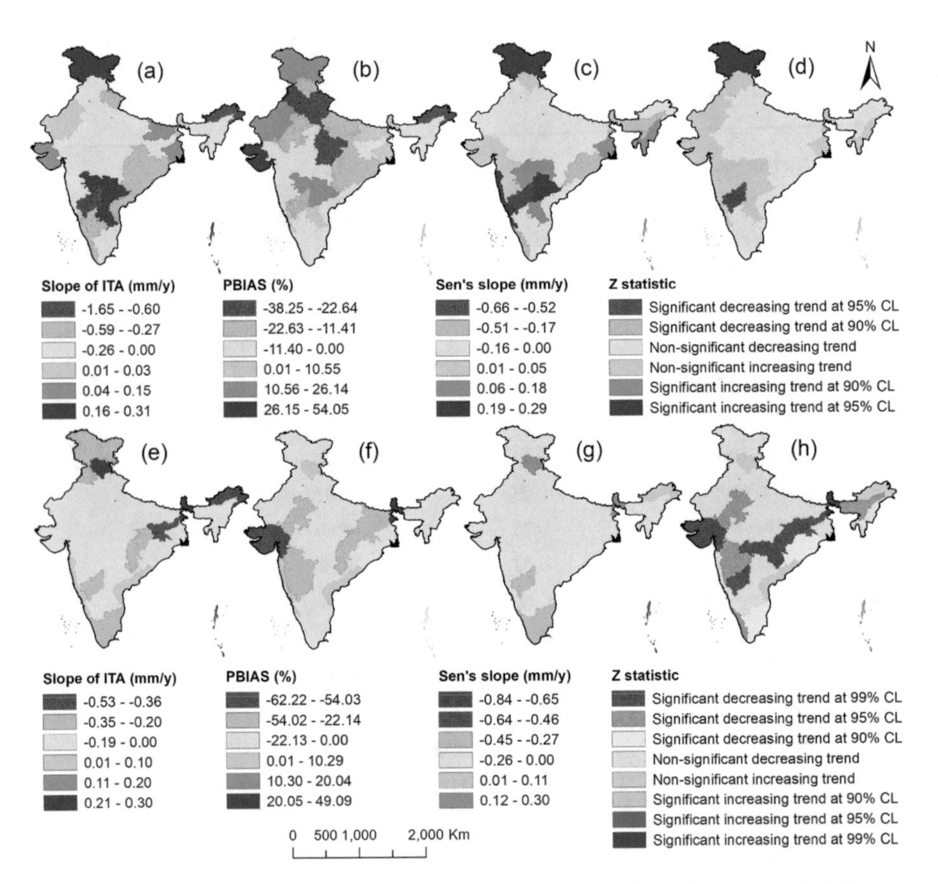

Fig. 7.6 Long-term (1901–2010) trends and magnitudes of the slope of seasonal rainfall post-monsoon (**a**)–(**d**) and in winter (**e**)–(**h**)

Comparison of ITA with *Z* Statistics and Sen's Slope

Figure 7.8 shows the comparison of trends between the ITA and the Z statistic, and the magnitudes of the ITA (B slope) and the Sen's slope of annual and seasonal rainfall. It is very important to test the reliability of the newly introduced method

Table 7.6 Details of winter rainfall trend of India (1901–2015)

Sl. No	Subdivisions	Slope IT	P_{BIAS}	Z statistics of MK/mMK	Sen slope (Q)	ITD
1	Lakshadweep	−0.46	−22.77	−2.78***	−0.51	▽
2	Arunachal Pradesh	−0.53	−16.57	1.11	0.07	△
3	Assam & Meghalaya	−0.13	−11.98	−2.06**	−0.18	▽
4	Naga Mani Mizo Tripura	−0.28	−22.14	−3.08***	−0.33	▽
5	Sub–Himalayan West Bengal & Sikkim	0.30	49.09	4.74***	0.24	△
6	Gangetic West Bengal	−0.05	−6.35	−0.09	−0.02	▽
7	Orissa	−0.13	−17.80	−1.80*	−0.13	▽
8	Jharkhand	−0.36	−36.37	−3.03***	−0.25	▽
9	Bihar	−0.16	−26.15	−1.39	−0.10	▽
10	Uttar Pradesh (East)	−0.13	−18.58	−1.19	−0.09	▽
11	Uttar Pradesh (West)	−0.15	−18.43	−1.33	−0.12	▽
12	Uttarakhand	−0.01	-0.39	−0.26	−0.08	▽
13	Haryana Delhi & Chandigarh	−0.12	−14.76	−0.85	−0.07	▽
14	Punjab	−0.20	−16.33	−1.20	−0.16	▽
15	Himachal Pradesh	0.26	7.10	1.41	0.30	△
16	Jammu & Kashmir	−0.27	−5.42	−0.10	−0.05	▽
17	West Rajasthan	−0.02	−11.67	−0.77	−0.01	△
18	East Rajasthan	−0.10	−30.85	−2.06**	−0.06	▽
19	West Madhya Pradesh	−0.02	−4.36	−1.35	−0.05	▽
20	East Madhya Pradesh	−0.12	−13.72	−1.36	−0.12	▽
21	Gujarat Region	−0.06	−54.03	−3.52***	−0.03	▽
22	Saurashtra & Kutch	−0.06	−62.22	−3.71***	−0.02	△
23	Konkan & Goa	0.01	9.04	−1.07	0.00	△
24	Madhya Maharashtra	−0.05	−22.65	−2.13**	−0.03	▽
25	Marathwada	−0.08	−25.13	−2.46**	−0.06	▽
26	Vidarbha	−0.11	−18.54	−2.63***	−0.15	▽
27	Chhattisgarh	−0.29	−35.56	−2.88***	−0.23	▽
28	Coastal Andhra Pradesh	0.03	6.52	0.26	0.00	△
29	Telangana	−0.01	−3.29	−0.07	0.00	△
30	Rayalseema	−0.08	−8.31	−0.88	−0.07	▽
31	Tamilnadu and Puduchery	−0.27	−12.22	−1.79	−0.35	▽
32	Coastal Karnataka	0.01	4.01	−0.46	0.00	△
33	North Interior Karnataka	−0.05	−20.68	−0.92	−0.02	▽
34	South Interior Karnataka	−0.07	−18.99	−1.18	−0.05	▽
35	Kerala	−0.27	−20.73	−2.51**	−0.27	▽
36	Andaman & Nicobar islands	−0.41	−9.43	−2.00**	−0.84	▽

(continued)

Table 7.6 (continued)

ITD innovative trend detection

*	The trend is significant at 10% significance or 90% confidence level
**	The trend is significant at 5% significance or 95% confidence level
***	The trend is significant at 1% significance or 99% confidence level
(shaded)	The trend was detected by modified Mann–Kendall test based on lag 1 autocorrelation

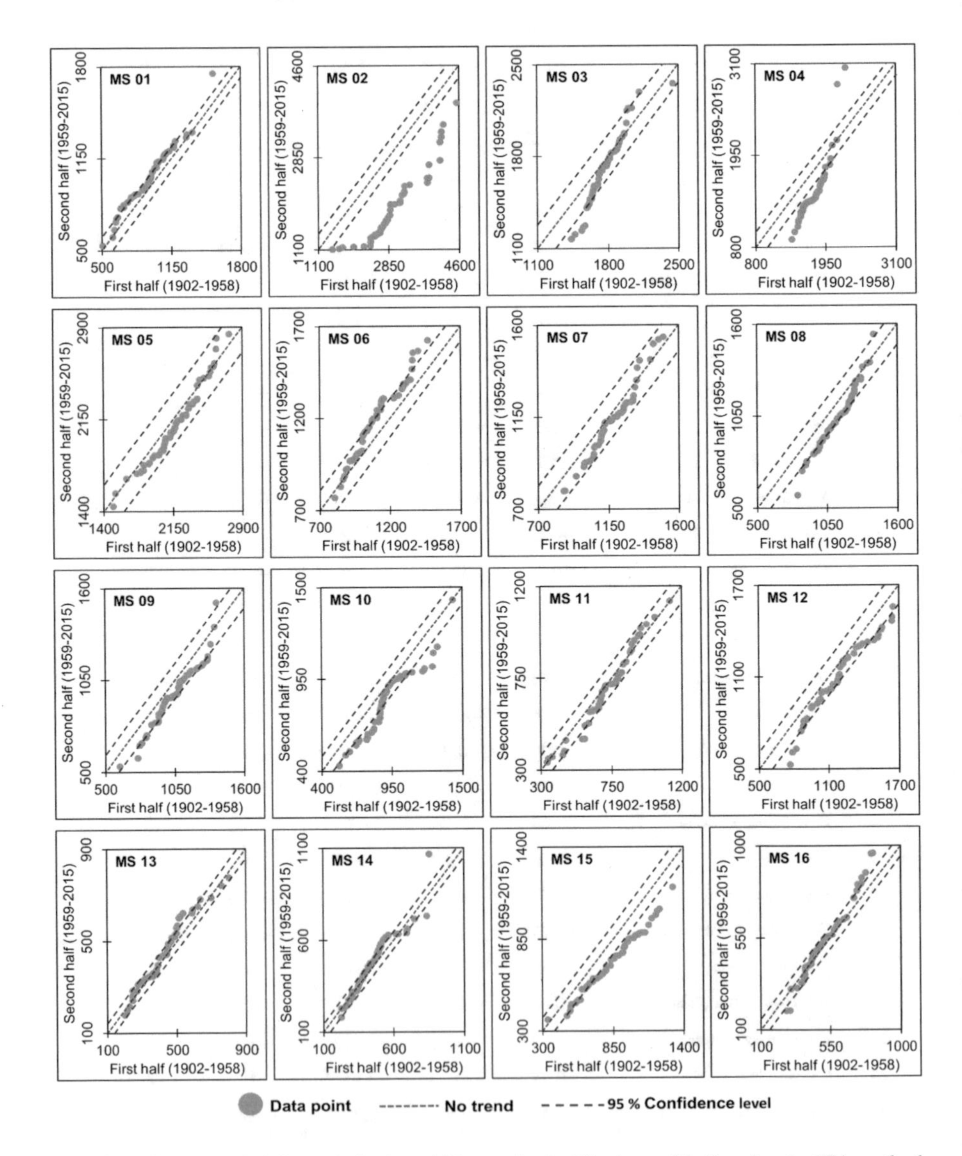

Fig. 7.7 Monsoon rainfall trend of selected Meteorological Stations of India using the ITA method

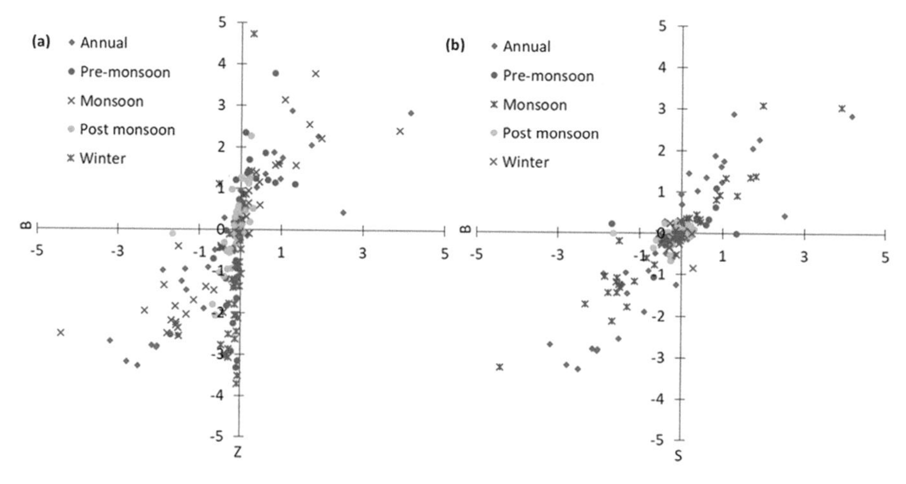

Fig. 7.8 Comparison of (**a**) Z statistic and ITA (B slope) and (**b**) Sen's slope and ITA (B slope)

with the conventional method for the acceptance of a new method. Henceforth, we have tried to compare the B slope of ITA with Z statistics of MK/mMK, and Sen's slope. To determine whether the results obtained from the three methods are significantly related, quadrantal analysis is performed. The slope of ITA and the Z statistic of MK/mMK and ITA and Sen's slope are shown in scatter plots in Fig. 7.8a, b. The results show most of the scatter (nearly 90%) is in the first and third quadrants, which indicates a strong agreement among the selected methods, and very few data points are scattered in the second and fourth quadrants, which would show a difference in agreement between the methods, and their magnitude is insignificant. The results indicate a strong statistical association of the B statistic of ITA with the Z statistic of MK/mMK tests and the β of Sen's slope estimator. Hence, the reliability of the results of these three trend methods affirms that the ITA is a consistent and effective method that is capable of efficiently analyzing the trends and magnitude of change in hydrometeorological data at low, medium, and high values from the graphical representations.

Conclusion

In the present study, we have analyzed the spatial variability and trends of annual and seasonal rainfall across India. By analyzing the trend of precipitation, it could be assumed that the results of ITA and P_{BIAS} showed a more or less similar trend for all the stations and time scales, and the Z statistic confirmed its reliability. The study also indicated that there is no evidence of a clear increasing or decreasing trend of annual rainfall in India as a whole, but at the regional level a decreasing trend is observed in recent decades in the central and western parts of India whereas a increasing trend is observed in the northern and southwestern parts. Overall decreasing trends are observed for the post-monsoon and winter seasons, and both positive and negative trends are observed for the pre-monsoon season. Monotonic trends in seasonal rainfall are very rare. In addition, the findings also revealed that some of the meteorological subdivisions have recorded a significant negative trend for long-term seasonal rainfall. Pre-monsoon (Mar–May), post-monsoon (Oct–Nov), and winter (Dec–Feb) seasons were dominated by a decreasing trend whereas decreasing and increasing trends were found equally for the monsoon (Jun–Sep) season. Even some of the subdivisions are showing no trends using the MK test, but innovative trend analysis detects a significant trend in those data points. However, for the better management of water resources and disaster mitigation, such robust analysis is very helpful and can provide good insight into the complex dynamic phenomena. However, more precise observation of this phenomenon is needed. The microlevel discrepancies in response to the climate change are recommended for further study. Also needed is the incorporation of other climatic parameters, especially temperature, air pressure, humidity, wind patterns and directions, and sunshine, with the changing dynamism of monsoon rainfall for more comphrehensive analysis.

References

Ahmad I, Zhang F, Tayyab M, Anjum MN, Zaman M, Liu J, Farid HU, Saddique Q (2018) Spatiotemporal analysis of precipitation variability in annual, seasonal and extreme values over upper Indus River basin. Atmos Res 213:346–360. https://doi.org/10.1016/j.atmosres.2018.06.019

Ay M, Kisi O (2015) Investigation of trend analysis of monthly total precipitation by an innovative method. Theor Appl Climatol 120(3–4):617–629

Banerjee A, Dimri AP, Kumar K (2020) Rainfall over the Himalayan foot-hill region: Present and future. Journal of Earth System Science 129(1):1–16. https://doi.org/10.1007/s12040-019-1295-2

Chang HI, Kumar A, Niyogi D, Mohanty UC, Chen F, Dudhia J (2009) The role of land surface processes on the mesoscale simulation of the July 26, 2005 heavy rain event over Mumbai, India. Glob Planet Change 67(1–2):87–103. https://doi.org/10.1016/j.gloplacha.2008.12.005

Changnon SA, Demissie M (1996) Detection of changes in streamflow and floods resulting from climate fluctuations and land use-drainage changes. Clim Change 32(4):411–421. https://doi.org/10.1007/bf00140354

Cui L, Wang L, Lai Z, Tian Q, Liu W, Li J (2017) Innovative trend analysis of annual and seasonal air temperature and rainfall in the Yangtze River Basin, China during 1960–2015. J Atmos Sol Terr Phys 164:48–59. https://doi.org/10.1016/j.jastp.2017.08.001

Das J, Bhattacharya SK (2018) Trend analysis of long-term climatic parameters in Dinhata of Koch Bihar district, West Bengal. Spat Inf Res 26(3):271–280. https://doi.org/10.1007/s41324-018-0173-3

Das J, Mandal T, Saha P (2019) Spatio-temporal trend and change point detection of winter temperature of North Bengal, India. Spat Inf Res 27(4):411–424. https://doi.org/10.1007/s41324-019-00241-9

Das J, Jha S, Goyal MK (2020a) On the relationship of climatic and monsoon teleconnections with monthly precipitation over meteorologically homogenous regions in India: wavelet & global coherence approaches. Atmos Res 238:104889. https://doi.org/10.1016/j.atmosres.2020.104889

Das J, Rahman ATMS, Mandal T, Saha P (2020b) Exploring driving forces of large-scale unsustainable groundwater development for irrigation in lower Ganga River basin in India. Environ Dev Sustain:1–21. https://doi.org/10.1007/s10668-020-00917-5

Das J, Rahman ATMS, Mandal T, Saha P (2020c) Challenges for sustainable groundwater management for large scale irrigation under changing climate in lower Ganga River Basin in India. Groundw Sustain Dev 11:100449. https://doi.org/10.1016/j.gsd.2020.100449

Das J, Mandal T, Saha P, Bhattacharya SK (2020d) Variability and trends of rainfall using non-parametric approaches: a case study of semi-arid area. Mausam 71(1):33–44

Dash SK, Kulkarni MA, Mohanty UC, Prasad K (2009) Changes in the characteristics of rain events in India. J Geophys Res Atmos 114(D10):D10109. https://doi.org/10.1029/2008JD010572

De Leo F, De Leo A, Besio G, Briganti R (2020) Detection and quantification of trends in time series of significant wave heights: an application in the Mediterranean Sea. Ocean Eng 202:107155. https://doi.org/10.1016/j.oceaneng.2020.107155

Farook AJ, Kannan KS (2016) Climate change impact on rice yield in India–vector autoregression approach. Sri Lankan J Appl Stat 16(3):161. https://doi.org/10.4038/sljastats.v16i3.7830

Gedefaw M, Yan D, Wang H, Qin T, Girma A, Abiyu A, Batsuren D (2018) Innovative trend analysis of annual and seasonal rainfall variability in Amhara regional state, Ethiopia. Atmosphere 9(9):326. https://doi.org/10.3390/atmos9090326

Gilbert RO (1987) Statistical methods for environmental pollution monitoring. Wiley, Hoboken

Girma A, Qin T, Wang H, Yan D, Gedefaw M, Abiyu A, Batsuren D (2020) Study on recent trends of climate variability using innovative trend analysis: the case of the upper Huai River Basin. Pol J Environ Stud 29(3):2199. https://doi.org/10.15244/pjoes/103448

Güçlü YS (2020) Improved visualization for trend analysis by comparing with classical Mann-Kendall test and ITA. J Hydrol 584:124674. https://doi.org/10.1016/j.jhydrol.2020.124674

Guhathakurta P, Sreejith OP, Menon PA (2011) Impact of climate change on extreme rainfall events and flood risk in India. J Earth Syst Sci 120(3):359. https://doi.org/10.1007/s12040-011-0082-5

Guntukula R, Goyari P (2020) Climate change effects on the crop yield and its variability in Telangana, India. Stud Microecon 8:2321022220923197. https://doi.org/10.1177/2321022220923197

Hamed KH, Rao AR (1998) A modified Mann-Kendall trend test for autocorrelated data. J Hydrol 204(1–4):182–196. https://doi.org/10.1016/S0022-1694(97)00125-X

Huang X, Zhou T, Turner A, Dai A, Chen X, Clark R, Jiang J, Man W, Murphy J, Rostron J, Wu B (2020) The recent decline and recovery of Indian summer monsoon rainfall: relative roles of external forcing and internal variability. J Clim 33(12):5035–5060. https://doi.org/10.1175/JCLI-D-19-0833.1

Islam MN, van Amstel A (eds) (2018) Bangladesh I: climate change impacts, mitigation and adaptation in developing countries. Springer, Cham. https://doi.org/10.1007/978-3-319-26357-1

Kendall MG (1975) Rank correlation measures. Charles Griffin, London 202:15

Khan TI, Islam MN, Islam MN (2018) Climate variability impacts on agricultural land use dynamics in the Madhupur tract in Bangladesh. In: Bangladesh I: climate change impacts, mitigation and adaptation in developing countries. Springer, Cham, pp 167–193. https://doi.org/10.1007/978-3-319-26357-1_7

Kisi O (2015) An innovative method for trend analysis of monthly pan evaporations. J Hydrol 527:1123–1129. https://doi.org/10.1016/j.jhydrol.2015.06.009

Kisi O, Ay M (2014) Comparison of Mann–Kendall and innovative trend method for water quality parameters of the Kizilirmak River, Turkey. J Hydrol 513:362–375. https://doi.org/10.1016/j.jhydrol.2014.03.005

Krishnan R, Sabin TP, Vellore R, Mujumdar M, Sanjay J, Goswami BN, Hourdin F, Dufresne JL, Terray P (2016) Deciphering the desiccation trend of the South Asian monsoon hydroclimate in a warming world. Clim Dyn 47(3–4):1007–1027. https://doi.org/10.1007/s00382-015-2886-5

Machiwal D, Jha MK (2012) Analysis of stream flow trend in the Susquehanna River basin, USA. In: Hydrologic time series analysis: theory and practice. Springer, Dordrecht, pp 181–200. https://doi.org/10.1007/978-94-007-1861-6_9
Mandal T, Das J, Rahman ATMS, Saha P (2020) Rainfall insight in Bangladesh and India: climate change and environmental perspective. In: Habitat, ecology and ekistics: case studies of human-environment interactions in India. Springer, Singapore. https://doi.org/10.1007/978-3-030-49115-4_3
Mann HB (1945) Nonparametric tests against trend. Econometrica J Econometric Soc 13:245–259. https://doi.org/10.2307/1907187
Mohorji AM, Şen Z, Almazroui M (2017) Trend analyses revision and global monthly temperature innovative multi-duration analysis. Earth Syst Environ 1(1):9. https://doi.org/10.1007/s41748-017-0014-x
Moriasi DN, Arnold JG, Van Liew MW, Bingner RL, Harmel RD, Veith TL (2007) Model evaluation guidelines for systematic quantification of accuracy in watershed simulations. Trans ASABE 50(3):885–900
Nikumbh AC, Chakraborty A, Bhat GS (2019) Recent spatial aggregation tendency of rainfall extremes over India. Sci Rep 9(1):1–7. https://doi.org/10.1038/s41598-019-46719-2
Nisansala WDS, Abeysingha NS, Islam A, Bandara AMKR (2020) Recent rainfall trend over Sri Lanka (1987–2017). Int J Climatol 40(7):3417–3435. https://doi.org/10.1002/joc.6405
Öztopal A, Şen Z (2017) Innovative trend methodology applications to precipitation records in Turkey. Water Resour Manag 31(3):727–737. https://doi.org/10.1007/s11269-016-1343-5
Phuong DND, Tram VNQ, Nhat TT, Ly TD, Loi NK (2020) Hydro-meteorological trend analysis using the Mann-Kendall and innovative-Şen methodologies: a case study. Int J Glob Warming 20(2):145–164. https://doi.org/10.1504/IJGW.2020.105385
Pour SH, Abd Wahab AK, Shahid S, Ismail ZB (2020) Changes in reference evapotranspiration and its driving factors in peninsular Malaysia. Atmos Res 246:105096. https://doi.org/10.1016/j.atmosres.2020.105096
Rahman MA, Yunsheng L, Sultana N (2017) Analysis and prediction of rainfall trends over Bangladesh using Mann–Kendall, Spearman's rho tests and ARIMA model. Meteorol Atmos Phys 129(4):409–424. https://doi.org/10.1007/s00703-016-0479-4
Rahman ATMS, Hosono T, Quilty JM, Das J, Basak A (2020) Multiscale groundwater level forecasting: coupling new machine learning approaches with wavelet transforms. Adv Water Resour 141:103595. https://doi.org/10.1016/j.advwatres.2020.103595
Roxy MK, Ghosh S, Pathak A, Athulya R, Mujumdar M, Murtugudde R, Terray P, Rajeevan M (2017) A threefold rise in widespread extreme rain events over Central India. Nat Commun 8 (1):1–11. https://doi.org/10.1038/s41467-017-00744-9
Sahoo SK, Ajilesh PP, Gouda KC, Himesh S (2020) Impact of land-use changes on the genesis and evolution of extreme rainfall event: a case study over Uttarakhand, India. Theor Appl Climatol 140:1–12. https://doi.org/10.1007/s00704-020-03129-z
Sarkar A, Chouhan P (2019) Dynamic simulation of urban expansion based on cellular automata and Markov chain model: a case study in Siliguri Metropolitan Area, West Bengal. Model Earth Syst Environ 5:1723–1732. https://doi.org/10.1007/s40808-019-00626-7
Sarkar A, Chouhan P (2020) Modeling spatial determinants of urban expansion of Siliguri a metropolitan city of India using logistic regression. Model Earth Syst Environ 6:2317. https://doi.org/10.1007/s40808-020-00815-9
Satishkumar K, Rathnam EV (2020) Comparison of six trend detection methods and forecasting for monthly groundwater levels–a case study. ISH J Hydraul Eng:1–10. https://doi.org/10.1080/09715010.2020.1715270
Sen PK (1968) Estimates of the regression coefficient based on Kendall's tau. J Am Stat Assoc 63 (324):1379–1389
Şen Z (2012) Innovative trend analysis methodology. J Hydrol Eng 17(9):1042–1046. https://doi.org/10.1061/(ASCE)HE.1943-5584.0000556

Sharma A, Goyal MK (2020) Assessment of the changes in precipitation and temperature in Teesta River basin in Indian Himalayan region under climate change. Atmos Res 231:104670. https://doi.org/10.1016/j.atmosres.2019.104670

Sinha P, Nageswararao MM, Dash GP, Nair A, Mohanty UC (2019) Pre-monsoon rainfall and surface air temperature trends over India and its global linkages. Meteorol Atmos Phys 131(4):1005–1018. https://doi.org/10.1007/s00703-018-0621-6

Sreelekha PN, Babu CA (2019) Is the negative IOD during 2016 the reason for monsoon failure over southwest peninsular India? Meteorol Atmos Phys 131(3):413–420. https://doi.org/10.1007/s00703-017-0574-1

Tabari H, Talaee PH (2011) Analysis of trends in temperature data in arid and semi-arid regions of Iran. Glob Planet Change 79(1–2):1–10. https://doi.org/10.1016/j.jhydrol.2010.11.034

Tabari H, Marofi S, Aeini A, Talaee PH, Mohammadi K (2011) Trend analysis of reference evapotranspiration in the western half of Iran. Agric For Meteorol 151(2):128–136. https://doi.org/10.1016/j.agrformet.2010.09.009

Varikoden H, Revadekar JV, Kuttippurath J, Babu CA (2019) Contrasting trends in southwest monsoon rainfall over the Western Ghats region of India. Clim Dyn 52(7–8):4557–4566. https://doi.org/10.1007/s00382-018-4397-7

Wang Y, Xu Y, Tabari H, Wang J, Wang Q, Song S, Hu Z (2020) Innovative trend analysis of annual and seasonal rainfall in the Yangtze River Delta, eastern China. Atmos Res 231:104673. https://doi.org/10.1016/j.atmosres.2019.104673

Chapter 8
Forest Phenology as an Indicator of Climate Change: Impact and Mitigation Strategies in India

Priyanshi Tiwari, Pramit Verma, and A. S. Raghubanshi

Abstract Forests are an integral part of the terrestrial ecosystem, maintaining the biodiversity, carbon flux, and ecosystem services and even supporting livelihoods. Rampant exploitation of forest resources has resulted in deforestation and the loss of associated benefits. It is estimated that deforestation accounts for 11% of global carbon emissions. Their role in carbon sequestration has helped international bodies to create programmes such as reducing emissions from deforestation and forest degradation (REDD+). The phenomena of climate change adversely affect the functioning of forest species: it changes the timing of their flowering and fruiting habits, and brings about changes in the ecophysiology of the species. However, such changes depend on the species and the climatic conditions to a large extent. Phenology, the temporal order of the annual cycle of plant functions, is quite sensitive to the changes in climate. In this chapter we have explored the relationships between climate change and its impact on tree phenology, and mitigation and adaptation strategies. Some species may tend to exhibit a certain amount of resilience against climate change.

Keywords Forest phenology · Climate change · Global carbon emission · Adaptation strategies in India

Introduction

Climate change is impacting forest health and the distribution of many species across the globe. The effect of this change is felt more by the poor who constitute about 70% of the rural populations and depend directly on forest resources for their

P. Tiwari
Department of Botany, Institute of Science, Banaras Hindu University, Varanasi, India

P. Verma · A. S. Raghubanshi (✉)
Integrative Ecology Laboratory (IEL), Institute of Environment & Sustainable Development (IESD), Banaras Hindu University (BHU), Varanasi, India

M. N. Islam, A. van Amstel (eds.), *India: Climate Change Impacts, Mitigation and Adaptation in Developing Countries*, Springer Climate,
https://doi.org/10.1007/978-3-030-67865-4_8

livelihood (Secretariat of the Convention on Biological Diversity 2009). With the current rate of biodiversity loss, it is to be expected that there is dire need for mitigation and adaptation measures (Priti et al. 2016). Human-mediated climate warming has resulted in changes in species habitat and distribution and in phenological changes, although seldom resulting in species extinction (Priti et al. 2016). Phenology has been described as one of the best described and well-publicised phenomena linking global warming to plant communities (Wolf et al. 2017), and describing this link between forest phenology and climate change, mitigation, and adaptation efforts is the focus of this discourse.

The forest type of a biogeographical region is a characteristic feature of that region, an outcome of habitat selection as well as gradual adaptation to climatic conditions required for its sustenance (Richardson et al. 2013). Research on plant phenology mainly describes the series of biological changes taking place in plants during their life cycle combined with seasonal progression, which is mainly caused by the combined effect of changes in the abiotic and biotic factors of a region (Lieth 1974). Abiotic factors such as rainfall pattern, sunlight availability, temperature, and soil type, which govern the climate of a region, also regulate the plant phenophases. Therefore, past records of phenological events guide us to notice the pattern of changes in seasonal phenomenon over a period of time. When recording the fruiting period, the flowering period of a particular ecosystem, we know the time of seasonal changes taking place in that ecosystem. The study of phenology provides us with information regarding the availability of resources in an ecosystem by the fruiting and flowering period, the resilience of vegetation to follow a particular phenological cycle in spite of environmental stresses or resource depletion.

Forests contain a high level of species diversity that displays a number of phenological events such as leaf drop, leaf flushing, flowering, and fruiting in relationship to time and space (Negi and Singh 1992; Singh and Singh 1992; Justiniano and Fredericksen 2000). The phenology of a forest mainly depends on the types of species present in the forest: the species found in a forest differ from one another in such matters as their season of onset, the falling patterns of leaves, and the flowering season. Hence, forest phenology constitutes the overall changes in the forest in different seasons and the combined effects of the types of plants species found in the forest. Local climate, which is the result of combined effects of the abiotic factors of a region, dominantly influences the plant growing season. Therefore, it is crucial to know which climatic factors govern particular phenological behaviour in different plant species (Wolkovich1 et al. 2013). Different types of forest exist in response to the differences in climate conditions on the Earth, each with its own phenological characteristics, and particular climatic factors largely regulate its growing season. The phenology of the boreal forests and temperate zone forests is mainly driven by temperature, regulating the timing of the onset and duration of the growing season as well as the time of foliage fall (Kramer et al. 2000). In contrast, the phenology of the tropical forest and Mediterranean coniferous forests is mainly controlled by water availability and precipitation patterns (Borchert et al. 2002; Kramer et al. 2000). The correlation between climate change and phenology can be established by building prolonged records of phenology of individual trees.

However, when there is no parallel relationship between the present climate and future climate, well-documented phenological data become useless in predicting future changes (Corlett and Lafrankie 1998).

The Intergovernmental Panel on Climate Change (IPCC), in its sixth report, predicted, by using climate models, that climate change will lead to an increase in mean temperature in most land and ocean regions, hot extremes in most inhabited regions, heavy rainfall in several regions, and the probability of precipitation shortage and drought in some regions. Changes in these factors could lead to disturbance of different phenophases such as leaf emergence, leaf area, life span of the foliage, growth period, abscission of foliage, and the onset of flowering of individual trees. Many researchers have shown that plant phenology might act as one of the most receptive and simply noticeable features found in nature that change with respect to climate change. Therefore, plant phenology data are widely being used as an indicator of climate change, because climate change (increase in average temperature and change in rainfall pattern) primarily affects the phenology of biological organisation (Stöckli et.al, 2003). Such changes are easily perceptible with a long-term phenological database.

The sensible features of phenological observation towards regional climate conditions and to climate change mean that phenological records become the most reasonable data for climate change and thus have recently emerged as a major area of research in ecology. Different methods had been chosen by scholars to discover the phenological patterns in leafing and other phenomena, some of which are ground-level observation, spatial photography observation. and satellite imagery. Ground-based observations actually gave rise to the science of phenology in the first place when people kept noticing the changes that plants underwent over time. These methods rely on volunteers to collect observations of the various phenophases of wild plants, fruit trees, and agricultural crops at numerous locations (Cleland et al. 2007). Ground-level observations become hard to follow in difficult terrains such as regions of alpine, arctic, tundra, and desert ecosystems (Richardson et al. 2013).

To study a large forest area and to estimate forest biophysical parameters, information collected by the amalgamation of forest resources inventory and remote sensing technique are two approaches (Krankina et al. 2004). Application of new technologies to study the phenology of plants has contributed to diversify the field and bring about a revival in its application to study major changes across Earth's systems. One such technology combines remote sensing and the Geographical Information Systems (GIS). With the help of satellite imagery, it is possible to collect a large number of phenological records across a large spatial and temporal scale. Changes such as leaf emergence, leaf fall, plant responses to environmental changes, and changes in the greening of the biosphere can be easily observed (Myneni et al. 1997). The most important index used in remote sensing data to measure the temporal changes taking place in vegetation is the normalised difference vegetative index (NDVI), which is based on the low reflectance of red colour (a green plant absorbs more of the red and blue region of electromagnetic radiation) and the strong reflectance of near-infrared radiation (Huemmrich et al. 1999). The NDVI has been related to canopy cover (Yoder and Waring 1994), leaf area index, and productivity (Prince et al. 1995). Some studies have also used NDVI data to assess regional phenology (the so-called green wave) (Moulinetal 1997) and to develop phenology models at a regional scale. Other indices are also used based on the objective of research such as the enhanced vegetation index (EVI), Green-Red Vegetation Index (GRVI), Leaf Area Index, Canopy Colour Index, and Leaf Strategy Index.

Other technologies, such as web cameras attached to drones (Fisher et al. 2007) and eddy covariance measurements (Gu et al. 2003; Baldocchi et al. 2005), are providing automated, continuous, and areally averaged measures of phenology (Noormets 2009). Web cameras allow continuous monitoring of a particular area of study.

Phenology modelling has emerged as a new branch of modelling to establish relationships between the climate and phenological patterns and to examine the reactions of plant phenology towards climate change: these models now have an eminent role in regional ecosystem simulation models and biosphere/atmosphere general circulation models (Chuine et al. 2000; Liu et al. 2019). A different global climatic model (GCM) is used to study the dynamics of climate and to provide different future projections of climate for a particular region. In the decade of the 1980s, a number of phenological models evolved (Chuine et al. 1999; Piao et al. 2019). The main attribute of these models is that they are result oriented: their prediction is based on previous experimental results, which have established the responses of plant phenology towards different climatic factors. These models, which are based on the data of a small confined area, might work for large spatial and temporal scale predictions. Hence, they can be used in making assumptions about changes in plant phenology caused by future climate change at a different scale (Chuine et al. 2000).

Hence, plant phenology, which has easily noticeable elements, is now widely used as an indicator of climate change that will have a primary role in the identification and assessment of climate change. In the fourth report of the IPCC in 2007, phenology was reported as a large part of the corroboration on climate change impacts (Rosenzweig et al. 2008).

The literature for this chapter was perused using the Web of Science website. The search words "climate change," "forest," and "phenology" were used in the first step to shortlist the literature. The search yielded 1961 documents in the past two decades starting from year 2000. The number of publications has steadily risen from about 20 articles in 2006 to more than 240 in the year 2018 (Fig. 8.1). Although the number of articles has increased, a multivariate and inclusive approach is needed. In the second step, articles were selected based on their abstract and general content. The rest of the chapter describes the climatic factors responsible for plant phenological changes, the impact of climate change on tree phenology, the consequences of a shift in the phenological cycle, and a few mitigation strategies.

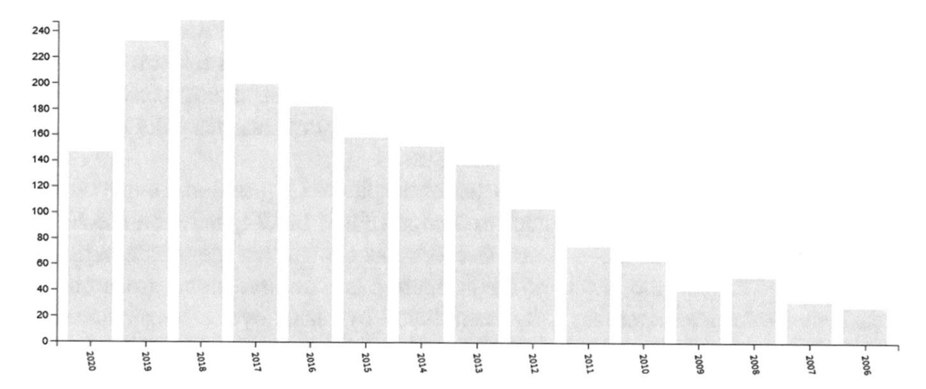

Fig. 8.1 Number of articles published during 2006–2020 based on the Web of Science results (August 1, 2020)

Climatic Factors Responsible for Plant Phenological Changes

Plant seasonal phenological processes have an important role in regulation of the hydrological cycle as well as the energy exchange process between terrestrial ecosystems and the surrounding atmosphere. Similarly, plant response towards climate change can help us in interpretation of climate. After the era of industrialisation when climatic factors were being disturbed by anthropogenic activity, an imbalance between these factors was created that may lead to changes in plant phenology. To study the extent of impact of these changes on forest phenology, first we need to know the relationship between climatic factors and phenology. In this chapter we mainly focus on only some important climatic factors such as carbon dioxide concentration, temperature, precipitation, and photoperiod.

Carbon Dioxide Concentration

Carbon dioxide, which is an essential gas in the process of photosynthesis in autotrophs, also regulates the phenology of plants and is observed to increase in the atmosphere because of anthropogenic activities mainly resulting from industrialisation. Statistical data show that before industrialisation (1750) carbon dioxide concentration was 250 ppm, and the post-industrialisation era witnessed an increase of carbon dioxide gas; in year 2000 it was elevated to 370 ppm, and further increased to 412 ppm in 2019. This elevation is observed more in the Northern Hemisphere because there is more human activity there compared to the Southern Hemisphere. Concentration of carbon dioxide in the atmosphere rises and falls in different seasons of a year and shows a Keeling curve. This variation is caused by the activities of plants on Earth: through photosynthesis and respiration, plants consume CO_2 gas during spring and summer, and release it in fall and winter seasons. After consumption of fossil fuels on a large scale in the era of industrialisation, the range of the carbon dioxide cycle is expanding. It had been estimated that forest trees hoard about 90% of all carbon fixed by the plant as plant biomass. After some research, the relationship between forest phenology and the response towards the continued rise in atmospheric CO_2 concentration is still ambiguous.

Plant phenophases are governed by the atmospheric CO_2 seasonal cycle on the Earth as it affects the net ecosystem exchange (NEE) of CO_2 between the Earth biosphere and the atmosphere. Some research has shown that the increasing concentration of CO_2 in the Northern Hemisphere has witnessed more green cover, considered to cause increasing CO_2 assimilation by plants over a longer growing season in response to global warming. Elevated CO_2 concentration has elongated the growing season of plant phenology by advancing the spring season and delaying leaf senescence. Remote sensing studies using the NDVI have supported this statement by observing increased NDVI values, which indicate geographically significant increases in the photosynthesis process (Myneni et al. 1997; Tucker et al. 2001). In the Boreal forest, which accounts for 13% of the carbon stored in the biomass and 43% of the carbon stored in soil, increased carbon assimilation and earlier leaf emergence have been shown, which notably increased ecosystem photosynthesis but had less effect on respiration (Black et al. 2000). Moderate increase in this gas leads to promotion of growth and productivity in plants having a C3 photosynthetic pathway, whereas negligible effects are seen in plants with a C4 pathway. In their study, Ahrends et al. revealed the association between forest phenology and gross primary productivity (GPP); they observed there is significant correlation between camera-based phenology and GPP and remarkable reliance of CO_2 exchange phenomenon during the growing stage of the tree species, mainly during spring season. As the respiration process is more likely to become acclimatised, compared to photosynthesis, therefore increase in carbon availability in the forest leads to more growth at increased temperatures by abating carbon loss then by assimilating carbon (Way and Oren 2010).

However, a large increase in CO_2 concentration leads to global rise in temperature, which in turn can increase respiration rate, increase plant life duration, accelerate nutrient mineralisation in soil, and increase the rate of evapotranspiration, thus affecting the phenology of plants. Such devices as open top chambers, free air carbon dioxide enrichment (FACE), and greenhouses are used more in recent research as controlled environment solutions to understand the impact of global climate change on plant growth and productivity.

Temperature

Temperature, which controls most of the enzymatic reactions in plant processes, also is crucial in governing the phenology of plants. In 1735, a pioneer phenologist, Rene Antoine Ferchault Reaumur, foremost confirmed the relationship between temperature and phenology by testing the relationship between phenology and the temperature of the preceding season (De Réaumur 1735). Temperature, which is an abiotic component of the environment, is found to primarily control phenological events in plants (Cleland et al. 2007; Chuine 2010). As global warming has the capability to increase global Earth temperature by 1° to 3 °C in upcoming years (Solomon et al. 2007), it has the potential to change regional weather patterns, which may reshape the timing, duration, and synchronisation of the events of forest phenology (Valdez-Hernández et al. 2010). Warming caused by an increase in greenhouse gases is not equally distributed over the globe but at higher altitudes will force more positive change in temperature than at equatorial regions, with mean annual temperatures in the Arctic region predicted to be warmer by 8.3 °C when greenhouse emission is at the highest, by year 2010 (Stocker 2014).

Increase in temperature caused by global warming may facilitate the poleward movement of species distribution; green vegetation may increase in Arctic and Antarctic regions as a result of the global rise in average temperature. Atmospheric temperature and soil temperature both are found to control forest community assemblage in the tropical moist deciduous forests, which differs based on the type of species in the forest (Bajpai et al. 2020). The emergence of new leaves and the end of the leafing season are governed by the environmental temperature. Liu et al. observed that warming preseason temperature was positively linked with the rate of leaf fall, except for arid or semiarid regions (Liu et al. 2016). Most experimental research on establishing a relationship between temperature and plant phenology has shown that elevation in environmental temperature results in earlier spring and later autumn seasons (Menzel et al. 2006), whereas decrease in temperature delays the timing of spring and accelerates the autumn season (Piao et al. 2019). Thus, the average increase of this abiotic factor increases the growing period of plants, which in turn will alter the carbon sequestration (carbon cycle) as well as the hydrological cycle of the biosphere. Increase in temperature not only elongated the growing season of plants but also changed the development trajectories, by increasing more biomass in the leaves compared to the roots. Also, at elevated temperatures, plants grow taller relative to specific stem diameter (Way and Oren 2010).

Temperature, which regulates plant development, was found to impact more importantly on some phenophases such as leaf bud break in spring night temperatures or the minimum temperature (Wielgolaski 1974), whereas other phenophases such as bud break and flowering are mainly dependent on day temperature or maximum temperature (Wielgolaski 2003). Various studies using different methods have been done in areas of plant phenology to find the appropriate threshold or basic temperature for the growth and development of different phenophases of perennial plants.

In temperate regions, the areas between latitudes 23.50°N and 66.32°S, the phenophase of plant vegetation is primarily dependent on atmospheric temperature (Wielgolaski 1999) because large differences in temperature variation and sunlight availability throughout the year compared to tropical regions, where plant phenophases are less affected by atmospheric temperature. It has been found that elevated temperatures magnify shoot height, stem diameter, and biomass in a deciduous forest; less response is recorded in an evergreen forest (Way and Oren 2010).

Temperature has an important role in flowering, Matthews et al. addressed that temperature had influenced the winter and spring flowering day of year, as warmer temperature minimum advanced the flowering whereas temperature maximum caused less significant effect on flowering day of year (Matthews and Mazer 2016). Most research studying the influence of temperature on flowering has revealed that temperature is a dominant and crucial factor for timing of flowering in plants (Templ et al. 2017). Most studies state that the strongest observed correlation in climatic factor and plant phenophase is between flowering phenology and temperature.

Average monthly temperatures of the month of onset season and the two preceding months of this season are observed to have more influence on most plant phenophases because these were found more significantly correlated. An experiment by Menzel et al. indicated that 19% of the phenophases studied showed the highest correlation with the mean temperature of month of onset, 63% of phenophases with preceding month temperature and the rest of the phenophases with temperatures 2 months earlier (Menzel et al. 2006).

Temperature also regulates seed dormancy. Low temperature activated plant stress responses and initiated plant endodormancy (Delpierre et al. 2016) whereas warm temperature broke ecodormancy (Hänninen 2016). Fruit ripening phenophases show strong correlation with increase in temperature or with warmer climate; agricultural plant fruit ripening was more correlated with warmer temperature than that found in wild plants (Menzel et al. 2006). Hence, temperature had an important role in regulating phenology in plants. However, temperature alone is not enough to explain all the variation in plant phenology in relationship to the environment.

Photoperiod

The duration of illumination each day or daylength period experienced by organisms (here, plants) is called photoperiod. It is another important abiotic element of climate that is crucial in stipulating the continuation of the seasons in a given region. Photoperiod had proved to be a critically regulating element in plant phenology (Flynn and Wolkovich 2018). The autumn phenological events are primarily regulated by photoperiodism in plants (Cooke et al. 2012). Photoperiod is found to usually regulate leaf senescence during the autumn season (Cooke et al. 2012). The dependence of autumn phenophases may be the result of modulation by low summer and autumn temperatures (Xie et al. 2015). Regulation of bud burst in plants by the influence of photoperiod varies among species (Miller-Rushing et al. 2008a). Receptors present just below the bud scales perceive light, and the genetic material maintaining leafout could be located in the young leaf primordial cells, which allows each bud to respond autonomously by its own circadian clock system during winter and to respond to daylength (Zohner and Renner 2015). However, the impact of photoperiod on spring phenophases such as leaf emergence is still being examined by phenologists (Chuine 2010). In temperate and boreal regions, the photoperiod coregulated leaf emergence with temperature (Fu et al. 2015).

Opler et al. (1976) stated that photoperiod and thermal accumulation seem insufficient to demonstrate the break of dormancy and the resulting synchronised flowering that is characteristic of many tropical trees, lianas, and shrubs, although short-day photoperiods are probably a major part of initial induction of the reproductive system.

A study to examine bud responses to photoperiod in adult trees growing outside revealed that dormancy break is controlled at the bud level, with light sensing (and probably also temperature sensing) occurring inside the buds. Leaf primordia only react to photoperiod during the late phase of dormancy break when warm days begin (Zohner and Renner 2015).

In another study it was investigated that longer chilling duration resulted in earlier bud burst and maximum bud burst occurred in less thermal time. Consequently, insufficient chilling is received during warmer winters, so bud burst in spring may be expected to be delayed. However, the photoperiod effect was found to be weaker for longer chilling durations; longer photoperiod may, at least in part, compensate for shorter chilling duration (Pletsers et al. 2015). After comparing other causes of change for phenology, more evidence of plants comes from changes observed in the spring (Rosenzweig et al. 2008).

Precipitation

Precipitation and soil water potential are other very important regulating factors in plant phenology. Sufficient nutrient availability during the growing season enhances plant tolerance and adaption to freezing stress, which thereby postpones autumn phenological events. Phenology in the tropical forest, which has more complex biodiversity than temperate and boreal regions, is majorly governed by precipitation patterns compared to temperature variability (Pau et al. 2011). Precipitation is crucial in leaf phenology in the tropical forest, compared to other abiotic elements, because the tropical climate is more regulated by rainfall-related events (Reich 1995). It was found that the extent of influence in phenological events by water and nutrient availability is less in temperate and boreal forests as compared to temperature and photoperiod (Jaworski and Hilzczanski 2013). In arid and semiarid regions, seasonal moisture may have a primary or synergetic function in vegetative and reproductive activity in plants (Beatley 1974; Kemp 1983).

In semiarid regions of California, Mazer et al. (2015) found that, in the phenophases of some species, precipitation had a central role in prognosticating the emergence of leaf phenology or flowering. This study also revealed a monthly synergetic relationship between temperature and precipitation that affected the onset of phenological transitions. The extent of stress experienced by tropical dry forests varies largely during the dry months in these areas as drought conditions are overcome in many tropical forest species by using stored trunk water, deeper roots, and soil moisture (Borchert 1994). The water stress resulting from regional drought may be the dominant contributor to widespread increases in tree mortality rates across tree species, sizes, elevations, longitudes, and latitudes (Peng et al. 2011). In the desert ecosystem, seed germination of annual species only becomes possible after a heavy rainfall of 15–25 mm, and completion of the plant life cycle depends on further rainfall (Kemp 1983).

Recent studies have also established a relationship between rainfall pattern and the advance of spring (Fu et al. 2014). Estiarte and Peñuelas (2015) observed that decreased water availability might regulate the effect of nitrogen presence on plant growth in arid and semiarid regions and thus affect plant phenology. Many studies have shown that long-term precipitation that exceeded plant demand (i.e., when precipitation is greater than evapotranspiration) in wet tropical forests can have a negative impact on forest growth because high water input is negatively correlated to the availability of other essential plant resources such as nutrients or light (Posada and Schuur 2011). Matthews and Mazer 2016 demonstrated that higher precipitation delayed flowering. However, more research is required to understand the mechanism by which water and nutrients interact with other abiotic factors such as temperature and photoperiod in determining plant phenophases.

Impact of Climate Change on Tree Phenology

It has been observed that multivariate and hierarchical modelling is needed for monitoring shifts in phenological changes (Qiu et al. 2020). It was also reported that, coupled with global warming, biodiversity (or lack of it) would also equally influence a change in phenological cycles, highlighting the importance of biotic interactions (Wolf et al. 2017). Extreme climate events such as drought, extreme temperature or precipitation, or availability of water and other resources might influence or trigger phenological changes. Phenology would be impacted by any activity triggering these changes, which would become more complex when these conditions occur in tandem with each other (Qiu et al. 2020). "The Buddhist quote can describe this phenomenon as "the wise will rejoice while the foolish will retreat (WND-1 2003), in other words, the more resilient and adaptive species would be able to survive and complete their phenological cycles, while more vulnerable ones would gradually wither.

Impact on Leaf Phenology Caused by Climate Change

Studying leaf phenological events help us to know seasonal pattern and evolution occurring on a temporal scale in different types of forest (Negi 2006). Climatic factors are central in regulating leaf phenology in plants. Changes such as leaf emergence, leaf fall, plant responses to environmental changes, and changes in the greening of the biosphere can be easily observed (Myneni et al. 1997). Different plant types differ greatly with varying environmental factors. For boreal and temperate vegetation, temperature and photoperiod factors mainly govern leaf phonology, whereas in tropical and subtropical vegetation precipitation has the most influence besides temperature and photoperiod. Earth climate changes caused by global warming lead to variation in leaf phenology. Impact on leaf phenology also varies with type of species found in a region. Some species have the capacity to adapt in a changing environment without any distinguishable change in leaf phenology whereas others indicate the climate change of a region by showing alterations in their leaf phenology. Modelling is now important in assessing the impact of climate change on leaf phenology and the productivity of terrestrial ecosystems.

Shifting of vegetation towards higher northern latitudes is a result of global heating of land surfaces. Many researchers note the northern pole is getting greener each year. Average lengthening of the growing season by about 11 days from increased global temperature has been detected in tree species in Europe from the beginning of the 1960s to the end of the twentieth century, related to the earlier onset of leaf buds and delayed leaf senescence (Menzel and Fabian 1999). A study on four woody plant species showed that the effect of temperature on leaf set was stronger at lower latitudes or altitude as well as in more continental climates, although the date of leaf colouring did not show a continuous pattern. In the same study, significant warming across China during the study period showed lengthening of the growing period of leaves: leaf emergence was advanced by 5.44 days from 1960 to 2009 and the leaf colouring date was delayed by 4.56 days during the same time (Dai et al. 2014).

Changes in the occurrence and timing of leaf phenophases may either accelerate or slow rates of climate change depending upon the vegetation at a particular geographical location. Leaf phenology, which regulates albedo, transpiration, exchange of molecules between vegetation and the atmosphere, and cloud formation, when faced with change in turn will impact the climate system. It is found that tropical forests lower the rate of global warming through evaporative cooling, whereas boreal forests, which have low albedo, will show positive climate forcing. The evaporative effect of temperate forests is still unclear as there is less information (Bonan 2008).

Impact on Reproductive Phenology Caused by Climate Change

Increase average temperature of the Earth is changing flowering phenophases in different species. Flowering in plants primarily depends upon temperature and precipitation, which is disturbed by climate change which in return affects flowering phenology. Many studies indicate that flowering dates are being advanced in flowering species. A study in Japan on the cherry tree found that flowering times in and around Osaka and other cities in Japan, with average temperature rise in urbanised cities, cause cherry plants to flower earlier than under the normal average temperature present in parks and outlying suburban areas (Primack et al. 2009). In Japan, the date of the cherry tree flowering season is celebrated as a large festival, providing one of the most important sources of data on the impacts of global warming on flowering phenology. From 1971 to 2000, cherry trees flowered an average of 7 days earlier in comparison to the average of all previous records. Flowering records at Mt. Takao from a large cherry botanical garden of Tokyo revealed that both among and within species, advanced flowering of trees is linked to their response against temperature variation (Primack et al. 2009).

Another study on North American species flowering date influenced by climate variation consisted of a series of earlier records of the dates of first flowering for more than 500 plant taxa in addition to recent observations from approximately 1852 to 2006. During this period the studied area had an increase in average temperature by 2.4 °C from global climate change, which shows that plants are now flowering 7 days earlier on average than they did in 1852. This study had shown that plant flowering times were largely influenced by mean temperatures in the 1 or 2 months just before flowering (Miller-Rushing et al. 2007).

It has also been found that flowering species blossoming during summer showed more year-to-year dissimilarity in flowering time than those species which blossom during spring, but mean monthly temperatures had strong influence on the flowering times of the spring-flowering species (Miller-Rushing et al. 2007).

The date of snowmelt is also advancing because of global warming; snowmelt also governs the flowering phenology as noted in a study at the Rocky Mountain Biological Laboratory (Colorado, USA), which makes this site ideal for examining phenological responses to climate change. This study examined the floral buds of *Delphinium barbeyi*, *Erigeron speciosus*, and *Helianthella quinquenervis*, which are sensitive to frost; from advancing of the growing season in recent years these species have faced frost kills more frequently during mid-June, interrupting their reproductive cycle (Inouye 2008).

In India, Gaira et al. (2014) investigated the influence of flowering phenology change on Himalayan species of *Rhododendron arboreum* by comparing herbarium records (1893–2003) and real-time field observations (2009–2011). Flowering phenology was strongly correlated with increase in seasonal temperature, mainly winter and post-monsoon, and annual mean maximum temperature. This research also predicted using the generalized additive model (GAM) that the advancement of flowering has been 88–97 days early during the past 100 years (Gaira et al. 2014).

It may also be possible that closely related species show different responses to climatic variation. Effects on community structure, ecosystem dynamics, and population processes and phenological changes can only be well understood by doing comparative studies that can adequately predict future phonological impact from climate change.

Changes in Forest Phenology Influencing the Environmental and Ecological Functions of Tree Species

Phenological changes caused by global climate change affect other aspects of plant life cycle. The correlation of a plant with other flora and fauna is also altered, which changes ecosystem-level, population-level, and community-level interactions. Changes in plant phenology that govern different activities and balance abiotic cycles may increase or decrease rates of global warming on Earth; vegetation change may vary the energy exchanges between the terrestrial ecosystem and the atmosphere. Changes in energy balance in the ecosystem will endanger the species that have synchronised life cycles with these phenophases, which eventually increase discrepancy between the species and increase the risk of extinction for affected species. With continued change in global climate and phenological response to these climatic changes, this risk of extinction is expected to further increase.

Change in Reproductive System

Reproduction in an organism aims to propagate its progeny. The occurrence of healthy progeny can only be determined by proper availability of a suitable environment for the maintenance of new offspring. Without proper resources, reduction in population and even extinction can occur. Reproduction is only accomplished when it occurs at the correct time; reproducing before the season or delaying reproduction from the appropriate season leads to unavailability of required resources. Change in reproductive season may lead to difficulty in finding mates or fulfilling the demands of growing offspring (Visser et al. 1998).

Some studies show that increase in carbon dioxide concentration may advance or delay the flowering season in different floral species. Elevated levels of CO_2 may induce self-compatibility, which may limit outcrossing fertilisation and lead to inbreeding depression (Bawa and Dayanandan 1998).

Effect on Plant–Pollinator Interactions

Plants depends upon biotic and abiotic factors for the process of pollination. Species who depend on particular species for pollination may unable to complete this process if those species are not available; hence, the presence of that species is crucial to completing the pollination process. The advance or delay of flowering caused by global increases in temperature and carbon dioxide may cause the absence of its pollinating species.

Increased drought may lower the population densities of bees and insects that are dependent on plant sap (nectar), and these insects act as a great pollinating medium for such flowering plants (Rouault et al. 2006; Le Conte and Navajas 2008). Mismatching in timing of the arrival of butterflies and flowering in plants in Britain caused by climate change will eventually affect the mutualistic relationship between plant and butterflies (Roy and Sparks 2000). Plants that depends on bird populations for pollination can also face difficulty in completing pollination as the bird species may have a specific nesting period (Miller-Rushing et al. 2008b).

Effect on Plant Seed Dispersal

Seed dispersal allows spreading of species population to a great distance. Edaphic factors and animals govern the seed dispersal process. To increase the probability of germination of seed and establishment of a new individual requires proper availability of resources, particularly water and temperature. When dispersal of seeds occurs earlier or later than the favourable season for seed germination, seed may be lost to predators or pathogens or will not have the proper medium for distant dispersal. Frankie et al. (1974) found that dry deciduous forest trees of central America disperse their seeds between wet and dry seasons to provide a favourable time for germination. However, change in climate alters the pattern of ripening of fruit and its timing for dispersal, which will decrease the probability of seed germination.

It is evident that plants which depend on animals for seed dispersal have a lesser probability of the dispersal because of differences in the timing of ripening of fruit and the arrival of animals with a major role in dispersion (Miller-Rushing and Primack 2008). These studies have shown that this discrepancy can cause an ecological mismatch, so that resources needed for the germination and survival of seed are no longer present (Waser and Real 1979).

Effect on Diversity of Plants

Floral species diversity is greatly affected by climate change. Many endemic species may no longer be able to sustain populations in their native habitat; they may migrate or become extinct because of altered temperatures and precipitation regimes (Thomas et al. 2004). Besides the extinction or migration of species, climate change will decelerate the rate of reproductive success, which decreases the population size of species in their native habitat. Global climate change affects plant species diversity by shifting populations northward. It has been recorded that species are migrating to higher latitudes or altitudes in response to climate change.

Effect on Tourism and the Economy

Tourism is crucial to the economic significance of a country and a region by providing large weightage in country gross domestic product (GDP). Tourism is mostly connected with environment, nature, and climate, so tourism can be called an economic sector that is highly sensitive to regional climate. Tourism can be directly affected by climatic change such as rising average temperatures of a region or indirectly by rising sea levels, loss of snow cover, change in precipitation pattern or impacts on the landscape. Thus, it becomes important to predict the consequences of climate change on future demand in tourism.

The changes in phenology, physiology, and diversity of the forests caused by climate change affect forest tourism as these changes lead to more frequent forest fires, storm damage, difference in precipitation patterns, and loss in species diversity (Hall et al. 2011).

Elevation in average temperature has caused winter seasons to shorten and summer duration to become longer. Snow cover is also affected, challenging local tourism. At higher altitudes, snow cover is being reduced and further decrease is predicted from ongoing global warming; reduced snow cover may lead to an appreciable decrease in future tourism for this natural attraction (Endler et al. 2010).

Other tourist destinations that are highly sensitive towards climate change are greatly affected, such as coral reef exploration, and the proliferation of jellyfish. It is also noticed that more effects of climate change might be experienced at lower altitudes than at higher altitudes.

In a study in Europe, Hanewinkel et al. (2013) found that change of temperature and precipitation is expected to decline an economically valuable species, Norway spruce (*Picea abies* Karst), which is one of the major commercial tree species in Europe. It is also predicted by models that Norway spruce may shift northwards and probably lose large parts of its present range in central, eastern, and western Europe in coming years (Hanewinkel et al. 2013).

Mitigation Strategies

Mitigation strategies towards reducing the effect of climate on forest health lie in analysing forest health by taking into account the multiple environmental, ecological, and social factors influencing it. Phenology would be the most important factor in developing mitigation efforts for the impact of climate change on forests. A keyword search on the Web of Science for "forest," "phenology," and "mitigation," in the abstract, keywords, or title of articles, produced only 24 results. Most of the literature originated in developed countries (Fig. 8.2), whereas less developed nations are prone to exploit their forest resources. It has been argued in the 2019 IPCC special report on Climate Change and Land that planting forest and protecting the existing forests, especially the dense forests, are crucial to limit the temperature rise below 2°C. In this section we have briefly highlighted the role of phenology as an indicator of climate change and the need to strengthen phenological resilience.

Phenology as an Indicator of Climate Change

The temporal variations in the green cover of forests show long-term changes in the phenological activity of the forests which can be used as an indicator of climate change impact (Chakraborty et al. 2018).

Species distribution in the forest can act as an indicator of climate change (Priti et al. 2016). It has been projected that tree species "face a significant decrease in suitable habitat area" from climate change (Dyderski et al. 2018). The ramification of the same for both ecosystem services and biodiversity would be very serious.

Plant phenology has been reported to be a "sensitive indicator of climate change" and essentially controls the seasonal variations of carbon budget and water and energy flow between the terrestrial and atmospheric systems of the ecosystem (Kosmala et al. 2016). Any change in the climate will result in a chain reaction affecting the material and energy flow across the forest ecosystem that is observable at the phenological level of the tree species.

Enhancing Phenological Resilience

Development of specific policy level and practical approaches is needed to enhance the phenological resilience of forests towards climatic disturbances; this is important

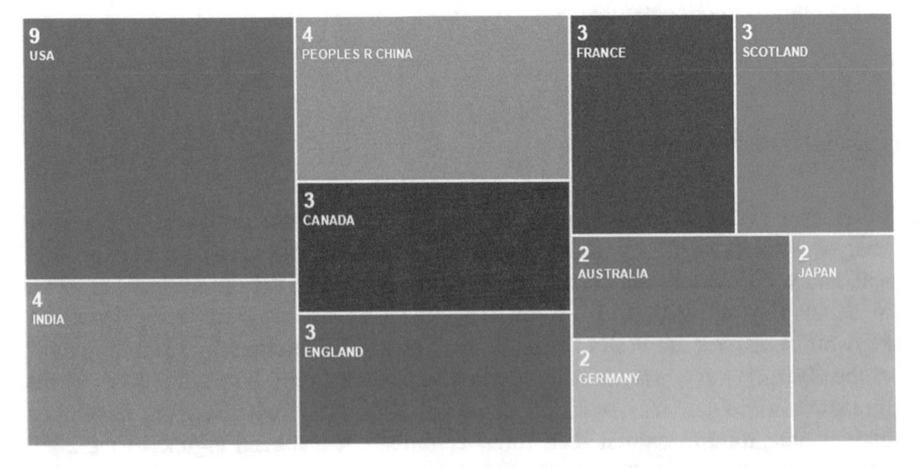

Fig. 8.2 Web of Science search results according to the country where the study was conducted (searched on August 1 2020)

to maintain the flow of ecosystem services. Phenological resilience is seen as the most apparent symptom of or strategy towards a climate-induced disturbance. These approaches may address the management of forests, including reforestation and protecting the existing forests, such as the United Nations REDD+ initiation, or, they may approach the monitoring aspect to be cautious of the changes taking place over the long term. Some of these approaches are mentioned here:

(a) The protection of forest biodiversity should be prioritised, especially in the core forest areas. Chakraborty et al. (2018) observed that the seasonal greenness over different types of forests in India suffered a gradual decrease from 2001 to 2014 that was highest over tropical moist deciduous forests; most of the high to medium negative trends were found in the core zones of forests. Significant among the protected areas exhibiting this decrease were the Simplipal wildlife sanctuary (Odisha), Rajaji National Parl (Uttarakhand), Achanakmar wildlife sanctuary, and the Sundarban (Chakraborty et al. 2018).
(b) Conservation of existing habitats, especially in protected and core areas, resuscitating the threatened species populations in existing areas, and identifying the more resilient species that can adapt to disturbances, would help in increasing forest resilience (Priti et al. 2016).
(c) Multicriteria and hierarchical modelling approaches that take into consideration socioeconomic and biological factors, including the response lag of certain species to environmental changes, would help in creating better strategies to protect forest species (Qiu et al. 2020; Hua et al. 2019).
(d) A multipronged approach aimed at educating the local stakeholders, and using traditional knowledge to monitor phenological changes, would work better in tandem with remote sensing and ground-based measurements. A recent example is a study conducted by Kosmala et al. (2016), wherein the authors used "citizen science in combination with near-surface remote sensing of phenology" to observe ground-based phenological changes. This approach helped in data processing and validation of satellite results. The volunteers' observed data was further used to improve the vegetation indices and computation algorithm to interpret satellite results more accurately (Kosmala et al. 2016).
(e) Carbon forestry and manure management has also been suggested to reduce the impact of climate change in forest areas as well as function in an adaptation strategy to manage forest resources (Singh et al. 2010).

Conclusions

Climate change is rapidly affecting the most pristine and vulnerable habitats of the Earth. Among these, some habitats are facing changes at more than global average rates. Gradually, this will result in loss of habitat and biodiversity. Extreme weather events, added to the direct human exploitation of biodiversity-rich land resources, place the phenology and revival of many forest tree species in jeopardy.

Towards this, availability of data and modelling approaches are important to create and implement policy level decisions. In situ observations and data collection may prove to be expensive and time consuming. The role of remote sensing technologies and citizen science would pave the way towards monitoring multiple spatial and temporal scales across large swaths of land. Satellite remote sensing can prove invaluable in monitoring and analysing large-scale and continuous phenological changes at the global scale.

References

Bajpai O, Dutta V, Singh R, Chaudhary LB, Pandey J (2020) Tree community assemblage and abiotic variables in tropical moist deciduous forest of Himalayan Terai eco-region. Proc Natl Acad Sci India B Biol Sci 2020:1–11

Baldocchi DD, Black TA, Curtis PS, Falge E, Fuentes JD, Granier A et al (2005) Predicting the onset of net carbon uptake by deciduous forests with soil temperature and climate data: a synthesis of FLUXNET data. Int J Biometeorol 49(6):377–387

Bawa KS, Dayanandan S (1998) Global climate change and tropical forest genetic resources. Clim Change 39(2–3):473–485

Beatley JC (1974) Phenological events and their environmental triggers in Mojave Desert ecosystems. Ecology 55(4):856–863

Black TA, Chen WJ, Barr AG, Arain MA, Chen Z, Nesic Z et al (2000) Increased carbon sequestration by a boreal deciduous forest in years with a warm spring. Geophys Res Lett 27 (9):1271–1274

Bonan GB (2008) Forests and climate change: forcings, feedbacks, and the climate benefits of forests. Science 320(5882):1444–1449

Borchert R, Rivera G, Hagnauer W (2002) Modification of vegetative phenology in a tropical semi-deciduous Forest by abnormal drought and rain. 1. Biotropica 34(1):27–39

Chakraborty A, Seshasai MVR, Reddy CS, Dadhwal VK (2018) Persistent negative changes in seasonal greenness over different forest types of India using MODIS time series NDVI data (2001–2014). Ecol Indic 85:887–903

Chuine I (2010) Why does phenology drive species distribution? Philos Trans R Soc B Biol Sci 365 (1555):3149–3160

Chuine I, Cour P, Rousseau DD (1999) Selecting models to predict the timing of flowering of temperate trees: implications for tree phenology modelling. Plant Cell Environ 22(1):1–13

Chuine I, Cambon G, Comtois P (2000) Scaling phenology from the local to the regional level: advances from species-specific phenological models. Glob Change Biol 6(8):943–952

Cleland EE, Chuine I, Menzel A, Mooney HA, Schwartz MD (2007) Shifting plant phenology in response to global change. Trends Ecol Evol 22(7):357–365

Cooke JE, Eriksson ME, Junttila O (2012) The dynamic nature of bud dormancy in trees: environmental control and molecular mechanisms. Plant Cell Environ 35(10):1707–1728

Corlett RT, Lafrankie JV (1998) Potential impacts of climate change on tropical Asian forests through an influence on phenology. Clim Change 39(2-3):439–453

Dai J, Wang H, Ge Q (2014) The spatial pattern of leaf phenology and its response to climate change in China. Int J Biometeorol 58(4):521–528

De Réaumur RAF (1735) Observations du thermomere. In: Memories Academie Royale Sciences Paris, pp 545–576

Delpierre N, Vitasse Y, Chuine I, Guillemot J, Bazot S, Rathgeber CB (2016) Temperate and boreal forest tree phenology: from organ-scale processes to terrestrial ecosystem models. Ann For Sci 73(1):5–25

Dyderski MK, Paź S, Frelich LE, Jagodziński AM (2018) How much does climate change threaten European forest tree species distributions? Glob Change Biol 24(3):1150–1163
Endler C, Oehler K, Matzarakis A (2010) Vertical gradient of climate change and climate tourism conditions in the Black Forest. Int J Biometeorol 54(1):45–61
Estiarte M, Peñuelas J (2015) Alteration of the phenology of leaf senescence and fall in winter deciduous species by climate change: effects on nutrient proficiency. Glob Chang Biol 21 (3):1005–1017
Fisher JI, Richardson AD, Mustard JF (2007) Phenology model from surface meteorology does not capture satellite-based greenup estimations. Glob Chang Biol 13(3):707–721
Flynn DFB, Wolkovich EM (2018) Temperature and photoperiod drive spring phenology across all species in a temperate forest community. New Phytol 219(4):1353–1362
Frankie GW, Baker HG, Opler PA (1974) Comparative phenological studies of trees in tropical wet and dry forests in the lowlands of Costa Rica. J Ecol 62:881–919
Gaira KS, Rawal RS, Rawat B, Bhatt ID (2014) Impact of climate change on the flowering of *Rhododendron arboreum* in central Himalaya, India. Curr Sci 106:1735–1738
Gu L, Post WM, Baldocchi D, Black TA, Verma SB, Vesala T, Wofsy SC (2003) Phenology of vegetation photosynthesis. In: Phenology: An integrative environmental science. Springer, Dordrecht, pp 467–485
Hall CM, Scott D, Gössling S (2011) Forests, climate change and tourism. J Herit Tour 6 (4):353–363
Hanewinkel M, Cullmann DA, Schelhaas MJ, Nabuurs GJ, Zimmermann NE (2013) Climate change may cause severe loss in the economic value of European forest land. Nat Clim Change 3(3):203–207
Hänninen H (2016) Boreal and temperate trees in a changing climate. In: Biometeorology. Springer, Dordrecht, Netherlands
Hua L, Wang H, Sui H, Wardlow B, Hayes MJ, Wang J (2019) Mapping the spatial-temporal dynamics of vegetation response lag to drought in a semiarid region. Remote Sens 11(16):1873
Inouye DW (2008) Effects of climate change on phenology, frost damage, and floral abundance of montane wildflowers. Ecology 89(2):353–362
Jaworski T, Hilszczański J (2013) The effect of temperature and humidity changes on insects development their impact on forest ecosystems in the expected climate change. For Res Pap 74 (4):345–355
Justiniano MJ, Fredericksen TS (2000) Phenology of tree species in Bolivian dry forests. Biotropica 32(2):276–281
Kemp PR (1983) Phenological patterns of Chihuahuan Desert plants in relation to the timing of water availability. J Ecol 71:427–436
Kosmala M, Crall A, Cheng R, Hufkens K, Henderson S, Richardson AD (2016) Season Spotter: Using citizen science to validate and scale plant phenology from near-surface remote sensing. Remote Sens 8(9):726
Kramer K, Leinonen I, Loustau D (2000) The importance of phenology for the evaluation of impact of climate change on growth of boreal, temperate and Mediterranean forests ecosystems: an overview. Int J Biometeorol 44(2):67–75
Le Conte Y, Navajas M (2008) Climate change: impact on honey bee populations and diseases. Revue Scientifique et Technique-Office International des Epizooties 27(2):499–510
Lieth H (1974) Purposes of a phenology book. In: Phenology and seasonality modeling. Springer, Berlin, Heidelberg, pp 3–19
Liu Q, Fu YH, Zhu Z, Liu Y, Liu Z, Huang M, Piao S (2016) Delayed autumn phenology in the Northern Hemisphere is related to change in both climate and spring phenology. Global Change Biol 22(11):3702–3711
Liu Q, Piao S, Fu YH, Gao M, Peñuelas J, Janssens IA (2019) Climatic warming increases spatial synchrony in spring vegetation phenology across the Northern Hemisphere. Geophys Res Lett 46(3):1641–1650

Matthews ER, Mazer SJ (2016) Historical changes in flowering phenology are governed by temperature × precipitation interactions in a widespread perennial herb in western North America. New Phytol 210(1):157–167

Mazer SJ, Gerst KL, Matthews ER, Evenden A (2015) Species-specific phenological responses to winter temperature and precipitation in a water-limited ecosystem. Ecosphere 6(6):1–27

Menzel A, Fabian P (1999) Growing season extended in Europe. Nature 397(6721):659

Menzel A, Sparks TH, Estrella N, Koch E, Aasa A, Ahas R, Chmielewski FM (2006) European phenological response to climate change matches the warming pattern. Global Change Biol 12 (10):1969–1976

Miller-Rushing AJ, Primack RB (2008) Global warming and flowering times in Thoreau's Concord: a community perspective. Ecology 89:332–341

Miller-Rushing AJ, Katsuki T, Primack RB, Ishii Y, Lee SD, Higuchi H (2007) Impact of global warming on a group of related species and their hybrids: cherry tree (Rosaceae) flowering at Mt. Takao, Japan. Am J Bot 94:1470–1478

Miller-Rushing AJ, Inouye DW, Primack RB (2008a) How well do first flowering dates measure plant responses to climate change? The effects of population size and sampling frequency. J Ecol 96(6):1289–1296

Miller-Rushing AJ, Primack RB, Stymeist R (2008b) Interpreting variation in bird migration times as observed by volunteers. Auk 125:565–573

Myneni RB, Keeling CD, Tucker CJ et al (1997) Increased plant growth in the northern high latitudes from 1981 to 1991. Nature 386:698–702

Negi GCS (2006) Leaf and bud demography and shoot growth in evergreen and deciduous trees of central Himalaya, India. Trees 20(4):416–429

Negi GCS, Singh SP (1992) Leaf growth pattern in evergreen and deciduous species of the Central Himalaya, India. Int J Biometeorol 36(4):233–242

Opler PA, Frankie GW, Baker HG (1976) Rainfall as a factor in the release, timing, and synchronization of anthesis by tropical trees and shrubs. J Biogeogr:231–236

Pau S, Wolkovich EM, Cook BI, Davies TJ, Kraft NJ, Bolmgren K et al (2011) Predicting phenology by integrating ecology, evolution and climate science. Glob Chang Biol 17 (12):3633–3643

Piao S, Liu Q, Chen A, Janssens IA, Fu Y, Dai J et al (2019) Plant phenology and global climate change: current progresses and challenges. Global Change Biol 25(6):1922–1940

Pletsers A, Caffarra A, Kelleher CT, Donnelly A (2015) Chilling temperature and photoperiod influence the timing of bud burst in juvenile *Betula pubescens* Ehrh. and *Populus tremula* L. trees. Ann For Sci 72(7):941–953

Posada JM, Schuur EA (2011) Relationships among precipitation regime, nutrient availability, and carbon turnover in tropical rain forests. Oecologia 165(3):783–795

Primack RB, Higuchi H, Miller-Rushing AJ (2009) The impact of climate change on cherry trees and other species in Japan. Biol Conserv 142(9):1943–1949

Priti H, Aravind NA, Shaanker RU, Ravikanth G (2016) Modeling impacts of future climate on the distribution of Myristicaceae species in the Western Ghats, India. Ecol Eng 89:14–23

Qiu T, Song C, Clark JS, Seyednasrollah B, Rathnayaka N, Li J (2020) Understanding the continuous phenological development at daily time step with a Bayesian hierarchical space-time model: impacts of climate change and extreme weather events. Remote Sens Environ 247:111956

Reich PB (1995) Phenology of tropical forests: patterns, causes, and consequences. Can J Bot 73 (2):164–174

Richardson AD, Keenan TF, Migliavacca M, Ryu Y, Sonnentag O, Toomey M (2013) Climate change, phenology, and phenological control of vegetation feedbacks to the climate system. Agric For Meteorol 169:156–173

Rosenzweig C, Karoly D, Vicarelli M, Neofotis P, Wu Q, Casassa G et al (2008) Attributing physical and biological impacts to anthropogenic climate change. Nature 453(7193):353–357

Rouault G, Candau JN, Lieutier F, Nageleisen LM, Martin JC, Warzée N (2006) Effects of drought and heat on forest insect populations in relation to the 2003 drought in Western Europe. Ann For Sci 63(6):613–624

Roy DB, Sparks TH (2000) Phenology of British butterflies and climate change. Global Change Biol 6:407–416

Secretariat of the Convention on Biological Diversity. (2009) Biodiversity, development and poverty alleviation: recognizing the role of biodiversity for human well-being, Montreal

Singh JS, Singh VK (1992) Phenology of seasonally dry tropical forest. Curr Sci:684–689

Singh SP, Singh V, Skutsch M (2010) Rapid warming in the Himalayas: Ecosystem responses and development options. Clim Dev 2(3):221–232

Solomon S, Manning M, Marquis M, Qin D (2007) Climate change 2007-the physical science basis: Working group I contribution to the fourth assessment report of the IPCC, vol 4. Cambridge University Press

Stocker T (ed) (2014) Climate change 2013: the physical science basis: Working Group I contribution to the Fifth assessment report of the Intergovernmental Panel on Climate Change. Cambridge University Press

Templ B, Templ M, Filzmoser P, Lehoczky A, Bakšienè E, Fleck S, Palm V (2017) Phenological patterns of flowering across biogeographical regions of Europe. Int J Biometeorol 61 (7):1347–1358

Thomas CD, Cameron A, Green RE, Bakkenes M, Beaumont LJ, Collingham YC et al (2004) Extinction risk from climate change. Nature 427(6970):145–148

Tucker CJ, Slayback DA, Pinzon JE, Los SO, Myneni RB, Taylor MG (2001) Higher northern latitude normalized difference vegetation index and growing season trends from 1982 to 1999. Int J Biometeorol 45(4):184–190

Valdez-Hernández M, Andrade JL, Jackson PC, Rebolledo-Vieyra M (2010) Phenology of five tree species of a tropical dry forest in Yucatan, Mexico: effects of environmental and physiological factors. Plant Soil 329(1–2):155–171

Visser ME, Van Noordwijk AJ, Tinbergen JM, Lessells CM (1998) Warmer springs lead to mistimed reproduction in Great Tits (*Parus major*). Proc R Soc Lond Ser B 265:1867–1870

Waser NM, Real LA (1979) Effective mutualism between sequentially flowering plant species. Nature 281:670–672

Way DA, Oren R (2010) Differential responses to changes in growth temperature between trees from different functional groups and biomes: a review and synthesis of data. Tree Physiol 30 (6):669–688

Wielgolaski FE (1974) Phenology in agriculture. In: Phenology and seasonality modeling. Springer, Berlin/Heidelberg, pp 369–381

Wielgolaski FE (1999) Starting dates and basic temperatures in phenological observations of plants. Int J Biometeorol 42(3):158–168

Wielgolaski FE (2003) Climatic factors governing plant phenological phases along a Norwegian fjord. Int J Biometeorol 47(4):213–220

WND-1 (2003) Gosho Translation Committee. The writings of Nichiren Daishonin. Soka Gakkai, Tokyo

Wolf AA, Zavaleta ES, Selmants PC (2017) Flowering phenology shifts in response to biodiversity loss. Proc Natl Acad Sci 114(13):3463–3468

Xie Q, Lou P, Hermand V, Aman R, Park HJ, Yun DJ et al (2015) Allelic polymorphism of GIGANTEA is responsible for naturally occurring variation in circadian period in Brassica rapa. Proc Natl Acad Sci 112(12):3829–3834

Zohner CM, Renner SS (2015) Perception of photoperiod in individual buds of mature trees regulates leaf-out. New Phytol 208(4):1023–1030

Chapter 9
Assessment of Stream Flow Impact on Physicochemical Properties of Water and Soil in Forest Hydrology Through Statistical Approach

Malabika Biswas Roy, Pankaj Kumar Roy, Sudipa Halder, Gourab Banerjee, and Asis Mazumdar

Abstract Forests affect the hydrology of watersheds in various ways with different hydrological parameters such as evapotranspiration, infiltration, intercepting cloud moisture, surface runoff, and groundwater. Sikkim, a part of northeastern India, a hilly area, is basically a critical terrain condition, and collection of secondary and primary data might be a challenging task to assess overall impacts, particularly on hydrological services. Three micro-watersheds, categorized as undisturbed, semidisturbed, and disturbed catchments covering a distance of 12 km at Geyzing, located in West Sikkim, were selected for this research work. Hydro-meteorological instruments were installed and recorded data for 4 years in three micro-watersheds to monitor the impact of forest cover on stream discharge and water quality. A composite rectangular weir was installed at field level to accommodate measuring stream discharge during the monsoon period as well as the pre-monsoon period. Water and soil quality parameters were analyzed during the monsoon as well as the pre-monsoon period. The overall impact of forests on the hydrological regime of a watershed is twofold. First, comparatively less discharge from a forest watershed during the rainy months shows that the forest controls excess runoff downstream. Second, forests induce infiltration, which leads to more uniform flow all year from the dense forest watershed stream. Conversions to forestland would have the potential to reduce erosion and subsequent sedimentation, as well as reduce levels of nutrients and pesticides in surface runoff and groundwater. These improvements in water quality can be a function of lower amounts of runoff and leaching as well as lower concentrations of potential pollutants.

M. B. Roy
Women's College, Kolkata, India

P. K. Roy (✉)
FISLM, School of Water Resources Engineering, Jadavpur University, Kolkata, India

S. Halder · G. Banerjee · A. Mazumdar
School of Water Resources Engineering, Jadavpur University, Kolkata, India

M. N. Islam, A. van Amstel (eds.), *India: Climate Change Impacts, Mitigation and Adaptation in Developing Countries*, Springer Climate,
https://doi.org/10.1007/978-3-030-67865-4_9

Keywords Surface runoff · Infiltration · Evapotranspiration · Watershed · Disturbed catchment · Stream discharge

Introduction

Forests affect the hydrology of watersheds in various and complex ways, such as increasing evapotranspiration, increasing infiltration, intercepting cloud moisture, and reducing the nutrient load of runoff. Runoff originating from a mixed forest produces high concentrations of ionic elements, organic matter, nitrogen, and phosphoric acids (Klimaszyk et al. 2015). Study on qualitative and qualitative analysis of runoff variation and its influencing factors has greater dominance on the ecological protection of a river basin (Shang et al. 2019). Degradation of the forest cover causes alternation of the hydraulic balance of the forest ecosystem. Such uncontrolled change in land use and land cover such as deforestation affects watershed hydrology in most of the tropical watersheds of developing countries across the world (Ewane and Lee 2020). Watershed deforestation induced surface runoff, interflow, base flow, and total runoff in a Bodrog basin of Northeastern Slovakia, which clearly visualizes an adverse effect of anthropogenic disturbance on natural ecology. Different parts of the trees such as the root system, stem, trunk, and leaves lead to the absorption of moisture from soil, ground, and soil surface cover, and converts it into biowater that ultimately helps develop a sound precipitation mechanism (Yu Yannian 1990). Tree canopy morphology is important in the determination of rainfall interception and hence ecological restoration accordingly (Yang et al. 2019). Various modeling techniques can be applied to estimate the water loss through forest degradation. Guzha et al. (2018) applied the Mann–Kendal test and Sen slope estimator to analyze the variation of peak discharge, mean annual discharge, and low flows in river basins of the East African region to observe the impact of land use and land cover change on these parameters.

Sometimes scientific clearing of forest and regrowth with herbicides increases the water yield of a basin and improves water quality (Hornbeck et al. 1993). Forest types also control the water yield of the basin. The capacity of forest hydrology in increasing the carbon capture and water storage under climate change depends on the interaction between hydrological processes and forest dynamics (Frene et al. 2019). Vildan and Michael (1996) applied fuzzy linear regression and estimated that water yield from coniferous type forests is higher than the eucalyptus type. Forest fire through natural and anthropogenic activities induces increased soil erosion rates, causes alteration of runoff generation, and provokes pollutant sources, which increase fluxes of sediment, nutrients, and water quality ingredients that contaminate water supplies (Smith et al. 2011). Forestation transformation and the horizontal migration of soil nutrients can exaggerate soil erosion activities, with more runoff and poor water recharge (Zhu et al. 2019). Dying of trees in a forest can also lead to the loss of canopy and changes in water and nutrient uptake, which further leads to observable changes in hydrology and the biogeochemical cycle of the ecosystem

(Mikkelson et al. 2013). Water quality improvements mainly occur in the areas where most of the excess precipitation moves across, in, or near the root zone of the riparian forest buffers, which enriches the water mineralogy (Lowrance et al. 1997). Wetland forests are also valuable, enriching the environment and societal economic development. Wetland forests are valued by society because of their ability to maintain or improve water quality, and forest practices such as timber harvesting produce unacceptable changes in surface water quality as well as quantity (Shepard 1994). Forested wetland also is important in flood risk management, water quality enhancement, and recreational benefits to society. Prevost et al. (1997) analyzed the soil and surface water quality for a 5-year-long term trend in a drainage experiment executed in a black spruce peat land of eastern Quebec, Canada, and found observable changes that can be handled through scientific management. Walega et al. (2020) made estimated direct storm runoff and peak flow rates using a modified SCS-CN model for a selected forest ecosystem in the southeastern Unites States. Harvesting, shifting agriculture, grazing, and invasion by alien plant species may all have profound effects on the functioning of the watersheds and runoff character of the basin. Liu et al. (2019) analyzed the runoff response of the basin based on flow path length and integral connectivity scale length considering the soil moisture conditions.

In India, research in forest hydrology was begun as early as the 1960s by the Central Soil & Water Conservation Research & Training Institute, Research Centre. The Himalayan region is facing extensive runoff from forest degradation, which is presently leading to increased landslides. The present study includes a reconnaissance survey at different locations, namely, Nambu and Rimbi, in three different areas such as disturbed by anthropogenic activities, undisturbed, and a pasture area of West Sikkim. In this study, an attempt has been made to assess the quantum of the stream flow and to check water quality with reference to the study area and its impact on forest plantation under the National Afforestation Programme, related to maintenance of forest hydrology. The effects of forest on rainfall, the impacts of various forest activities (thinning, selection felling, clear-felling) on stream flow, and the soil and water impacts of reforesting degraded or agricultural areas have been analyzed. Similarly, effects of forest management and conversion of land use on water quality and ways to minimize any adverse impacts are covered with forest plantation. Water draining from undisturbed forest watersheds generally has the highest quality, particularly with regard to beneficial uses including drinking water, aquatic habitat for native species, and contact recreation. The role of forests for providing water supplies of the highest quality has been one of the driving factors for establishment of forest reserves and for the development of forest management practices designed to protect the high water quality. Advanced forest management activities such as road construction, logging operations, site preparation for regeneration of forest tree species, and fertilization of existing forests have been shown to alter water quality, primarily by causing changes in sediment loads, stream temperature, dissolved oxygen, and dissolved nutrients, particularly nitrogen.

Description of Study Area

West Sikkim is a district of the Coloured Stone Carving at Tashiding Monastary of the Indian state of Sikkim. Its capital is Geyzing, also known as Gyalshing. The district is a favorite among trekkers because of the high elevation. Other important towns include Pelling and Jorethang. The district is enriched with a wide variety of fauna and flora. Because most of the district is hilly, it enjoys a temperate climate. Above 3800 m the slopes are mainly covered with rhododendron forests. The economy is mainly agrarian, although most of the land is unfit for cultivation owing to the precipitous and rocky slopes. The region has developed many power projects and enjoys almost uninterrupted electricity. Roads, however, are in poor condition because of the frequent landslides. Thus, an attempt has been made to assess the quantum of the stream flow and its impact on water quality status in the defined study area where advanced forest conservation activities are going on to preserve the forest hydrology.

The selected study area is located at the western part of Sikkim, India. The total geographical area of West Sikkim is about 116 km^2, and the study area is demarcated by three different sites (Fig. 9.1) from which the samples were collected. The detailed description of the study area is also shown in Table 9.1. The density of the population is found to be 836 km^2 whereas the population of West Sikkim as per the 2011 census is about 109,513. In the present study, sites are classified into disturbed, semidisturbed, and undisturbed with a catchment area of 3.5 km^2. About 60–65% of the area is covered by forest. Rangit River is one of the main rivers flowing across the region, which originates from Rimbi and Nambu Khola and is discharged into the Teesta River.

Figure 9.2 indicates the movement of a conjectural stormwater flow path in different surface layers and the subsurface profile with the flow direction in West Sikkim. Figure 9.3 is an idealized section showing soil formation found where soil moisture is high on the hill slopes of West Sikkim for plantation activities. Different forest canopies in Sikkim are also important in increasing soil moisture content. Figure 9.4 shows the elevation profile of the study area, which is higher near the areas of the reserved forest. In recent days changes in land use and land cover have altered the topsoil cover, triggering more soil erosion that alters the soil and water quality of the selected sampling sites. No or few physical studies have highlighted estimation of the surface runoff and its impact on soil and water quality.

Materials and Methods

Data Collection

A nonrecording rain gauge station and pan evaporimeter were installed in the selected catchment. Rainfall and evaporation data have been recorded from May

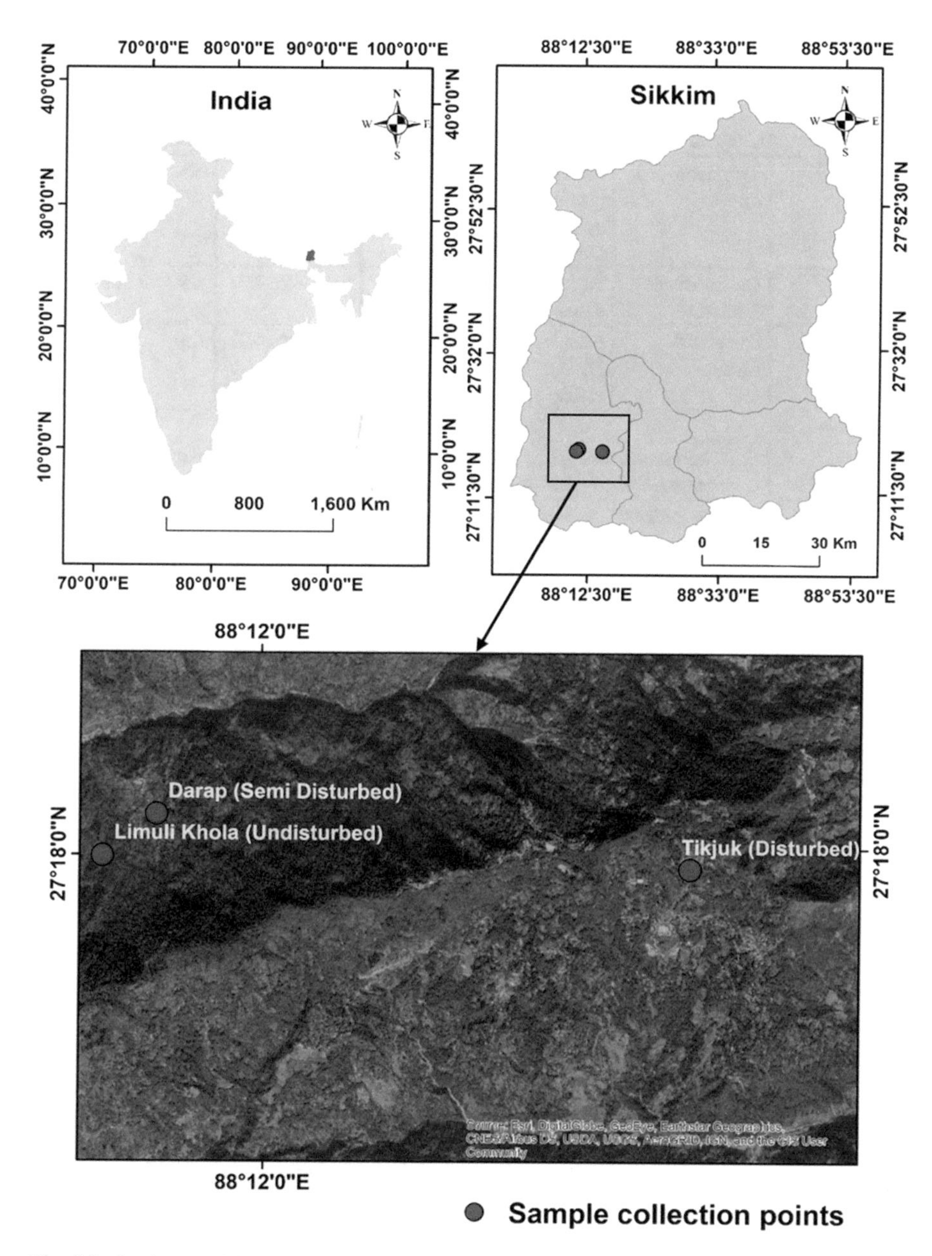

Fig. 9.1 Study area delineated in West Sikkim

2008 to December 2018. Historical discharge data of the Rangit River have been collected from the Central Water Commission (CWC) from July 2003 to December 2017. The rate of infiltration has been calculated using a double ring infiltrometer (Fig. 9.5) at three different sites in three different seasons: pre-monsoon, monsoon,

Table 9.1 Description of the study area

Sample collection points	Character	Location	Latitude	Longitude	Area (km^2)	Elevation (m)
Darap	Semidisturbed forest	Forest check post, Darap	27°18′15.4″N	88°11′15.5″E	0.5	1567
Limuli Khola	Undisturbed forest	Near waterfall	27°17′59.7″N	88°10′52.4″E	1.5	1600
Tikjuk	Disturbed forest	Near DFO office	27°17′52.8″N	88°14′59.4″E	1.5	1834
Tree category						
Natural regeneration	*Elaeocarpus sikkimensis*, *Engelhardtia* spp., *Ichnocarpus dasycarpus*, *Castonopsis indica*, *Betula alnoides*, *Beilschamiedia* spp.					
Native	*Alnus nepalensis*, *Acer oblongum*, *Bambusa nutans*, *Casearia glomerata*, *Castonopsis indica*, *Castonopsis tribuloides*, *Duabanga sonneratiodes*, *Ficus hookerii*, *Juglans regia*, *Litsea polyantha*, *Machilus* spp., *Magnolia campbellii*, *Michelia excelsa*, *Nyssa sessiliflora*, *Prunus nepalensis*, *Quercus* spp.					
Exotic	*Cryptomeria japonica, Juniperus* spp. *Cupressus* spp.					

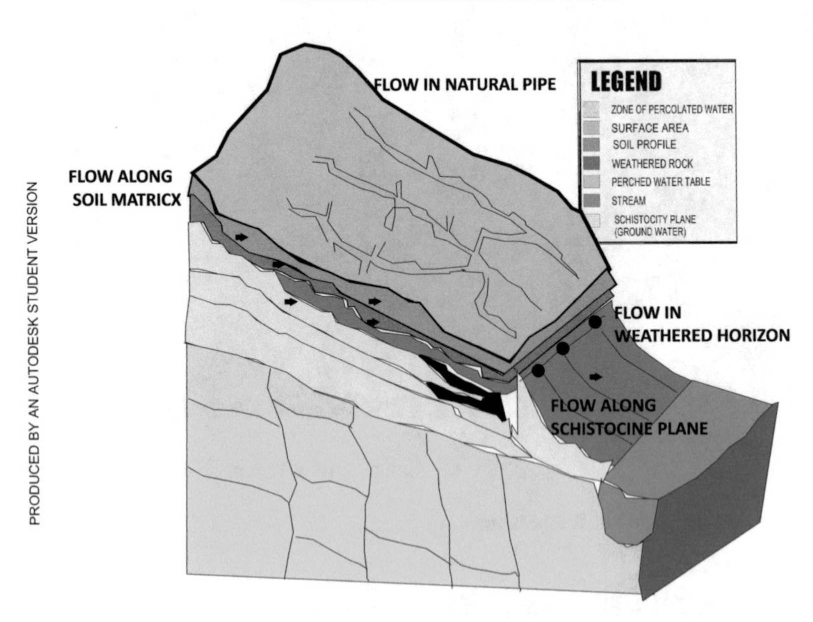

Fig. 9.2 Conjectural stormwater flow model

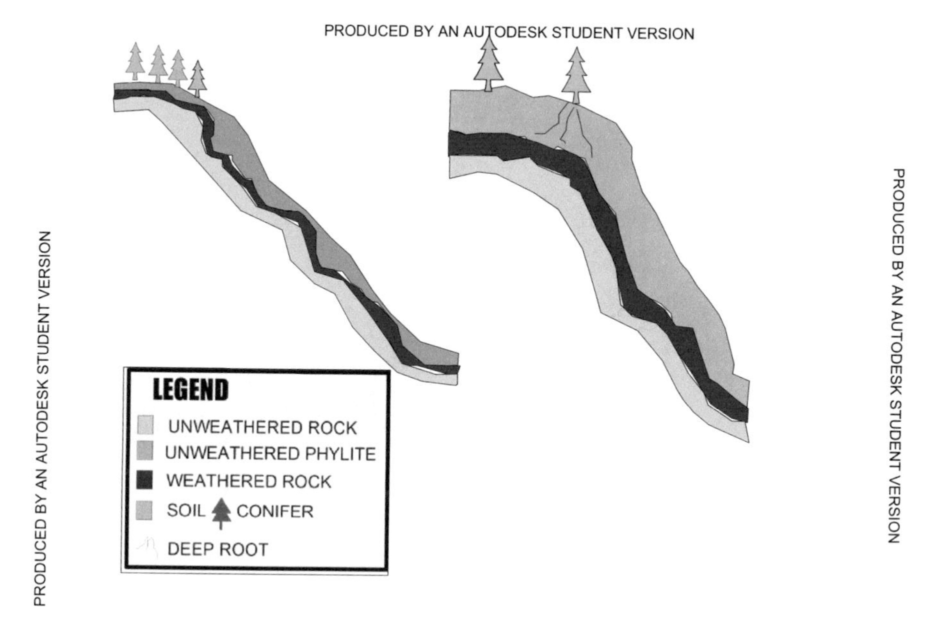

Fig. 9.3 Idealized section showing formation of soil on the hill slopes of West Sikkim

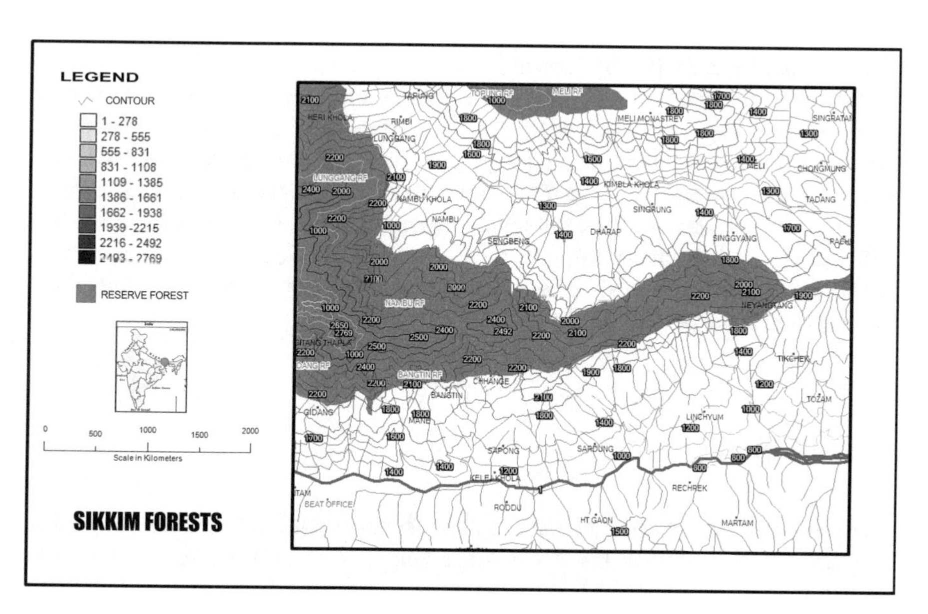

Fig. 9.4 Contour map of West Sikkim. (Source: Forest Environment and Wildlife Management Department, Government of Sikkim)

Fig. 9.5 Rain gauge, evaporimeter, and double ring infiltrometer installed at site in West Sikkim

and post-monsoon. Soil moisture at depths of 15 cm and 30 cm has also been measured using a soil tensiometer in similar seasons.

Soil and Water Sampling

Quantitative soil and water samples have been collected from the 0 to 30 cm soil layer on a seasonal basis from different locations: disturbed by anthropogenic activities, undisturbed area, and the pasture area of a forest area situated near a natural canal from the nearby forest.

Different physicochemical parameters of the soil (pH, EC, moisture content, water-holding capacity, bulk density, soil texture, C:N ratio, total carbon, soil organic carbon, total nitrogen, total potassium, total phosphorus, total sodium, sodium absorption ratio, total calcium, potassium oxide, phosphoric acid, aluminum content, silica) and water [pH, turbidity, suspended solids (SS), conductivity, alkalinity, acidity, ammoniacal nitrogen, COD, BOD, TC, FC, phosphorus] have been analyzed on a seasonal basis in the laboratory of the School of Water Resources Engineering, Jadavpur University.

Data Analysis

Analysis of data was carried out using SPSS software for all soil chemical properties. The aim was to estimate the water-conserving effects of the forest and also to investigate the probability of exceedance and equality of annual runoff and its components in river flow in forested catchments compared to the areas where forest coverage was poor. The following steps were applied to complete the research.

Procedure

- Three micro-watersheds were selected, namely, disturbed, semidisturbed, and undisturbed, separated at a distance of 12 km. There is no significant variation in climatic parameters and geological conditions.
- Meteorological observations have been undertaken in three micro-watersheds and sampling sites to measure rainfall, evaporation, temperature, humidity, and wind velocity.
- A meteorological observatory consists of dry and wet bulb thermometer, rain gauge, pan evaporimeter, and anemometer.
- Composite weirs (1200 V notch with rectangular weir) have been used at three different sites to measure stream discharge for the three micro-watersheds during the monsoon period as well as during the pre-monsoon period (March–April). V notches have been used to measure the very low discharge (Fig. 9.6).
- Water and soil quality parameters have also been analyzed during the monsoon and pre-monsoon periods (Fig. 9.7). Principal component analysis was applied to find the correlation among the parameters.

Fig. 9.6 Discharge measured using the composite weir installed at site

Fig. 9.7 Water quality measured at site by using a rugged field kit

Results and Discussion

Analysis of Stream Flow

From the study, humus thickness was found to vary from 8 to 10 cm. It is apparent that the infiltration capacity values of soils are subject to wide variation depending on several factors. Infiltration capacity can be the result of alterations in land use and land cover. The values varied from 1.2 to 1.5 cm/h for the selected sites at Darap and Limbuni Kholsa, whereas at the other sites these values increased by as much as tenfold because of good grass cover or vegetation cover. In almost the entire watersheds of the study area, stream flow increases as forest cover decreases, and vice versa, because evaporation from the forest is significantly greater than that from grasslands or crops. Although transpiration rates (soil water uptake by plants) under conditions of ample soil water do not differ much between forests and non-forest vegetation, rates of evaporation from vegetation wet by rain (rainfall interception) are much higher from tall, aerodynamically rough surfaces such as forests. In addition, the deeper roots of trees allow continued water uptake where plants with more shallow roots cannot take up water during prolonged rainless periods. Because rainfall interception totals are higher in wet years, the impact of forest clearance (i.e., after interception ceases) increases with mean annual rainfall. Apart from rainfall, the magnitude of the changes in annual stream flow is also affected by forest type and the slope aspect. Generally, the largest changes are observed in the case of clearing conifers, owing to their dense evergreen habit and high interception, followed by native (but not exotic) forest species, then deciduous hardwoods which are leafless in winter or dry seasons. Rainfall and runoff data have been analyzed for May 2008 to December 2018. Flow in the stream was minimum during March–April and maximum during July–August. Discharge of stream from the disturbed forest micro-watershed (Tikjuk area) remained negligible for a period of 3 months from March to May whereas discharge of the stream from the undisturbed micro-watershed remained low but almost steady (0.0028 m^3/s) during the same period. Also, the discharge of streams from the semidisturbed micro-watershed remained low (0.0015 m^3/s) compared to the undisturbed area. Maximum daily average discharges in the disturbed and semidisturbed catchment were recorded on 24 August 2010 (0.053 m^3/s and 0.015 m^3/s, respectively), whereas maximum daily average discharge of 0.01 m^3/s was recorded on 12 July 2010 in the undisturbed catchment. In terms of runoff per unit area, maximum runoff was 57 mm and 13 mm in disturbed and semidisturbed catchments, respectively, on 24 August 2010, and 9 mm in the undisturbed catchment on 12 July 2010. Runoff coefficients during the monsoon period (June to September) were 0.55, 0.25, and 0.20 for disturbed, semidisturbed, and undisturbed micro-watersheds, respectively.

Standard deviation of daily discharge was 25% higher in the disturbed (0.006 m^3/s) than the undisturbed (0.005 m^3/s) catchment (Fig. 9.8). These results indicate that runoff was more uniform in the undisturbed catchment. Stream flows from the undisturbed catchment are perennial whereas streams from disturbed and semidisturbed forests were intermittent and flowed for only 7–8 months. Stream flow during the post-monsoon season is supported by subsurface flow in a hilly watershed, known as the base flow of the stream. Rainfall infiltrates the ground through intergranular pore spaces, openings, fissures, fractures, joints, bedding planes, etc., and reappears downslope as springs and seepage. The hydrographs of three different micro-watersheds (Fig. 9.5) revealed the quick response to rainfall by stream discharge from the micro-watersheds that results from the small size and steep slope of both catchments, undisturbed and disturbed. The recession parts of the hydrographs of the two streams differ. Discharge declined slowly in the undisturbed catchment during post-monsoon months whereas it declined more rapidly in the disturbed catchment; the stream dries up in summer months. Streams flowing through dense forest sustained discharge during non-monsoon months with input from subsurface flow. Runoffs in the three different catchments were maximum during August and minimum during May. Total rainfall received in the undisturbed, disturbed, and semidisturbed micro-watersheds was 2950 mm, 2805 mm, and 2801 mm, respectively, generating runoff of 1470 mm, 1646 mm, and 1517 mm, respectively, during the year. Total runoff in disturbed and semidisturbed catchments was found to be 57% and 54%, respectively, whereas 49% was recorded from the undisturbed catchment.

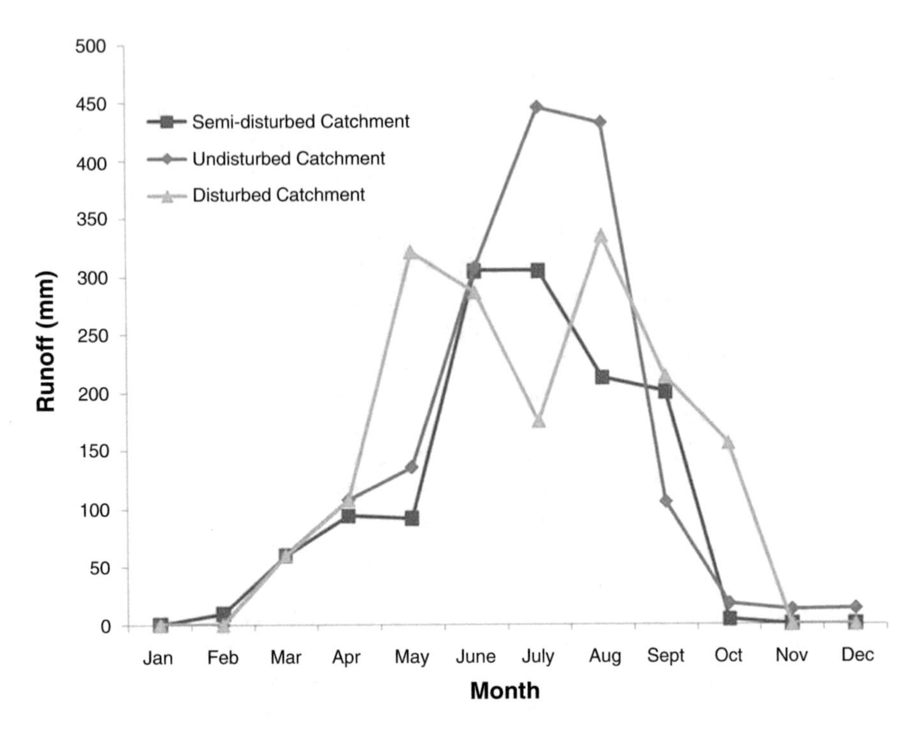

Fig. 9.8 Monthly variation of stream discharge (2008–2018) in three different micro-watersheds

Impact of Runoff on Water and Soil Quality Analysis

Dissolved oxygen (DO) is a critical water quality parameter for characterizing the health of an aquatic system. It is a measurement of the oxygen dissolved in water that is available to fish and other aquatic life. The DO content of water results from the photosynthetic and respiratory activities of the flora and fauna in the system and the mixing of atmospheric oxygen with waters through wind and stream current action. The optimum level of dissolved oxygen, 7–8 mg/l, is said to be good for agricultural prospects. A DO content greater than 6.0 mg/l is favorable for plants for their maturation (Table 9.2). pH values between 7 and 8 are optimal for supporting a diverse aquatic ecosystem. A pH range between 6.5 and 8.5 is generally usable. The range of pH values of water taken from the field is suitable for the plants because it lies between 7 and 8, which is safe for human requirements. Biochemical oxygen demand (BOD) is a measure of the amount of oxygen consumed by the respiration of microorganisms while feeding on decomposing organic material such as algae and other dead plants. Excessive nutrients (phosphates and nitrates) can cause algal blooms, and their eventual decomposition can cause massive fish kills if BOD drastically lowers the DO levels. The BOD should not be so great as to lower the dissolved oxygen to an unacceptable level (6.0 mg/l or less). The range of BOD, 0.5–1.9, of the water taken from the field is suitable for plants. These very low values

Table 9.2 Calculated mean of hydrological parameters of three different selected micro-watersheds

Sample no.	Parameter	Undisturbed	Disturbed	Semidisturbed
1	Total rainfall (mm)	2950	2878	2801
2	No. of rainy days	110	130	128
3	Total runoff (mm)	1470 (49%)	1646 (57%)	1517 (54%)
4	Standard deviation (m^3/s)	0.005	0.006	0.006
5	Rainfall (June–October)	92%	73%	80%
6	Runoff (June–October)	89%	71%	68%
7	Runoff (November–April)	13.2%	10%	11%

of BOD indicate that people of this area use a sanitary system and not open defecation. Fecal coliform bacteria indicate the likely presence of water-borne pathogenic (disease-causing) bacteria or viruses, including *Escherichia coli*, which are present in the intestinal tracts of all warm-blooded animals, including humans. Fecal coliform levels are measured in fecal colonies (FC) per 100 ml. Fecal coliform levels in freshwater should not exceed an average count of 100 colonies. A fecal coliform level less than 50 ml MPN/100 is optimal. For marine waters, fecal coliform levels should not exceed 14 MPN/100. The health standard for drinking water is 0 MPN/100, for swimming 200 MPN/100, and for partial body contact (boating) 1000 MPN/100. A maximum fecal coliform of 100 MPN/100 ml as found in a semi-disturbed catchment is safe for plant use but in the disturbed catchment value of 800 MPN/100 ml indicated anthropogenic activities. Soil of the undisturbed catchment has a favorable amount of nitrogen for plant growth whereas a disturbed catchment has slightly more. Therefore, mixing of soil from undisturbed, disturbed, and semidisturbed catchments is needed to get a homogeneous nitrogen concentration in the soil. The soil at the disturbed catchment has a little more nitrogen than required for plants and semidisturbed catchments have a favorable amount of nitrogen for plant growth. Therefore, these two soils also need to be mixed properly (Table 9.4).

Net impacts on water quality depend on prior land use and crop management, current forest management practices, soil type, local hydrology, and climate. In general, conversions to forestland have the potential to reduce erosion and subsequent sedimentation, as well as reduce levels of dissolved nutrients and pesticides in surface runoff and groundwater. These improvements in water quality are a function of the lesser amounts of runoff and leaching as well as a lower concentration of potential pollutants that are expected to result from the conversion of forestland. Water draining from undisturbed forested watersheds is generally of the highest quality, particularly with regard to beneficial uses including drinking water, aquatic habit for native species, and contact recreation. Our results show consistent patterns of relatively high water quality draining forested catchments, undisturbed in comparison to other land uses such as agriculture or urbanization (disturbed or semidisturbed catchment). Recognition of the relative role of forests for providing water supplies of the highest quality has been one of the driving factors for

Table 9.3 Estimated values of water quality analysis

Sample no.	Parameter	Undisturbed catchment	Disturbed catchment	Semidisturbed catchment
1	pH	7.12	7.10	7.12
2	BOD, mg/l	<1.00	1.9	1.20
3	COD, mg/l	3.00	7.00	4.00
4	Turbidity, NTU	4.50	5.40	4.50
5	Total hardness (as $CaCO_3$), mg/l	16.00	16.00	16.00
6	Alkalinity ($CaCO_3$), mg/l	12.00	16.00	12.00
7	Ammoniacal nitrogen (as N), mg/l	0.012	0.14	0.033
8	Phenolphthalein acidity ($CaCO_3$), mg/l	1.00	1.00	1.00
9	Total suspended solids (SS), mg/l	3.20	6.50	5.40
10	Conductivity (μmhos/cm)	38.50	45.80	42.80
11	Dissolved oxygen (as O_2), mg/l	6.00	4.70	6.10
12	Total coliform organisms, MPN/100 ml	2.20	3×10^3	1.20×10^3
13	Fecal coliform, MPN/100 ml	<1.10	0.80×10^3	0.10×10^3

establishment of forest reserves and for development of forest management practices designed to protect this high quality. Some forest management activities such as road construction, harvesting, site preparation for regeneration of forest tree species, and fertilization of existing forest have been shown to alter water quality, primarily by causing changes in sediment loads, stream temperature, dissolved oxygen, and dissolved nutrients, particularly nitrogen (Tables 9.3 and 9.4).

Application of Principal Component Analysis in Water and Soil Quality Analysis

In this study, principal component analysis (PCA) has been applied to determine the correlation among soil and water quality parameters. In this analysis, 11 water quality parameters (Table 9.5) and 22 soil quality parameters have been used for the correlation matrix, which has been calculated using IBM SPSS Statistics 23 software. The extraction method of component transformation matrix has been calculated using PCA, and Varimax with Kaiser normalization has been applied for the rotation method. In water quality, pH shows negative correlation with the BOD, COD, alkalinity, ammoniacal nitrogen, phenolphthalein acidity, total suspended solids, and conductivity and positive correlation with dissolved oxygen. The correlation among other parameters is given in Table 9.5. The loading matrix (Table 9.6)

Table 9.4 Estimated values of soil quality analysis

Sample	Parameters	Unit	Disturbed catchment	Undisturbed catchment	Semidisturbed catchment
1	pH	–	5.24	4.93	5.03
2	Electrical conductivity	μs/cm	60	46	37
3	Moisture content	%	7.7	2.5	6.1
4	Water-soluble aggregates	%	0.026	0.022	0.017
5	Water-holding capacity	kg/kg	0.392	0.346	0.378
6	Bulk density	gm/cm^3	1.39	1.41	1.37
7	Soil texture	–	Sandy loam	Sandy loam	Sandy loam
(a)	Sand	%	54.5	53.8	50.6
(b)	Silt	%	29.3	28.1	29.2
(c)	Clay	%	16.2	18.1	20.2
8	Carbon:nitrogen ratio (C:N)	wt. basis	11.2	12.3	11.7
9	Total carbon (C)	g/kg	45.6	50.4	99.5
10	Soil organic carbon	gm/kg	10.7	2.8	17.1
11	Total nitrogen (N)	mg/kg	957	228	1,457
12	Mineral nitrogen	mg/kg	216	48	248
13	Total potassium (K)	mg/kg	9,237	6,222	6,640
14	Total phosphorus (P)	mg/kg	428.1	227.9	404.8
15	Total sodium (Na)	mg/kg	1,398	1,319	1,524
16	Sodium absorption ratio	–	11.1	13.2	14.2
17	Total calcium (Ca)	mg/kg	3,164	2,807	2,971
18	Na_2O	mg/kg	1,884	1,778	2,054
19	K_2O	mg/kg	11,127	7495	7,999
20	P_2O_5	mg/kg	1,013	539	958
21	Al_2O_3	mg/kg	38,164	47,590	42,443
22	SiO_2	mg/kg	77.7	75.1	64.4

shows the first two components with eigen values greater than 1 that together contribute 100% of the total variance explained. In component 1, the highest correlation is found among pH, BOD, COD, turbidity, alkalinity, ammoniacal nitrogen, phenolphthalein acidity, and DO whereas in the second component correlation is high among turbidity, total hardness, and total suspended solids. The same procedure has been applied for analysis of soil quality parameters wherein the first two components with eigen values greater than 1 together contribute 96.56% of the total variance explained (Table 9.7). In the first component, pH, moisture content, water-holding capacity, sand, clay, total nitrogen (N), mineral nitrogen, total

Table 9.5 Computation of intercorrelation matrix of water quality parameters

Water quality parameters	Intercorrelation matrix										
	pH	BOD	COD	Turbidity	Total hardness	Alkalinity	Ammoniacal nitrogen	Phenolphthalein acidity	Total suspended solids (SS)	Conductivity	Dissolved oxygen
pH	1.000	−0.977	−0.971	−0.999	0.001	−0.996	−0.988	−0.945	−0.756	−0.810	0.998
BOD	−0.977	1.000	1.000	0.977	0.212	0.977	0.998	0.993	0.877	0.916	−0.962
COD	−0.971	0.999	1.000	0.971	0.240	0.971	0.996	0.996	0.891	0.927	−0.953
Turbidity	−0.990	0.977	0.971	1.000	0.001	0.998	0.988	0.945	0.756	0.810	−0.998
Total hardness	0.002	0.212	0.240	0.000	1.000	0.998	0.153	0.327	0.655	0.586	0.064
Alkalinity	−0.999	0.977	0.971	0.999	0.000	1.000	0.988	0.945	0.756	0.810	−0.998
Ammoniacal nitrogen	−0.988	0.998	0.996	0.988	0.153	0.988	1.000	0.984	0.847	0.890	−0.976
Phenolphthalein acidity	−0.945	0.993	0.996	0.945	0.327	0.945	0.984	1.000	0.929	0.958	−0.922
Total Suspended solids	−0.756	0.877	0.891	0.756	0.655	0.756	0.847	0.929	1.000	0.996	−0.712
Conductivity	−0.810	0.916	0.927	0.810	0.586	0.810	0.890	0.958	0.0996	1.000	−0.771
Dissolved oxygen	0.998	−0.962	−0.953	−0.998	0.064	−0.998	−0.976	−0.922	−0.712	−0.771	1.000

Table 9.6 Principal component loading matrix of water quality parameters

Water quality parameters	Component	
	1	2
pH	−0.996	−0.089
BOD, mg/l	0.955	0.298
COD, mg/l	0.945	0.326
Turbidity, NTU	0.996	0.089
Total hardness (as $CaCO_3$), mg/l	−0.089	0.996
Alkalinity ($CaCO_3$), mg/l	0.996	0.089
Ammoniacal nitrogen (as N), mg/l	0.971	0.241
Phenolphthalein acidity ($CaCO_3$), mg/l	0.912	0.410
Total suspended solids, mg/l	0.694	0.720
Conductivity (μmhos/cm)	0.755	0.656
Dissolved oxygen (as O_2), mg/l	−0.998	−0.025
Eigen value	8.614	2.386
Total factor covariance (%)	78.310	21.690
Cumulative % of total factor covariance	78.310	100

Table 9.7 Principal component loading matrix of soil quality parameters

Soil quality parameters	Component	
	1	2
pH	0.904	−0.427
Electrical conductivity	0.398	−0.917
Moisture content	0.999	−0.021
Water-soluble aggregates	0.219	−0.976
Water-holding capacity	0.999	−0.017
Bulk density	−0.691	−0.723
Soil texture	−0.070	−0.998
Sand	0.978	0.208
Silt	−0.253	0.968
Clay	−0.983	0.186
Carbon:nitrogen ratio (C:N)	0.158	0.987
Total carbon (C)	0.733	0.680
Soil organic carbon	0.764	0.645
Total nitrogen (N)	0.907	0.420
Mineral nitrogen	0.805	−0.593
Total potassium (K)	0.984	0.177
Total phosphorus (P)	0.590	0.807
Total sodium (Na)	−0.468	0.884
Sodium absorption ratio	0.959	−0.282
Total calcium (Ca)	0.589	0.808
Na_2O	0.805	−0.593
K_2O	0.984	0.177
P_2O_5	−0.983	0.185
Al_2O_3	−0.054	−0.999
SiO_2	0.904	−0.427
Eigen value	13.499	10.501
Total factor covariance (%)	56.246	43.754
Cumulative % of total factor covariance	56.246	95.56

potassium (K), sodium absorption ratio, Na_2O, K_2O, P_2O_5, and SiO_2 show the highest correlation whereas in the second component electrical conductivity, water-soluble aggregates, bulk density, soil texture, silt, carbon:nitrogen ratio, total phosphorus (P), total sodium (Na), total calcium (Ca), and Al_2O_3 have the highest correlation.

Conclusion

In this study, the impact of forest degradation has been analyzed using stream flow estimation and water and soil quality analysis. The overall impact of forests on the hydrological regime of a watershed is twofold. First, comparatively less discharge can be observed from forest watershed (undisturbed catchment) during the rainy months, showing the forest control over excess runoff downstream. Annual runoff from a dense forest was lower than that from a disturbed catchment. Second, forests induce infiltration, which leads to more uniform flow year round from the undisturbed forest watershed. Net impacts on water quality depend on prior land use and crop management, current forest management practices, soil type, local hydrology, and climate. In general, conversions to forestland have the potential to reduce erosion and subsequent sedimentation, as well as to reduce levels of dissolved nutrients and pesticides in surface runoff and groundwater. These improvements in water quality are a function of the lower amounts of runoff and leaching as well as lower concentrations of potential pollutants that are expected to result from the conversion of forestland. Further, PCA has been applied to ascertain correlation among the parameters, and significant correlation has been observed. Such studies will incur valuable information to stakeholders and policy makers to conserve the forest watersheds in a sustainable way through physical exploration.

Acknowledgments The authors express their sincere gratitude to The Ministry of Environment, Forest and Climate Change, Government of India, for providing a research grant to the National Afforestation and Eco-Development Board, Regional Center, Jadavpur University for the years 2008–2010. Afterward, the work has been continued until 2018 under the collaboration of School of Water Resources Engineering, Jadavpur University. The authors are also thankful to the Forest Environment and Wildlife Management Department, Government of Sikkim for providing data and their continuous support. The authors are also grateful to Central Water Commission, Government of India for providing discharge data. Last but not the least, the authors acknowledge the water quality laboratory and digital library of the School of Water Resources Engineering, Jadavpur University, for allowing to conduct the water and soil quality tests and to access all the GIS and statistical software.

References

Ewane BE, Lee HH (2020) Assessing land use/land cover change impacts on the hydrology of Nyong River Basin, Cameroon. J Mt Sci 17:50–67. https://doi.org/10.1007/s11629-019-5611-8

Frene C, Dörner J, Zuniga F, Cuevas JG, Alfaro FD, Armesto JJ (2019) Eco-hydrological functions in forested catchments of Southern Chile. Ecosystems. https://doi.org/10.1007/s10021-019-00404-7

Guzha AC, Rufino MC, Okoth S, Jacobs S, Nobrega RLB (2018) Impacts of land use and land cover change on surface runoff, discharge and low flows: evidence from East Africa. J Hydrol Reg Stud 15:49–67

Hornbeck JW, Adams MB, Corbett ES, Verry ES, Lynch JA (1993) Long-term impacts of forest treatments on water yield: a summary for northeastern USA. J Hydrol 150:323–344. ISSN: 0022-1694

Klimaszyk P, Rzymski P, Piotrowicz R et al (2015) Contribution of surface runoff from forested areas to the chemistry of a through-flow lake. Environ Earth Sci 73:3963. https://doi.org/10.1007/s12665-014-3682-y

Liu J, Engel BA, Wang Y et al (2019) Runoff response to soil moisture and micro-topographic structure on the plot scale. Sci Rep 9:2532. https://doi.org/10.1038/s41598-019-39409-6

Lowrance R, Altier L, Newbold J et al (1997) Water quality functions of riparian forest buffers in Chesapeake Bay Watersheds. Environ Manag 21:687. https://doi.org/10.1007/s002679900060

Mikkelson KM, Bearup LA, Maxwell RM et al (2013) Bark beetle infestation impacts on nutrient cycling, water quality and interdependent hydrological effects. Biogeochemistry 115:1. https://doi.org/10.1007/s10533-013-9875-8

Prevost M, Belleau P, Plamondon AP (1997) Substrate conditions in a treed peat land: responses todrainage. Écoscience 4:543–554

Shang X, Jiang X, Jia R, Wei C (2019) Land use and climate change effects on surface runoff variations in the Upper Heihe River Basin. Water 11:344

Shepard JP (1994) Effects of forest management on surface water quality in wetland forests. Wetlands 14:18. https://doi.org/10.1007/BF03160618

Smith HG, Sheridan GJ, Lane PNJ, Nyman P, Haydon S (2011) Wildfire effects on water quality in forest catchments: a review with implications for water suppl. J Hydrol 396:170–192. ISSN: 0022-1694

Vildan S, Hall MJ (1996) The effects of afforestation and deforestation on water yields. J Hydrol 178:293–309. ISSN: 0022-1694

Walega DM, Amatya P, Caldwell D, Marion S (2020) Assessment of storm direct runoff and peak flow rates using improved SCS-CN models for selected forested watersheds in the Southeastern United States. J Hydrol Reg Stud 27:100645

Yang B, Lee DK, Heo HK, Biging G (2019) The effects of tree characteristics on rainfall interception in urban areas. Landsc Ecol Eng 15(3):289–296. https://doi.org/10.1007/s11355-019-00383-w

Yannian Y (1990) Hydrological effects of forests. The hydrological basis for water resources management (Proceedings of the Beijing symposium, October 1990). IAHS Publ. no. 197

Zhu X, Lin J, Dai Q, Xu Y, Li H (2019) Evaluation of forest conversion effects on soil erosion, soil organic carbon and total nitrogen based on 137Cs tracer technique. Forests 10:433

Chapter 10
Impact of Climate Extremes of El Nina and La Nina in Patterns of Seasonal Rainfall over Coastal Karnataka, India

A. Stanley Raj and B. Chendhoor

Abstract This research investigates the influence of the El Nino Southern Oscillation (ENSO) on the rainfall characteristics of coastal Karnataka. An attempt has been made to study the behavior of rainfall in every season during an ENSO along the coast of Karnataka. Rainfall data from 1901 to 2017 have been used for studying the patterns. Rainfall variations occur with El Nino and La Nina events, which are based on sea surface temperature (SST), taking place in the Pacific Ocean along the Equator. The Southern Oscillation is like a pendulum, in which El Nino takes place in one side (west) and La Nina takes place along the other side (east), and vice versa. Its major cause is the trade winds that move from east to west, especially in Ecuador, Peru, and South America. By correlating the rainfall data and El Nino, the effects along the coast of Karnataka can be identified. As El Nino and La Nina have major impacts on agriculture, food supply, and the economy of the country, it is necessary to study the patterns and influence of rainfall. This study shows that La Nina has a major impact on coastal Karnataka, bringing more rainfall than El Nino.

Keywords El NiNo · La NiNa · Karnataka · Rainfall · Seasonal variations

Introduction

Karnataka, a district that lies along the southwestern part of the Indian Peninsula, is classified into Northern Karnataka, Central Karnataka, Southern Karnataka, and Coastal Karnataka based on geographical location. Coastal Karnataka, which is the primary site of this research, comprises three districts, Dakshina Kannada (12.87°N

A. Stanley Raj (✉)
Department of Physics, Loyola College, Chennai, Tamil Nadu, India
e-mail: stanleyraj@loyolacollege.edu

B. Chendhoor
Centre for Geo Technology, Manonmaniam Sundaranar University, Tirunelveli, Tamil Nadu, India

M. N. Islam, A. van Amstel (eds.), *India: Climate Change Impacts, Mitigation and Adaptation in Developing Countries*, Springer Climate,
https://doi.org/10.1007/978-3-030-67865-4_10

74.88°E), Udupi (13.3389°N 74.7451°E), and Uttara Kannada (14.6°N 74.7°E), where Mangalore, Udupi, and Karwar form the headquarters of these districts, respectively (Vinaya Kumar et al. 2017). The coast lie to the west of Karnataka and forms the northern part of the Malabar coast. Kanara, Canara, and Karavali are other popular names for coastal Karnataka or the Karnataka coast. Kanara is surrounded by coastline to the west, which is flooded by the Arabian Sea, a vast landscape of Western Ghats to the east, Konkan to the north, which forms the western coastline of India, and to the south lie the Kerala Plains. Kanara stretches from the village of Talapady to the south to the village of Majali to the north, covering a distance of 320 km and a width of 64 km in the south. The Western Ghats act as a geographical barrier to the southwest monsoon (June–September) which accounts for 90% of annual rainfall over the west coastal belt. The Sharavathi, the Mandavi, the Kalinadi, the Gangavalli, the Aghanashini, the Varahi, the Chakra Nadi, the Barapole, and the Netravathy are the west-flowing rivers of Karnataka that eventually merge with the Arabian Sea, forming a vast delta region. The most common type of soil found in the coastal region of Karnataka is alluvial soil, which is composed of sand, clay, and silt formed by the deposition of alluvium and sedimentary particles carried down by rivers and deposited over the estuaries. Alluvial soils are rich in organic nutrients, so cashew, coconut, banana, paddy, and other crops are grown in this type of soil. Agriculture is the primary occupation of the state: about 65% of the people depend on agriculture activities for their livelihood. Of 190.5 lakh hectares of land, only 125.93 lakh hectares is suitable for agriculture [https://shodhganga.inflibnet.ac.in]. Fishing is also a major occupation next to agriculture. A temperate climate prevails along the coastal regions of Karnataka, with an average temperature of 33° in the summer and 20° in the winter. It is mostly humid in the coastal regions. This chapter paper discusses the relationship between El Nino, Southern Oscillation, and sea surface temperature with that of the summer monsoon rainfall of coastal Karnataka.

India is the second most populous country in the world with a rich heritage and culture. India has diverse geographical features that make the study of its physical features more fascinating. Indian rainfall shows drastic variations from region to region and year to year Normand (1953).The Indian summer monsoon rainfall (June–September) is very important for disaster management planning, economic development, and the hydrology of our country (Guhathakurta and Rajeevan 2008). Several studies have analyzed rainfall on annual as well as seasonal time scales in peninsular India (Pramanik and Jagannathan 1954; Parthasarathy and Dhar 1978; Parthasarathy 1984; Mooley and Parthasarathy 1984; Parthasarathy et al. 1993). Linear trend analysis has been made for each of the 306 stations of IITM and its spatial pattern studied (Kumar et al. 1992). Nonparametric statistical tests such as the Mann–Kendall test and Sen's slope estimator were used for trend analysis of seasonal and annual rainfall of coastal Karnataka (Vinaya Kumar et al. 2017). Coastal Karnataka receives heavy rain because of the orographic effect (Soman and Kumar 1990). The monsoon season is responsible for 84–90% of the annual rainfall. The post-monsoon season contributes 6.2–9.5% of the annual rainfall, the winter season accounts for 0.2–0.6% of annual rainfall, and the summer season contributes 2.5–5.5% of annual rainfall (Vinaya Kumar et al. 2017). Research shows that decrease in southwest monsoon rainfall and increase of post-monsoon rainfall were observed (Krishnakumar et al. 2009; Soman et al. 1988). Climate change is a primary element in manipulating the fishery and agricultural productivity of coastal Karnataka (Vinaya Kumar et al. 2017). The Uttara Kannada district shows an increasing trend for monsoon rainfall, and the Udupi district (−1.9 mm year^{-1}) and Dakshina Kannada (−7.1 mm year^{-1}) show a decreasing trend in monsoon rainfall over the period of 1980–2013 (Vinaya Kumar et al. 2017). The relationship between El Nino and summer monsoon rainfall has been studied for the years 1875–1979 (Rasmusson and Carpenter 1983). The Southern Oscillation Index (SOI) was established by Wright (1977). Increase in sea surface temperature (SST) along the eastern Pacific Ocean is shown to be related to the deficits of southwest monsoon rainfall over India (Rasmusson and Carpenter 1983; Kiladis and Diaz 1989; Mooley and Paolino 1989). Correlation between El Nino, Southern Oscillation, and summer monsoon rainfall over India has been meticulously observed (Sikka 1980; Rasmusson and Carpenter 1983; Parthasarathy and Pant 1985; Mooley 1997).The El Nino Southern Oscillation has a strong impact on the frequency of winter monsoon rainfall (Revadekar and Kulkarni 2008). Daily rainfall data have been used to monitor the substantial increase in frequency and magnitude of intense rainfall events, and a major decreasing trend in the frequency of rainy days of the monsoon season has been observed from 1951 to 2000 (Goswami et al. 2006). For the past two decades, the relationship between ENSO and summer monsoon rainfall has been diminishing (Kripalani and Kulkarni 1997; Krishna Kumar et al. 1999).

Methodology

The rainfall data from 1901 to 2019 have been utilized (https://www.indiastat.com) to correlate with the El Nino years, obtained from the Golden Gate Weather Services to determine the impact of ENSO along coastal Karnataka. The El Nino Southern Oscillation (ENSO) is a seasonal variability in sea surface temperature and atmospheric air pressure overlying the tropical Pacific Ocean. ENSO consist of three phases: El Nino (warm ocean current), the La Nina (cold ocean current), and the neutral phase. Centuries ago, fishermen along the South American coast first discovered coastal water to be warming more than normal around Christmas time, so they termed it El Nino, which means 'boy child' in Spanish. Later, in the 1980s, the opposite phase of El Niño was identified, the cooler than normal ocean temperatures, which scientists named La Niña, which means 'girl child' in Spanish. El Nino and La Nina phases last as long as 9–12 months, developing during the months of March to June (spring), intensifying during the months of November to February (late autumn or winter), and weakening during March to June (spring or early summer). In rare conditions El Nino and La Nina might last more than a year. The longest El Nino phase was 18 months, and for La Nina 33 months. The increase in sea surface temperature (SST) along the Equator occurs because this area is exposed to more sun rays than the regions lying along the Tropic of Cancer and the Tropic of Capricorn.

According to NOAA, the Oceanic Nino Index (ONI) is a key element in monitoring and classifying the relative intensities of El Nino and La Nina. The normal value of the ONI is 0.5: if the value exceeds (+0.5) higher it shows that the central and eastern tropical Pacific Ocean is warmer than normal, indicating El Nino conditions. If the value lies (−0.5) lower it shows that the region is cooler than normal, indicating La Nina conditions. The ONI values are measured for a consecutive 3 months average of SST in the central and eastern tropical Pacific Ocean, which lies between 120° and 170° west. Scientists have named the region as the Niño 3.4 region. The average SST of each month is measured in the Nino 3.4 region and is averaged with previous and succeeding months; then the consecutive 3 months average is compared with a 30-year average. The obtained difference from the average temperature is the ONI value for those 3 months.

ENSO intensity is classified based on SST (https://ggweather.com):

- Weak (with a 0.5 to 0.9 SST anomaly)
- Moderate (1.0 to 1.4)
- Strong (1.5 to 1.9)
- Very strong ($\geq$2.0)

Walker Circulation

Walker circulation is the ocean-based air circulation system that affects weather on the Earth. The change in surface pressure and temperature over the eastern and western tropical Pacific Ocean results in Walker circulation. In recent studies from NASA, Walker circulation is the result predominantly of the subsurface current, which originates from the deep ocean, leading to the accumulation of warm water in the Western Pacific Ocean.

Neutral Phase

There are certain times where El Nino and La Nina fail to exist and the temperature of ocean surface water remains close to average SST (Fig. 10.1). Sometimes the ocean may appear as a prevailing El Nino or La Nina but the atmosphere does not accompany the changes, and vice versa. Neutral conditions refer to the winds, temperature, rainfall, and convection (rising air) across the tropical Pacific regions being closer to their long-term averages. Generally, the trade winds blow from the eastern to western Pacific along the surface while in the atmosphere (troposphere) the winds blow in the opposite direction, from western to eastern Pacific, forming a looping pattern over the Pacific Ocean called the Walker circulation. These trade

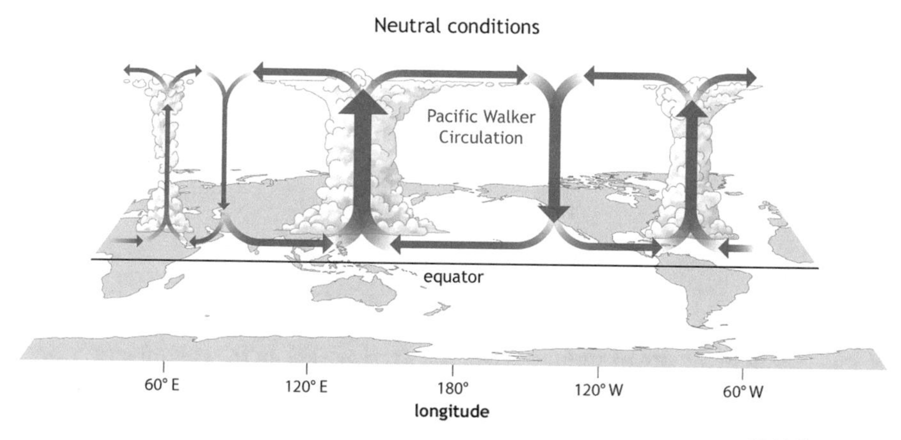

Fig. 10.1 Neutral conditions and Walker circulation in the Pacific Ocean. (Courtesy of NOAA)

winds push the warm pool of ocean water from the tropical central Pacific to the tropical western Pacific. Thus, the ocean water towards the western Pacific remains warmer. Cool ocean water prevails at central and eastern Pacific, with an atmospheric pressure difference between western and eastern Pacific. Because the ocean water is warm along the Western Pacific the hot air molecules tend to rise up into the atmosphere, creating a low-pressure zone; when hot air rises above this it becomes cooler and forms clouds that later precipitate as rain. Normal rainfall is observed in both the Western and Eastern Pacific Ocean during the neutral phase. El Nino and La Nina each has its own impact on Walker circulation. As warm ocean water is less dense than cold ocean water, it tends to stay at the surface. The thermocline is a thin boundary line that divides the upper warm layer from the cold deep water, below which the temperature tends to decreases with increase in depth.

El Nino

Warming of the ocean surface above the average sea surface temperature (SST) in the central and eastern tropical Pacific Ocean is caused by the weakening of trade winds. The warm pool of ocean water is gradually shifted towards the central and eastern Pacific, thus breaking the looping pattern of Walker circulation. As mentioned earlier, the Eastern Pacific Ocean holds the warm ocean water, which in turn shifts the pressure from high to low in the Eastern Pacific. Evaporation will occur at a much faster rate followed by the formation of clouds, so that heavy rainfall occurs over the Western coast of South America. During stronger El Nino years more floods, a lower daytime temperature, and further cyclonic conditions may prevail while the regions lying along the Eastern Pacific, that is, New Zealand, Australia, and Indonesia, experience drought conditions that severely affect agricultural activities. Warming of ocean water in the Eastern Pacific causes a huge loss for the fishing industries warm water kills all the microorganisms, making it difficult for the smaller fishes to survive, so the fish start moving towards cold ocean water. In olden days, predictions were based on these phenomena to find out whether it is a El Nino or La Nina year.

La Nina

With cooling of the ocean surface below the average SST in the central and eastern tropical Pacific Ocean (or Nino 3.4 region) in La Nina, these trade winds start to blow more strongly from east to west, thus shifting the warm pool of ocean water towards the west. A shift in atmospheric pressure is caused by the Southern Oscillation. La Nina is an intensified form of neutral phase wherein floods exist along Australia and the regions lying around the Western Pacific and severe drought conditions prevail on western coast of South America. With the return of cold ocean water in the Eastern Pacific, fish tend to return to the coast, which is favourable for the fishing industries. Because El Nino and La Nina are opposite events, the countries lying along the tropical Eastern Pacific experience El Nino conditions, then the western part experiences the La Nina conditions, and vice versa. Atlantic hurricanes are formed during La Nina conditions. In much simpler words, if floods exist in South America, then it is the El Nino phase. Or, if floods occur in Australia and in other Western Pacific regions, then it is the La Nina phase.

Indian Ocean Dipole

The dipole is the difference in SST between the eastern part (Bay of Bengal) and the western part (Arabian Sea) of the tropical Indian Ocean. The Indian Ocean dipole is classified into two types: positive and negative. When the SST of the Arabian Sea is greater than that of the Bay of Bengal, it signifies a positive Indian dipole. When the SST of the Arabian Sea is less than that of the Bay of Bengal, a negative Indian Ocean dipole (IOD) forms (Fig. 10.2). The positive phase of IOD is more likely to correspond with the El Nino conditions whereas the negative phase corresponds with La Nina conditions. IOD occur during the months of May to December, lasting up to 2 to 7 months (Bureau of Meteorology). In the positive phase of IOD, SST is higher west of the Indian ocean compared to the east. Similarly, in the negative phase the eastern part of the Indian Ocean remains warmer than the west. The correlation between the IOD and Indian summer monsoon rainfall (ISMR) is not clear (Saji et al. 1999).

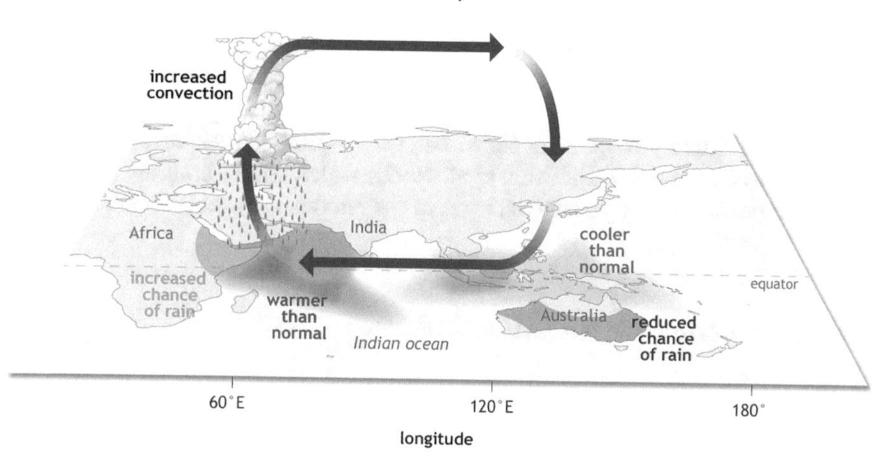

Fig. 10.2 The Indian Ocean dipole and the movements of warmer and cooler air. (Courtesy of NOAA)

Types of ENSO Indexes

Air Pressure Indexes

The Southern Oscillation Index (SOI) is the oldest method to indicate ENSO. It is measured through the atmospheric pressure differences at sea level between Tahiti and Darwin, Australia. The contrast of pressure at the two ends shows the atmospheric components of ENSO, noticed in the early 1900s by Walker and Bliss (1932). During El Nino, low pressure is created in Tahiti and high pressure is created at Darwin, and this represents a negative Southern Oscillation Index. Similarly, during La Nina pressure is reversed at Tahiti and Darwin and the index changes to positive. One drawback was that the Tahiti and Darwin stations were established south of the Equator (Tahiti at 18°S, Darwin at 12°S) but the ENSO effect acts mostly along the Equator.

This is where the Equatorial Southern Oscillation Index is taken into consideration as it is used to find the sea level pressure difference between two large regions, that is, Indonesia and the eastern tropical Pacific, which lies centered on the Equator (5°S to 5°N).

Sea Surface Temperature Indexes

Because ENSO is largely dependent on the ocean, sea surface temperature (SST) is the important factor in determining ENSO (Bjerknes 1969; Rasmussen and Carpenter 1982; Wyrtki 1985). Nino 1 and Nino 2 are the regions chosen for measurements, later combined into Nino 1 + 2. Nino 3 and Nino 4 are the regions where data are acquired consistently by passing ships. Later the region 3.4 was recognised as the representative for ENSO (Barnston et al. 1997), which covers both the Nino 3 and Nino 4 regions. The temperature changes of the Nino 3.4 region are shown in ONI.

Outgoing Longwave Radiation Indexes

In this satellite-based data collection, the outgoing longwave radiation (OLR) emitted from the cloud tops is observed with the advance very high-resolution radiometer (AVHRR) instrument. With this we could detect the areas that are rainier or drier than normal over the tropical Pacific Ocean. Negative OLR indicates the greater convection that leads to greater cloud coverage, usually of El Nino conditions. Greater convective activity over the central and eastern tropical Pacific shows that the cloud tops are higher and colder thus emitting less infrared (IR) radiation into space. The positive OLR is exactly the opposite of the negative index. The OLR are not greatly related to ENSO but act as a key element in linking the remote climate teleconnection outside the tropical Pacific (Chiodi and Harrison 2013).

Wind Indexes

A wind index is the measure of change in air flow in the upper and lower part of the Walker circulation.

Hadley Circulation

The Equator is that part of Earth that receives more sun radiation than either the Northern or Southern Hemisphere. When overheating occurs at the Equator the heat needs to be evenly distributed towards nearby locations, which is possible through the convection process. Because the Equator is hot, the hot air molecules rise, forming a low-pressure zone; when the air is further away from the ground, the heat source begins to cool down. As the heat source begins to cool the air wants to fall down, but it cannot fall directly down as the warm air is rising, so it is pushed to the sides; then the warm air actually cools even further so that it begins to fall back

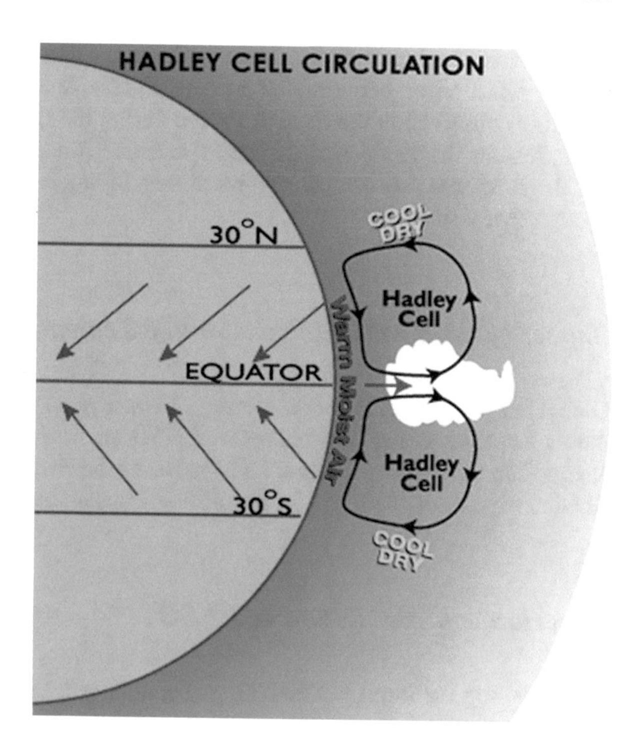

Fig. 10.3 Hadley Cell circulation. (Source: https://www.windows2universe.org/earth/Atmosphere/hadley_cell.html)

towards the Earth at 30°N and 30°S from the Equator where piling air on top of more air causes increased pressure, forming high pressure zones at 30°N and 30°S. As the high pressure always wants to flow towards low pressure, the wind moves back to the Equator, thus forming a convection process. The curve in the wind path is caused by the Coriolis effect, which results from the Earth's rotation, which causes all the moving particles such as air in the Northern Hemisphere to deflect to the right and those in the Southern Hemisphere to the left. The Coriolis force is absent at the Equator. During El Nino, overheating of the upper atmosphere causes the air flow towards the poles to become much stronger, which in turns affects the Hadley circulation pattern. During El Nino, jet streams extend all the way to North America and bring more than average storms to the Southern region of the United States (Fig. 10.3).

Impacts of ENSO on Hurricanes

During El Nino more hurricanes are observed in the central and eastern Pacific Ocean and fewer in the Atlantic Ocean because of the difference in strength of the vertical wind shear. During El Nino, the vertical wind shear is weaker at the Pacific Ocean and stronger at the Atlantic. Greater atmospheric stability is observed near the Atlantic Ocean. Similarly, in La Nina fewer hurricanes are observed in central and eastern Pacific and more in the Atlantic Ocean because of the switch in strength of vertical wind shear between the Eastern Pacific and the Atlantic. Atmospheric stability is observed in the Eastern Pacific during the La Nina phase. Vertical wind shear denotes the fluctuation in wind speed and direction between approximately 5,000 and 35,000 ft above the ground. A developing hurricane can be torn apart by strong vertical wind shear.

Impact of ENSO on Global Average Temperature

Global temperature is increased during El Nino years whereas it tends to decrease during La Nina years. The phases of ENSO are neither creating nor eliminating energy from our climate system; La Nina buries the Earth's prevailing heat while the El Nino exposes it. This type of reshuffling is called an internal climate variability.

Importance in Prediction of ENSO

On predicting the impact of ENSO, precautionary measures could be taken before the influence of extreme weather, which could cause a life-threatening situation for people in the coastal regions. Unnecessary damage could be avoided if safety measures are provided in anticipation of prevailing flood or drought conditions. Fishery, agriculture, energy, transportation, water and energy resources, and healthcare units are the sectors most affected by ENSO.

Discussion

As obtained from the three-dimensional (3D) models, Figs. 10.4 and 10.5 show that both El Niño and La Niña have impacts on rainfall along coastal Karnataka. A graph is plotted with the El Niño and La Niña intensities, which are classified into weak, moderate, strong, and very strong based on the Oceanic Nino Index (ONI), which is the measure of deviations from the normal sea surface temperature. The 3D model was plotted using Matlab software.

ENSO Effect on Peak Rainfall

A graph plotted between El Niño years and peak rainfall gives a positive linear regression of 0.082 (Fig. 10.6). The positive sign indicates that rainfall increases with an increase in El Niño intensity, and vice versa. Similarly, another graph is

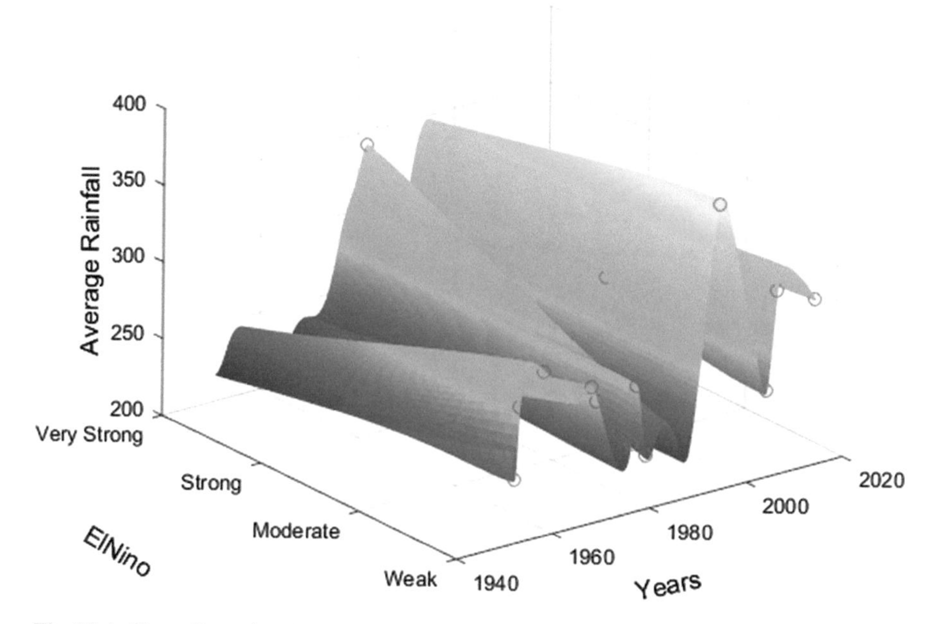

Fig. 10.4 Three-dimensional model for average rainfall for El Nino years

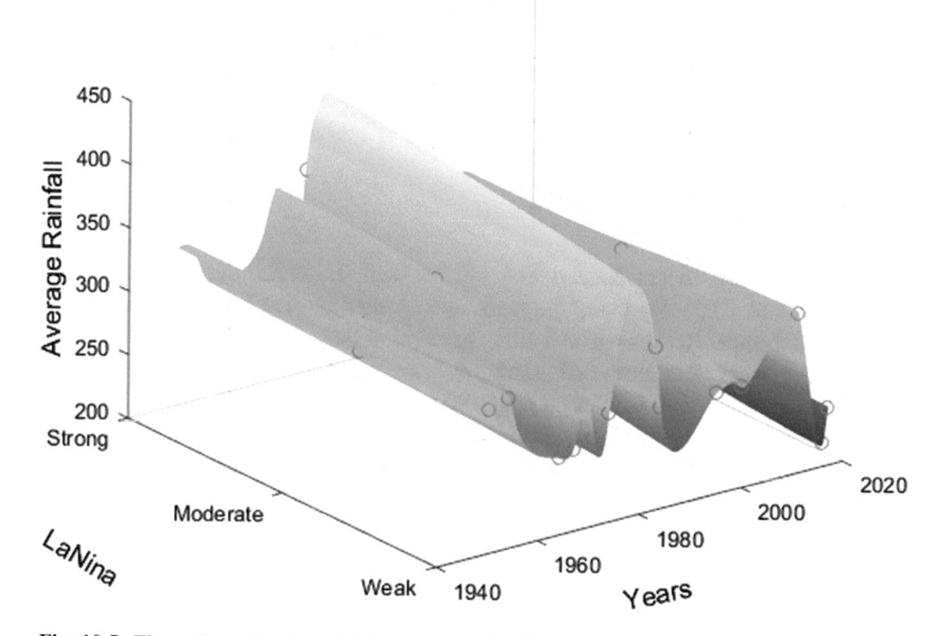

Fig. 10.5 Three-dimensional model for average rainfall for La Niña years

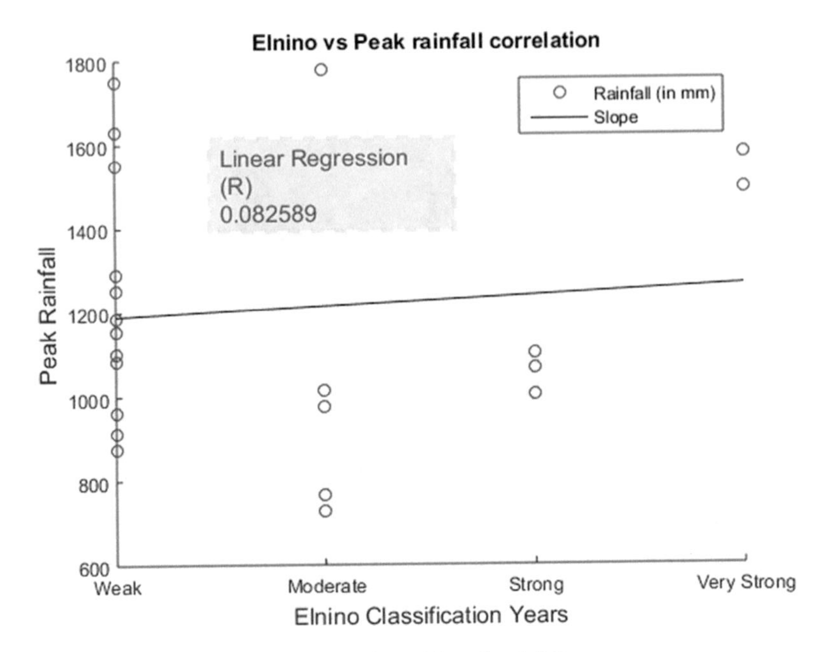

Fig. 10.6 Correlation of El Nino classification with peak rainfall

drawn considering the La Niña years and peak rainfall that shows a positive linear regression of 0.02 (Fig. 10.7). The positive regression indicates that rainfall increases as the La Niña intensity increases, and vice versa. Peak rainfall is calibrated by the month of the year with the highest amount of rainfall. By using the least squares method, the linear regression (*R*) is obtained.

ENSO Effect on Deficit Rainfall

Graphs are drawn to correlate deficit rainfall with El Niño and La Niña years (Figs. 10.8 and 10.9). By using the least squares method, the linear regression (*R*) is obtained. Figure 10.8 provides a clear description that increases in intensity as El Niño events occur with more deficit rainfall, with an *R* of 0.01. Figure 10.9 shows a negative regression, but it is negligible when compared to the El Niño regression value. These graphs show that peak rainfall occurs in El Niño years in comparison with La Niña years. Deficit rainfall is classified by the month of the year that had the least amount of rainfall.

ENSO comprises both El Niño and La Niña, which have negative and positive impacts, respectively, along coastal Karnataka. A negative impact signifies deficit rainfall, which in turn produces drought conditions along the inner parts of coastal Karnataka, whereas the positive impact signifies excess rainfall, which in turn produces floods if the dams and reservoirs are not properly maintained. However, El

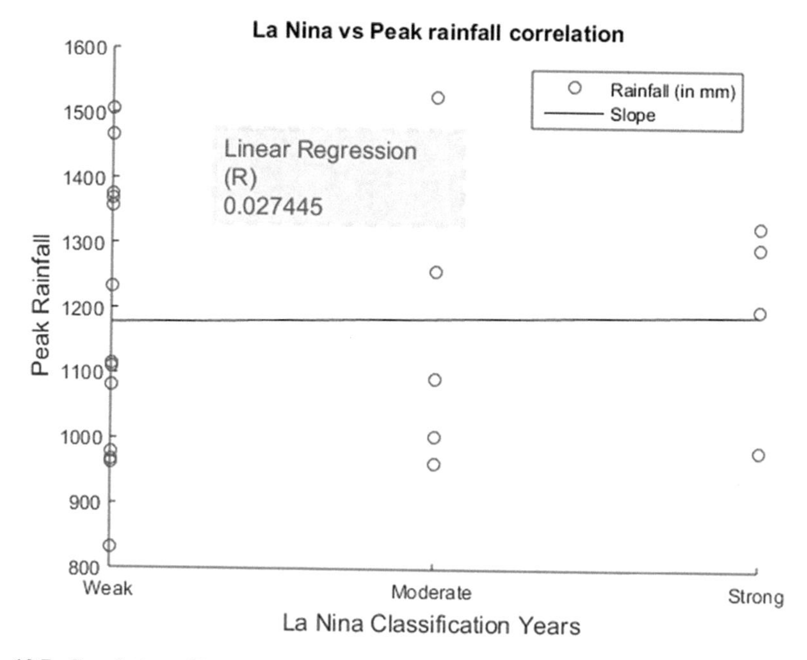

Fig. 10.7 Correlation of La Nina classification with peak rainfall

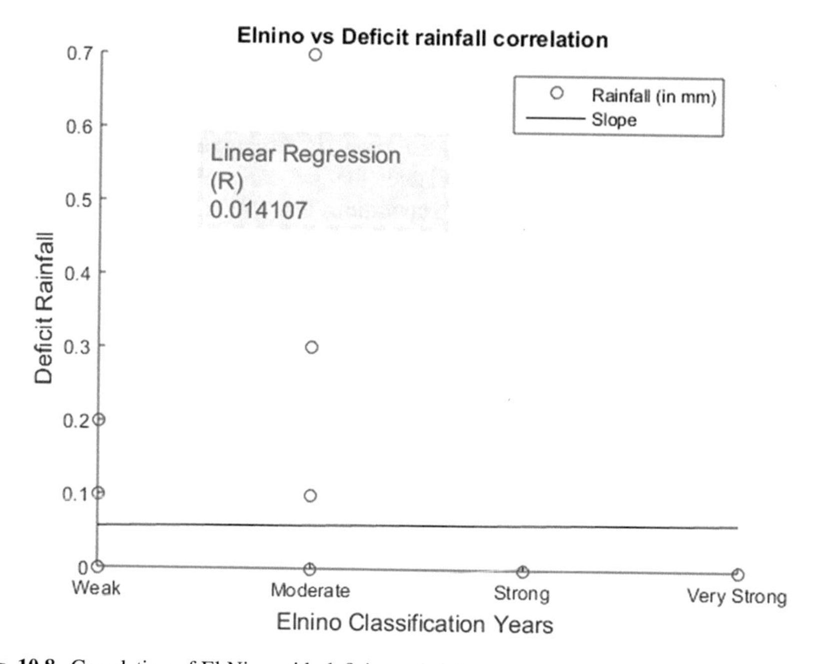

Fig. 10.8 Correlation of El Nino with deficient rainfall

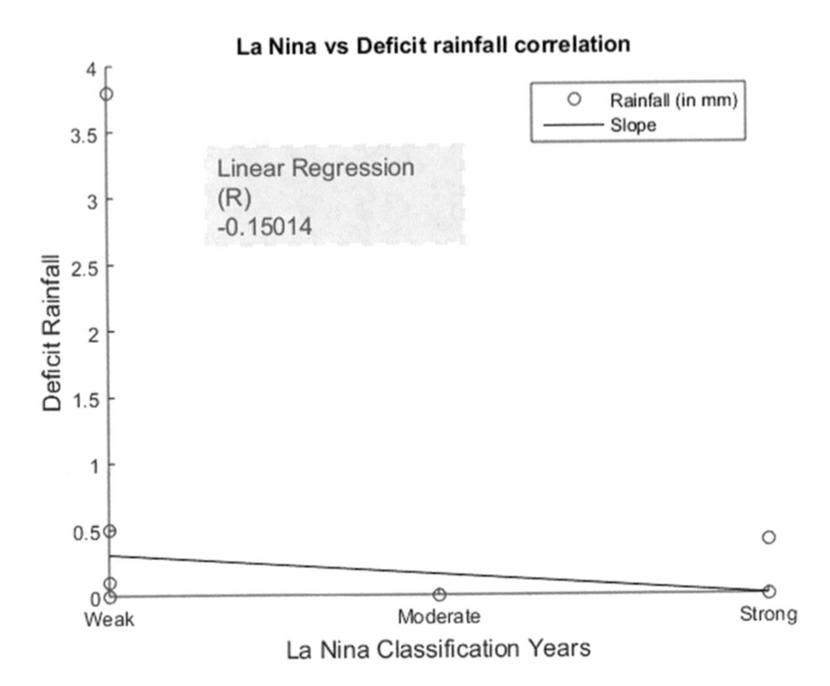

Fig. 10.9 Correlation of La Nina with deficient rainfall

Niño is very helpful for Karnataka as the state economy relies on agricultural activity, and an adequate water supply is required for the cultivation of rice, jowar, bajra, maize, ragi, small millets, pulses, castor, cotton, and sugarcane. Flood occurs only when the water in the reservoirs is not properly discharged at regular intervals. With early prediction of ENSO available, the government can take necessary measures related to the drought or flood conditions that will prevail in the upcoming year. In drought conditions, available water needs be stored to last for months, and agricultural activities can be shifted to preceding or succeeding months as per the rainfall. During La Niña years, the fishery is severely affected along coastal Karnataka by the warm ocean current. Similarly, in El Niño years, the fishery gains because of the cold ocean current. Warm water is less dense than cold water, and fishes do not survive in warm water because all the microorganisms are destroyed, thus causing a shortage in their food supply. Hence, the fish move towards cold water.

References

Barnston AG, Chelliah M, Goldenberg SB (1997) Documentation of a highly ENSO-related SST region in the equatorial Pacific. Atmos Ocean 35:367–383

Bjerknes J (1969) Atmospheric teleconnections from the Equatorial Pacific. J Phys Oceanogr 97:163–172

Chiodi AM, Harrison DE (2013) El Niño impacts on seasonal U.S. atmospheric circulation, temperature, and precipitation anomalies: the OLR-event perspective. J Climate 26:822–837
Goswami BN, Venugopal V, Sengupta D, Madhusoodanam MS, Xavier PK (2006) Increasing trends of extreme rain events over India in a warming environment. Science 314:1442–1445
Guhathakurta P, Rajeevan M (2008) Trends in the rainfall pattern over India. Int J Climatol 28:1453–1469
https://ggweather.com/enso/oni.htm
https://www.climate.gov/news-features/understanding-climate/climate-variability-oceanic-ni%C3%B1o-index
https://www.ncdc.noaa.gov/teleconnections/enso/indicators/soi/
Kiladis GN, Diaz HF (1989) Global climatic anomalies associated with extremes in the Southern Oscillation. J Climate 2:1069–1090
Kripalani RH, Kulkarni A (1997) Climatic impact of El Nino/La Nina on the Indian monsoon: a new perspective. Weather 52:39–46
Krishna Kumar K, Rajagopalan B, Cane MA (1999) On the weakening relationship between the Indian monsoon and ENSO. Science 284:2156–2159
Krishnakumar KN, Prasada Rao GSLHV, Gopakumar CS (2009) Rainfall trends in the twentieth century over Kerala, India. Atmos Environ 43:1940–1944
Kumar KR, Pant GB, Parthasarathy B, Sontakke NA (1992) Spatial and sub-seasonal patterns of the long-term trends of Indian summer monsoon rainfall. Int J Climatol 12:257–268
Mooley DA (1997) Variation of summer monsoon rainfall over India in El Nino. Mausam 48:413–420
Mooley DA, Paolino DA (1989) The response of the Indian monsoon associated with the change in sea surface temperature over the eastern south equatorial Pacific. Mausam 40:369–380
Mooley DA, Parthasarathy B (1984) Fluctuations in all-India summer monsoon rainfall during 1871–1978. Clim Change 6:287–301
Normand C (1953) Monsoon seasonal forecasting. Q J R Meteorol Soc 79:463–473
Parthasarathy B (1984) Inter-annual and long-term variability of Indian summer monsoon rainfall. Proc Indian Acad Sci (Earth Planet Sci) 93:371–385
Parthasarathy B, Dhar ON (1978) Climate fluctuations over Indian region. Rainfall: a review, Research report no. RR-025. Indian Institute of Tropical Meteorology, Pune
Parthasarathy B, Pant GB (1985) Seasonal relationships between Indian summer rainfall and the southern oscillation. J Climate 5:369–378
Parthasarathy B, Kumar KR, Munot A (1993) Homogeneous Indian monsoon rainfall: variability and prediction. Proc Indian Acad Sci (Earth Planet Sci) 102:121–155
Pramanik SK, Jagannathan P (1954) Climate change in India. I. Rainfall. Indian J Meteorol Geophys 4:291–309
Rasmussen EM, Carpenter TH (1982) Variations in tropical sea surface temperature and surface wind fields associated with the Southern Oscillation/El Niño. Mon Weather Rev 110:354–384
Rasmusson EM, Carpenter TH (1983) The relationship between eastern equatorial Pacific sea surface temperatures and rainfall over India and Sri Lanka. Mon Weather Rev 111:517–528
Revadekar JV, Kulkarni A (2008) The El Nino Southern Oscillation and winter precipitation extremes over India. Int J Climatol 28:1445–1452
Saji NH, Gowami BN, Vinayachandran PN, Yamagata T (1999) A dipole moment in the tropical Indian Ocean. Nature 401:360–363
Sikka DR (1980) Some aspects of large scale fluctuations of summer monsoon rainfall over India in relation to fluctuations in the planetary and regional scale circulation parameters. Proc Indian Acad Sci (Earth Planet Sci) 89:179–195
Soman MK, Kumar KK (1990) Some aspects of daily rainfall distribution over India during the south west monsoon season. Int J Climatol 10:299–311
Soman MK, Krishna Kumar K, Singh N (1988) Decreasing trend in the rainfall of Kerala. Curr Sci 57:7–12

Vinaya Kumar HM, Shivamurthy M, Govinda Gowda V, Biradar GS (2017) Assessing decision-making and economic performance of farmers to manage climate induced crisis in coastal Karnataka (India). Clim Change 142(1):143–153. https://doi.org/10.1007/s10584-017-1928-x

Walker GT, Bliss EW (1932) World weather. V. Mem R Meteorol Soc 4:53–84

Wright PB (1977) The southern oscillation patterns and mechanisms of the teleconnections and the persistence, Report HIG-77-13. Hawaii Institute of Geophysics, Honolulu

Wyrtki K (1985) Water displacements in the Pacific and the genesis of El Niño cycles. J Geophys Res 90:7129–7132

Chapter 11
Geothermal Energy and Climate Change Mitigation

Kriti Yadav, Anirbid Sircar, and Apurwa Yadav

Abstract The energy sector in the world needs major changes for reduction in greenhouse gas emission and climate change mitigation. It has been believed that with the current rate in increase of greenhouse gas emissions a rise in sea levels will occur with changes in global climate patterns. These effects hinder adaptation for climate change mitigation, so it is important to incorporate and understand the models implemented in the energy sector for decarbonisation. According to this perspective, the geothermal resource is contemplated as being one of the most productive energy sources that can be adopted for control of greenhouse gas emissions. The purpose of this study is to understand the global status of geothermal energy and its important role in climate change mitigation. In this chapter it is seen how geothermal energy has been used for centuries for heat extraction and power generation in geothermal countries such as the USA, New Zealand, Iceland, Kenya, and the Philippines, with minimal greenhouse gas emission with respect to fossil fuels. In the present study, the GeoAdam method of geothermal climate change mitigation is also discussed and modified in terms of scenario setup, assessment stage, and implementation stage. This method can be used for the climate change mitigation process in India with geothermal energy. The life cycle assessment of a system, which is an important factor to understand and control greenhouse gas emission, is reviewed in detail here. The present study provides a complete guide for climate change mitigation with geothermal energy.

Keywords Geothermal · Climate change · Greenhouse gas · Power generation · Environmental impact

K. Yadav (✉) · A. Sircar
Centre of Excellence for Geothermal Energy, Pandit Deendayal Petroleum University, Gandhinagar, Gujarat, India
e-mail: kriti.yphd15@spt.pdpu.ac.in

A. Yadav
Silver Oak College of Engineering and Technology, Ahmedabad, Gujarat, India

M. N. Islam, A. van Amstel (eds.), *India: Climate Change Impacts, Mitigation and Adaptation in Developing Countries*, Springer Climate,
https://doi.org/10.1007/978-3-030-67865-4_11

Introduction

In the year 2015 the Paris Agreement was signed, which established a landmark in combating climate change effects, with international consensus to reduce global temperatures to 2 °C with a goal of 1.5 °C reduction by 2100 (UNFCCC 2015). Although the 2 °C goal was deemed sufficiently aggressive by the IPCC to avoid catastrophic climate changes, the current trend points to rising temperatures of as much as 3 °C by 2100 (Masson-Delmotte et al. 2018). It is clear to the research community that greater efforts are needed. The production of electricity is at the forefront of attention as it is the greatest source of greenhouse gases and thus can cut carbon emissions almost entirely and faster than other manufacturing sectors. Through the development of arresting climate change, renewables are expected to have a major effect on greenhouse gas (GHG) emission, and geothermal energy has significant advantages over other renewables. This energy source is basically omnipresent (in the subsurface of the Earth, the temperature of formation rises by depths created by the environmental geothermal gradient), it is independent of global climate transition and thus a means of electricity production, and it is affordable (with an increased mean power cost as minimal as USD 0.07/kWh) (IRENA 2019). The important aspect is that it uses well-known thermodynamic methods to produce electricity.

As with other techniques, geothermal resources have been harnessed from high-caliber and less obvious potentials (Schechinger and Kissling 2015). Geothermal system technology pumps heated geothermal fluid from deep aquifers at depths between 1 and 4 kilometers (km). With increase in enthalpy of water, the consistency of an aquifer improves, basically from liquid reservoirs to steam reservoirs by improving water vapour concentrations. In comparison, petrothermal system technology seeks to provide geothermal water where traditional reserves do not occur by the development of an "engineered reservoir." These types of systems, also referred to as enhanced geothermal systems (EGS), are drawing considerable interest as this theoretically enables geothermal resources to be harnessed anywhere, although the technology is financially viable only in the case of significant geothermal gradients. In the year 2018 geothermal power production was calculated as about 630 petajoules, with approximately half of this in electricity generation (89.3 TWh) and half as other applications in the form of heat. However, the overall capacity is projected to be nearly 21 GWe from 2015 to 2020 (Bertani 2016). Single flash systems currently account for nearly 40% of the global installed capacity; dry steam and double flash plants follow with almost 30%, whereas binary technologies used in petrothermal and hydrothermal techniques add around 14% (Bertani 2016).

In recent decades, considerable attempts have been made to identify the ecological consequences of enhanced geothermal energy systems for the reduction of GHG emissions and to assess the life cycle of geothermal plants (Frick et al. 2010; Lacirignola and Blanc 2013; Pratiwi et al. 2018; Sullivan et al. 2010, 2011; Treyer et al. 2015). However, geothermal resources still dominate the current generation of electricity and are expected to grow in future.

For this aim, it is therefore essential to examine the ecological effects of conventional geothermal technologies (i.e., hydrothermal systems) with the life cycle, such as done by Bravi and Basosi (2014) and Parisi et al. (2019) for conditions found in Italy and for generic conditions by Sullivan et al. (2010), 2011).

Reduction in GHG emissions and climate change mitigation are the major issues of our time. However, development of geothermal resources has been proved one of the key solutions for reduction of CO_2 emission and can be a substitute for fossil fuels. The eco-friendly nature of geothermal resources has been found most advantageous to add to a renewable energy mix. The application of geothermal energy in direct and indirect forms has, on the basis of its enthalpy, already been discussed by various researchers (Rybach 2010a). It has been estimated that by 2050 the geothermal development will reduce CO_2 by 100 Mt/year, with electricity generation by geothermal resources and around 300 Mt/year by direct uses (Rybach 2010b). India is a developing country in the field of geothermal exploration, exploitation, and power production. In India, organisations such as GSI, NGRI, and CEGE are working on its development. In this study the application of geothermal energy in various forms is described. The present work also narrates the development of geothermal sources throughout the world, with their power generation capacity. This chapter suggests some of the geothermal energy mitigation plans and flow that can be adapted by the Indian geothermal industry. The chapter also discusses how to instigate a life cycle assessment plan of an enhanced geothermal system for the reduction of GHG emission.

Applications and Status of Geothermal Energy in the World

The three important features of any geothermal energy reservoir are a water source, an impervious rock cap, and a heat source. Structural faults provide the means of movement for hot water and steam to disperse in the cap rocks, which creates hot springs, fumaroles, and geysers.

These geothermal resources can be utilized for numerous direct and indirect applications (Basaran and Ozegener 2013). The present researchers analysed in depth the discovery and development status of geothermal resources in India and around the world (Sircar et al. 2015). In India, exploration activities for geothermal resources have been conducted in four major locations: in the Himalayan provinces (Puga geothermal field, Chhumthang geothermal field, Manikaran geothermal field, Beas valley, Sutluj and Spiti valley, Tapoban geothermal field), the Son-Narmada Lineament (Tatapani geothermal field, Anhoni- Sanmoni Area, Unkeshwar geothermal field), West coast geothermal provinces, and the geothermal field present in Cambay region (Dholera, Gandhar, Unai, Lasundra, Tuwa, and Chhabsar geothermal provinces). Various geophysical activities have been conducted in these areas such as magnetotelluric, magnetic, and gravity. For the first time in India a three-dimensional (3D) magneto-telluric survey was conducted for geothermal exploration in Unai (Sahajpal et al. 2015). Geothermal energy can be used in two major ways:

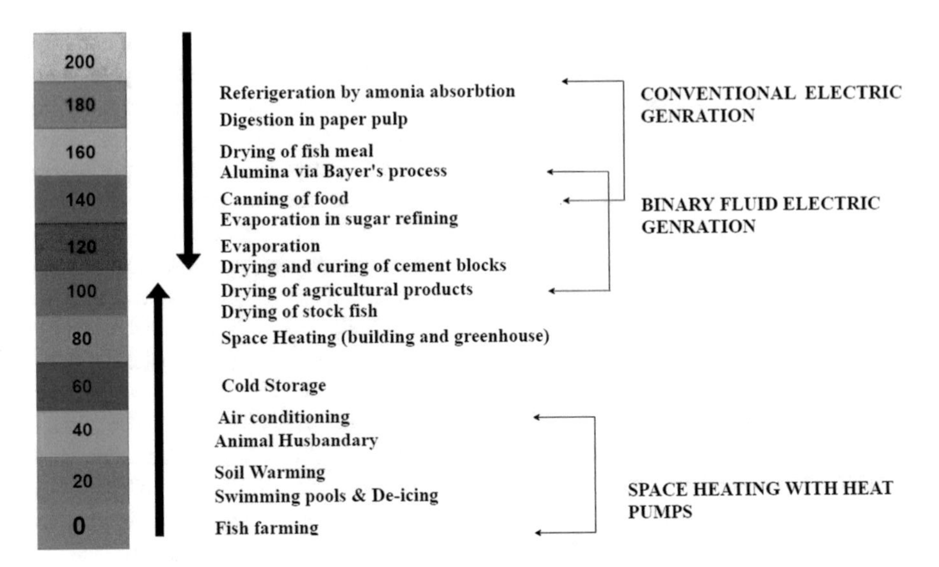

Fig. 11.1 Geothermal utilization at different temperatures. (Modified after Lindal 1973)

direct and indirect. Direct application of geothermal energy involves such needs as agriculture, aquaculture, space heating of buildings, crop drying, milk pasteurisation, and balneology. The indirect use of geothermal resources involves district heating and cooling, power generation, and numerous industrial applications (Yadav and Sircar 2019).

The geothermal fluid temperature utilised for direct application is minimal compared to that used for commercial electricity generation (Gudmundsson and Lund 1985; Lund and Boyd 2015). The optimal temperature range for geothermal plant outputs for various uses is shown in Fig. 11.1 (Lindal 1973).

The direct utilization of geothermal resources is generally limited to geothermal water with low to medium enthalpy. High-enthalpy geothermal energy can be directly used for heating and drying purposes. The process of refrigeration can take place at temperatures above 120 °C.

Kenya

Detailed work on most of the identified prospects in the central sector of the Kenyan rift has been completed. The Eburru, Paka, Suswa, Menengai, Korosi, Longonot, and Lake Bogoria volcanic fields have been studied. Electric power is currently being generated at Olkaria, with 209 MWe installed, and exploration drilling has been undertaken at Eburru. KenGen has commissioned a pilot plant of 2.5 MWe. The Geothermal Development Company (GDC), is currently undertaking detailed studies at Silali, Paka, Barrer, and Korosi for drilling exploration wells. It is estimated that more than 7000 MWe can be generated from such prospects in the Kenya rift (Figs. 11.2 and 11.3). In 2018, Kenya experienced a modest improvement to its geothermal resources with the Olkaria III power plant of 11 MW phased

Fig. 11.2 Olkaria geothermal plant

Fig. 11.3 Foam drilling at a geothermal field in Kenya

expansion stage, increasing the plant capacity to 150 MW (IRENA 2018). The total production of the country was 0.68 GW at the end of 2018 (IRENA 2018). Geothermal development in Africa is centred towards the East African Rift Basin, spreading from Djibouti mostly on the southern Gulf of Aden through Ethopia, Kenya, and Tazania to Malawai, where the broad Tanganyika and Malawi Lakes outline much of the rift features that make the region rich in geothermal potential (Wood and Guth 2019). Numerous ventures were assigned in Ethiopia and Kenya in 2018 (Peter Omenda 2018). Nevertheless, geothermal potential development in Eastern Africa was already hampered by considerable cost and business risks as opposed to other regions that are rich in geothermal resources. Other barriers include insufficient exploration funding and grant support, insufficient qualified human resources, and a lack of consistent, coherent policy and regulations across the country.

Iceland

The geothermal resources in Iceland are located in the Mid-Atlantic Ridge. Geothermal energy accounts for about 69% of Iceland's major energy demand. In the twentieth century, geothermal energy was used for space heating, followed by other similar uses. The district system is by far the most significant usage, providing 90% of all room space heating used in the region. Many direct applications include warm water in balneology, snow melting, agricultural uses, greenhouse effect, and aquaculture. In the villages and cities of the country, approximately 30 geothermal space heating systems are operational. In addition, about 200 small networks are running in rural areas. In Reykzavik the geothermal space heating network is the largest in the country, including more than 87861 people or more than 67% of the population in Iceland. In the nation there are more than 225 balneology facilities of which 150 are heated by geothermal water. The geothermal resource is the main industrial usage of a seaweed drying process in Thorverk, situated in west Iceland. Many industrial uses include the use of geothermal resources to extract salt from the oceans. The common uses include geothermal greenhouse warming and aquaculture. Geothermal heat pumps are utilised to complement northern Iceland's district heating network. Iceland has an established capacity for geothermal production in the magnitude of 755 MW.

The Teistareykir 90 MW geothermal electricity plant was established in 2018, with the second of two installed 45 MW phases (National Power Company of Iceland Landsvirkjun 2018). Iceland has also seen the initial output of electricity from a collection of low-enthalpy, binary cycle power systems (Charlotte Becker 2019). Iceland was historically a prominent generator of high-enthalpy geothermal energy. This construction was therefore incorporated to use low-enthalpy energy of about 100 °C (Varmaorka 2018). Iceland's 753 MW of geothermal energy capacity including thermal cogeneration for space heating and district water accounted for about 31% of the country's power production in 2018 (Energy Statistics in Iceland 2017).

Italy

Italy generally uses geothermal potential for electricity generation, and all systems are situated in Tuskuny in the key prominent areas of Mt. Amiata and Larderllo-Travale. Geothermal generation capacity as built in 2016 is 940 MW. The positive outcomes of the deep drilling and a careful and aware management of resources through reinjection systems and chemical enhancement helped to increase steam generation in the Larderllo-Travale, considering the long and rigorous history of mining. During the time span 2009–2013, five new units were built. After several years Enelgreen Power had obtained approval to restart drilling and development operations in the Mt. Amiata field. In 2009–2013, 31 geothermal wells were drilled in the entire geothermal area totalling approximately 100,000 meters (m), and 26 AMIS systems were constructed and are currently in service. Enel constructed the first hybrid geothermal-biomass electricity station in Italy in 2015 with increase in production from 12 MW to 17.5 MW of power for all plants in the region.

New Zealand

In the 1950s and 1960s, New Zealand led the world in the development of water-dominated geothermal resources. The harmony of growth in New Zealand deteriorated in the later period of the twenty-first century given the large and small coastal Maui gas field that offered a conveniently exploitable energy source. Mostly with reduction of gas and hydrocarbon alternatives, New Zealand has encouraged geothermal growth in the country. At present the geothermal potential of New Zealand is about 970 MWe, producing between 12% and 15% of the total production based on the season of year. By comparison, direct implementations are the best practical choice for uses. Geothermal heat pump construction is recorded in the coldest areas of the southern island, such as in Queenstown, to large homes in the warmer north, as in Oakland.

Although the most general use is for balneology, lesser amounts are used for heating water, and sometimes in direct application for frost prevention and agriculture. The variety and significance of demand grows for the higher enthalpy tools available in the volcanic region of Taupo, which involves greenhouse heating, prawn cultivation, powdered milk, and a special growth of tourism.

In November 2008, Wairakei installed the first turbo-generator. Throughout the years, the plant has undergone several changes, the current alteration being the construction of a 14 MWe net or matched binary cycle system. For reinjection activities, the geothermal separated water below to the area close to the facility is widely used. At the initial Wairakei location, the gross effective generating capacity has become 176 MWe.

The next major development is the Tauhara field, which has a 20 MW steam supply. A 23 MW Ormat binary power plant was commissioned in 2010. The other two binary plants are Ohaaki and Moki. The acceleration in the development process has brought improvement in the capacity of the Kawerau field, which is subject to the longest commercial development. In Rotokawa and Ngatamariki, good prospects are identified, and drilling has shown the potential for electricity generation at higher capacity. The earliest drilling at Rotokawa was undertaken in 1965 whereas in

Ngatamariki the drilling started in 1985. In Nagatamariki field, the 80 MWe Ormat plant was commissioned in 2013 in partnership with local Maorilantras.

Modest growth has been observed in New Zealand in its geothermal potential in subsequent years, mostly because of the stable consumption of electricity and limited demand for any extra generating power. The country mounted a 25 MW binary cycle network at Te Ahi O Maui in 2018. Unlike others of its kind, this relatively small power plant can drain large amounts of geothermal water from the geothermal source: three extraction wells and two injection wells draw and pump as much as 15,000 tons of water per day (Eastland Group 2018). Initially considered a 22 MW plant, it was subsequently raised to 25 MW, which lowered project costs per megawatt to 5.45 million USD (REN21 2019). In 2018, geothermal energy contributed 17% of electricity generation in New Zealand (OECD/IEA 2019).

Mexico

Throughout Mexico, geothermal power is often used exclusively for power generation, as its direct applications are undergoing growth and are mostly limited to balneology and swimming. Mexico's potential operates in four geothermal areas, namely Prieto, Los Humerus, Cerro, Los Azufres, and Las Tres Virgenes. In 2016, the total geothermal electric power in Mexico was estimated to be 1005 megawatts (MW).

The amount of power generated from geothermal energy is 2.71%, compared to 40.70% from hydrocarbons. Cerro Prieto is the largest and most influential geothermal region in action in Mexico, situated in the northern region of Mexico, with 13 condensing systems installed in 1973. The geothermal area of Cerro lies in tectonic basins formed between two effective strike slip faults that belong to the fault zone of San Andreas. Continental crust thinning in the area has induced thermal phenomena. The geothermal waters are found in 2400-m-thick sedimentary formations. In three decades more than 350 geothermal wells have been drilled in Cerro Prieto, to depths greater than 4400 m. The second largest geothermal region active in Mexico is Los Azufres, situated in the central part of the nation and about 250 km from the cities of Mexico, occurring inside the pine forest of the Mexican volcano bed physiographic region. In 1982, the very first power system in this region was constructed, and there are approximately 14 active systems with a combined capacity of 1888 MW.

Indonesia

More than 200 volcanoes are clustered across Indonesia's eastern region of Java, Sumatra, the islands, and Bali, regarded as the "Ring of Fire." Indonesia has the largest geothermal energy potential opportunities in the world from the large accumulation of geothermal systems with high temperatures. Indonesia reportedly contains 256 possible geothermal areas. Geothermal projects are currently run from seven locations with a total output capacity equal to 1380 MW, including Dieng, Darajat, Gunung Salak, Kamojang, Wayang Windu, Sibayak, and Lahen Dong. With development of Pertamina Geothermal Energy (PGE), the subsidiary of Pertamina State Oil and Gas Company and National Research Institute, the direct use of geothermal energy in Indonesia is beginning to expand, with the introduction of the Kamojang Mushroom Harvesting venture. Other uses of geothermal energy are the production of white copra and palm sugar; this does not include the numerous spas and balneology facilities. Several other organizations are progressing in research of the direct uses of geothermal for the purification of Akarwangi (the key raw ingredient for perfume). The use of geothermal heat pumps in the nation is not intensive.

Philippines
At present the country is next to the USA with an established capacity for geothermal production of the order of 1915 MW. Geothermal energy contributes about 14% of the total electricity requirement for the country. A road map that has been prepared by the National Power Corporation (NPC) depicts the utilization of geothermal resources between 2013 and 2030. Between 2013 and 2015, installation of 100 MW is planned, with an employment capacity of 2006 employees. Research is continuing to optimise and improve the geothermal power plant efficiency and energy conversion. Important work on policy changes and development of all the asset-generating NPCs has taken place during the past 5 years. The Makben, Laguna/Batangas/Quezon, are the Philippine geothermal development corporation geothermal energy plants, of Ink PGPC. In 1975, a maximum of 130 wells was drilled in these regions.

Croatia
In 2018 Croatia installed its first geothermal power plant, placing sixth in the world for new energy. The Velika Ciglena farms of the region produce 17.5 MW power from a medium-enthalpy geothermal resource. The thermal reservoir was discovered in 1990 during an inadequate search for oil resources (Turboden 2018b). Croatia aims to develop geothermal energy further in an attempt to increase the proportion of renewable energy sources in its energy mix (Igor Ilic 2019). In recent times, the geothermal industry in the Philippines has seen minimal development but in 2018 the country installed its first new geothermal pant of 12 MW with the extension of Maibarara-2 (Lenie Lectura 2018). The Philippines concluded 2018 with the net operating capacity of 1.93 GW (Power Supply and Demand Highlights 2017).

United States of America
The geothermal resources are used in all forms of electricity generation and for direct usages in the US. Primary energy use includes balneology, greenhouses and fish farming facilities, room heating and cooling, snow melting, forestry, commercial applications, and heat pumps. The nation's 88% annual energy consumption is for geothermal heat pump, followed by the highest use of aquaculture and balneology. During the previous 5 years with managing the profits and losses the direct usage of geothermal resources has remained unchanged; heat pumps are mounted with an estimated growth of 8% with 1.4 million (of 12 kW) in service (Boyd et al. 2015).

The United States continues to be the world leader by a broad margin in installed geothermal power efficiency. The nation placed around 58 MW in three locations in 2018 with a net annual operating capacity of 254 GW (EIA 2019). One introduction was the McGinnis Hills geothermal complex of the 48 MW third phase in Nevada, which seems to use two binary units of the new generation, increasing plant efficiency and availability, whereas three units would have been needed with earlier technology (Ormat 2018b). In Nevada too, a Chinese facility developer recharged Wabuska with the power capacity of 4.4 MW (REN21 2019).

A 14 MW binary power station has been installed in New Mexico to recharge an existing 4 MW system with upgraded and more effective binary infrastructure (Turboden 2018a). In 2018, geothermal energy generated 16.7 TWh power in the regions of New Mexico, a significant 5% increase over 2017, representing 0.4% of net generating capacity.

Denmark

Throughout Denmark, district heating is used to heat 63% of all homes. The space heating infrastructure is provided with electricity from a wide range of heating systems based on a variety of fuels. The amount of renewables is about 50%, and a small percentage of this comes from geothermal energy systems. The yearly heat retrieved from geothermal energy and utilised for space heating was approximately 300 TJ in 2012. About 90% of financing in space heating using geothermal energy is personal. In Denmark three geothermal space heating systems are in action, and the first has been in operating condition for 30 years.

Operations at various levels are currently underway, ranging from pilot projects for plant setup to basic geology studies related to geothermal development. In the power system in 2050 the Danish Government has a political goal of clean energy. Even scientific research of geothermal sources has grown as one of several projects working for this aim. In year 2014 the team of Danish Energy Agency produced a white paper for the development and implementation of space heating geothermal systems in the country.

A GIS database has been established with current and performance-controlled data appropriate for geothermal potential evaluation, with the goal of concentrating the preparation of new geothermal energy plants on instructional designs and speeding up the process of developing a heating facility. The Danish political research center funds multiple active geothermal research studies: Denmark's geothermal power potential and reservoir resources, temperature field "usage," and "heat storage in hot water bodies" models, both running until 2015, and the development of a network of geothermal activities in Denmark. Danish geothermal companies are also participating in cooperation with collaborators in 12 nations in the EU Intelligence Energy Europe Program project "GeoDH." This project is expected to construct the first power station in Copenhagen with 11 geothermal wells, some of which could be equipped for greater period heat storage.

Such a facility should be required to obtain around 64 MW of hot water from 1000 m^3/h (Rogen et al. 2015).

Pakistan

Pakistan is among the favourable developing nations with great geothermal possibilities that exist on the tectonic belt. Geothermal occurrences are analysed in

Pakistan in the form of mud volcanos, geysers, and hot springs (Zaigham et al. 2009, Ahmed and Rashid 2010; Bakr 1965; Kheli et al. Kheli et al. 1979; Bukhari 2000). The northern field, Chagai, Karachi, etc., contains several geothermal possibilities. Geothermal springs have strong brine temperatures in the northwest region of Baluchistan, although these springs have moderate brine temperatures in south Baluchistan. The geothermal springs in the Indus delta and Western Sind region are of low and medium enthalpy. Similarly, the geothermal springs in southwestern and northern Punjab areas have low enthalpy. Such geothermal reserves are located along Main Mantle Thrust (MMT), Main Boundary Thrust (MBT), and Main Karakorum Thrust (MKT), involved in the creation of the Indian and Eurasian plate collisions (Bakht 2000).

India

India boasts seven major geothermal regions. However, although many geothermal tasks are underway in India, no power station has ever been installed. One such initiative is the outcome of a partnership between Norway and the Himalayan provinces of India. The objective of this cooperation is to determine multiple indirect and direct uses of the geothermal resources in India. Two demonstration systems have been used for this research to investigate the space heating usages of these low- and medium-enthalpy aquifers, which have improved the living standards of the native community. Because the region has a very limited energy supply, about 3 h/day, and the winter temperature drops to below 20 °C, the area population practically depends on hydrocarbon fuel and coal for heating purposes.

Researchers working in this field have evaluated the resource capacity and heat load for warming restaurants and hotels and have succeeded in installing heating systems that sustain the indoor air temperature at around 20 °C. Solar farms are utilized as a resource to sustain the uninterrupted supply of heat pumps. These types of initiatives are important in increasing the life span and overall standard of living for communities in these areas that are marked by fuel deprivation and relative isolation as well as the potential for geothermal resources (Richter 2016).

The proposal to create 10,000 MW of electricity by 2030 has been declared by India in partnership with the best geothermal electricity production companies including New Zealand, Mexico, Philippines, and the USA. Tattapani, Chhumathang, Puga, Manikaran, Cambay basin, Ratnagiri, and Rajgir are among the explored geothermal locations in India. The initiative forms part of the government's commitment to expand the proportion of renewables to 350 GW by 2030 (Richter 2016).

Indian Geothermal Projects

In 1973, under the wings of the Geological Survey of India, a geothermal study was actively undertaken and about 340 active geothermal spots were identified. About 11 potential geothermal regions have been identified so far. Almost all of these seem to have potential at temperatures around 100 °–120 °C whereas others seem to have aquifers at a depth of 1–3 km with measured 200 °–250 °C enthalpies.

In the northwest Himalayas, the most promising fields are Puga-Chumathang (Ladakh district, J&K), Parbati valley, and Manikaran field in Himachak Pradesh. In Central India, in the fields, namely, the regions of the Madhya Pradesh area of Tattapani, there is a plan to set up a 20 MWe binary farm. The huge availability of alternative energy resources such as coal has evidently hindered the growth of production of geothermal resources (Kumar et al. 2008).

As well as the aforementioned ongoing projects in India, the National Geophysical Research Institute (NGRI), Hyderabad, has performed magneto-telluric studies in the Chhatisgarh Tattapani geothermal regions to classify geological surface features to determine their energy and thermal capacity. Depending on the findings of these inquiries, the construction of an electricity generation network is planned. NGRI has carried out similar studies in Puga valley for power generation through geothermal energy. In the states of Himachal Pradesh, Jammu, and Kashmir, and parts of Uttarakhand, which experience extreme cold weather for longer periods of time, the feasibility of heat pumps for space heating inside houses should be studied (Shah and Dutt 2014).

Major work in India related to application of geothermal energy has been carried out mainly in Gujarat. According to a GSI report, 17 hot springs were identified in Gujarat. Integrating the geological studies, geochemical studies, and remote sensing of all the locations, six locations—Unai, Tulshishyam, Tuwa, Dholera, Chabsar, and Gandhar—were identified for further investigation. The government of Gujarat (GoG) has established the Centre of Excellence for Geothermal Energy (CEGE), which is dedicated to the research and development of geothermal energy in Gujarat. CEGE has carried out an extensive exploration survey in various parts of Gujarat to understand the geothermal potential present in these regions. A three-dimensional (3D) magneto-telluric survey has been conducted for the first time in India for geothermal prospect identification. CEGE has drilled three shallow wells, of 1000 ft, for exploiting the shallow depth geothermal prospects on the basis of results obtained from the exploratory survey. The shallow geothermal well drilled by CEGE produces water at 4–5 l/s at 45 °C.

However, this geothermal aquifer is of relatively low enthalpy compared to influential thermal springs all over the globe. The geothermal space heating and cooling system based on heat pumps, which is the first of its type in India (Fig. 11.4), is designed in such a way that it can achieve the desired parameters. The system is established at Swaminarayan Temple (Fig. 11.5a, b), Dholera. The hot water produced from the wells acts as the driving force for the heat pump-based space heating and cooling system. During gatherings and other temple activities, the output from the conditioning side of the plant is used for comfort conditioning to assemble in the Hall at the temple. The output from the heating part of the device is utilised as heat input for scaled-up electricity production in the organic Rankine cycle (ORC). As an additional advantage, the plant provides conditioning of 32 TR capacity for the temple Sabhamandap.

A geothermal-based heat pump is a system that can use low-enthalpy storage energy. Geothermal based heating and cooling plant consists of heat exchanger at

Fig. 11.4 Space heating and cooling system set up at Dholera, Gujarat, India. (Photograph by the authors)

Fig. 11.5 (**a**, **b**) Swaminarayan Temple at Dholera. (**c**) Application of geothermal energy in the form of balneology

both cooling and heating side of the system. The heat pump is made up of four elements, which include condenser, evaporator, compressor, and valve for expansion. The outcome from the evaporator side is used for comfort conditioning of Sabhamandap of the temple whereas the output from the condenser side is used for direct applications such as power generation of 20 kW, milk pasteurisation, and honey processing.

The geothermal heating and cooling plant contributes to energy-efficient replacement of traditional air conditioners, water coolers, and water heaters. In this plant, electricity utilization is decreased, and for the long term both economic and societal advantages can be accomplished.

Power Generation

The geothermal resource is exploited for power generation and for several thermal purposes including industrial utilization and space heating. In 2018 the total global power output from geothermal energy is estimated to be about 630 PJ (IEA 2018).

Regardless of the complexities of data gathering, estimates of geothermal resources utilization are more reliable than those for electricity. Several geothermal installations produce both heat and electricity for various thermal utilizations. An additional 0.5 GW of advanced geothermal electricity generation resources came on board in 2018, bringing the total worldwide production to about 13.3 GW (Hondo 2017). Turkey and Indonesia persisted as the leaders in new facilities, taking into account roughly two thirds of the new generation capacity.

In 2018, New Zealand, Iceland, the United States, Croatia, Kenya, and the Philippines added resources (Hondo 2017). The nations with major resource potential for electricity production at the end of 2018 were Indonesia, the USA, Turkey, Philippines, Mexico, New Zealand, Italy, Iceland, Japan, and Kenya.

In 2018, Turkey completed multiple geothermal projects, increasing generation capacity to 1.3 GW by 21% (IRENA 2018). Turkey is the fourth largest nation in the world for geothermal combined power generation with almost 1 GW of capacity installed during 2013 and 2018 (REN21 2017).

The 65.5 MW Unit 2 at the Kizildere III was the largest single unit built in 2018, thereby becoming the largest geothermal power plant in Turkey (165 MW) (Zorlu Energy Group 2018). Certain completed projects involved the Baklaci 19.4 MW, the Pamukoren 32 MW Unit 4, and the Kale 25 MW 3S throughout the year (IRENA 2018). A final expansion, the 30 MW Alsehir Unit 3, joined the Turkish array in November. Geothermal plants in Turkey generally use binary-cycle engineering, as do all other plants in the nation (Tevfik Kaya 2017). In comparison, most emerging geothermal power stations around the globe are using flash or dry-steam techniques that are suitable for high-enthalpy energy.

Worldwide, the binary-cycle technique has been the largest growing engineering method in recent decades because of the growing use of comparatively low-enthalpy resources (US National Renewable Energy Laboratory 2019). Indonesia persisted to generate its geothermal system capacity 140 MW to the Philippines by a decent margin to position second worldwide for generation capacity, and finished with 1.95 GW at the end of 2018 (ESDM 2018a). In the regions of North Sumatra, the ultimate 110 MW unit of the Sarulla plant was constructed in year 2018, after completion of the two systems in 2017 (Ormat 2018a). The plant in Sarulla is the nation's first integrated cycle geothermal system, incorporating the techniques of two organizations to implement both water vapour and brine collected from the geothermal area to enhance plant efficiency (Hondo et al. 2017). Indonesia also commenced operation of its 30 MW power generation Karaha Unit 1 in 2018 (Pertamina 2018). During the latter half of 2018, the Indonesian government stated that it did not meet its targets for aggressive geothermal development, primarily because of contractor delays in drilling (PABUM News 2018). However, the nation's geothermal potential

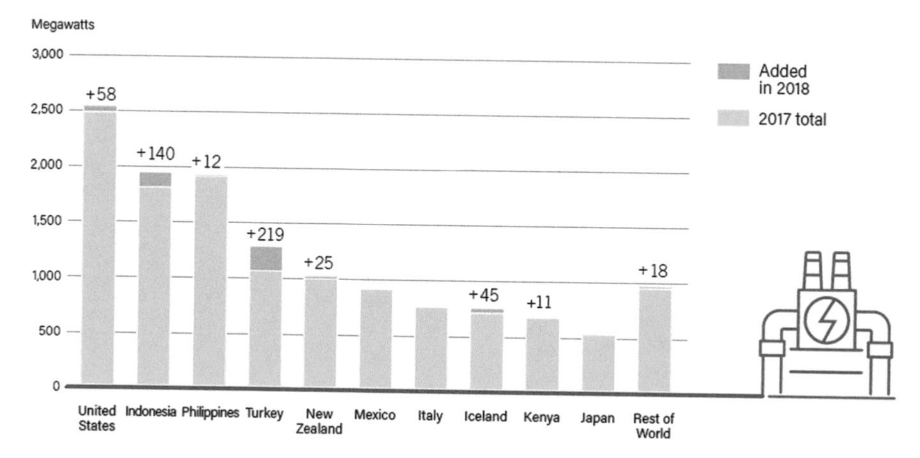

Fig. 11.6 Geothermal power capacity and additions for the top 10 countries and the rest of the world in 2018 (REN21 2019)

Table 11.1 Geothermal energy production of countries up to May 2018 (Richter 2018)

Country	Energy production (MW)
Kenya	676
Iceland	755
Italy	944
New Zealand	980
Mexico	951
Indonesia	1925
Philippines	1868
United States of America	3591

is calculated at 29 GW; only about 7% of that was developed because of high resource impact and potential drilling costs (World Bank 2018). Geothermal potential supplies about 5% of the nation's electricity demand (ESDM 2018b). Many of the issues faced by the geothermal sector in Indonesia are common, such as commercial threats from large project timeframes, exploration and resource risks, and generally high production costs (Bloomberg NEF (BNEF) and Business Council for Sustainable Energy 2018). Thanks to these industry-based obstacles, several nations that lead geothermal power use, such as the United States, Italy, and New Zealand, have not seen rapid growth in recent years. New Zealand and the US both finalized new projects in 2018 (Fig. 11.6).

The eight countries listed next indicate that renewable energy is the future of the world, as most of their power production is from geothermal energy. Table 11.1 shows the established capacity for geothermal production in megawatts (MW).

The power installed capacity from a geothermal plant globally is listed in Table 11.2. The 2020 forecast is based on an accurate estimate at the organizational stage of all current sites: a clear transition in the current linear continuing demand should be required. Table 11.2 displays information from all nations currently

Table 11.2 Total worldwide installed capacity from 1950 up to end of 2015 and short-term forecast

Year	Installed capacity (MWe)	Produced energy (GWh)
1950	200	
1955	270	
1960	386	
1965	520	
1970	720	
1975	1180	
1980	2110	
1985	4764	
1990	5834	
1995	6832	38,035
2000	7972	49,261
2005	8933	55,709
2010	10,897	67,246
2015	12,635	73,549
2020	21,443	

producing geothermal electric power, with the 2010 and 2015 revised figures of annual production and generated energy annually, the rise after 2010 in both absolute terms, and the aforementioned short-term projection for generation capacity up to 2020.

USA

In the United States, significant geothermal energy plants are situated in California, Utah, Hawaii, and Nevada, with current installations in Alaska, Idaho, New Mexico, Oregon, and Wyoming.

These regions currently have constructed electricity potential of 3450 MWe with total of 2542 MWe running, generating about 16,600 GWh per annum. California has two significant geothermal fields, Imperial Valley and The Geysers. In Alaska the world's largest first minimal temperature binary cycle has been established, using Chena geothermal springs 74 °C hot fluid with three phases for a total of 730 kWe. In Nevada, the very first integrated solar PV and geothermal project was carried out at Stillwater, where a geothermal system of 48 MWe is completely integrated with 26 MWe of PV panels and 17 MWh of solar heat transfer coefficient, with an extra production of 2 Mwe. In this region the 2.0 cents/kWh production tax incentive has been applied and the renewable portfolio has a growth rate of 3.6% per annum.

However, in the United States geothermal energy is a small contributor for electric power generation and capacity, which has an estimated total power generation contribution of 0.48% (Boyd et al. 2015).

China

In China geothermal energy is mainly used for direct applications such as balneology and drying of fruits and vegetables. However, electricity generation in China is still in the nascent stage. Several private and government organizations have signed an agreement on "Guidelines for Supporting Geothermal Power Generation and Use" to improve the exploration and exploitation of geothermal power in China. By building the Yangyi power station, the nation's growth capacity will be doubled (Zheng et al. 2015). The total installed capacity in China is 28 MWe, with an electricity production capacity of about 155 GWh/year by geothermal energy.

India
In India various exploration surveys have been carried out by GSI and other organisations, but geothermal energy is still in the development phase. The Centre of Excellence for Geothermal Energy (CEGE), PDPU, has carried out exploration work in three major areas of Gujarat, namely, Dholera, Unai, and Gandhar, to identify the geothermal potential in these regions. In Dholera two wells has been drilled by CEGE to exploit geothermal energy. In the Swaminarayan Temple at Dholera a heating and cooling system is installed by CEGE and its output is used as a source for the organic Rankine cycle to produce electricity of 20 kW at pilot scale.

Iceland
Iceland is a major hub of geothermal energy in the world. Geothermal power contributes nearly 68% to Iceland's total energy consumption with energy utilised for house heating for about 90%. Geothermal energy in Iceland fulfils 29% of the power demand of the country. The nation's total generating capacity is greater than approximately 665 MWe with an annual production of almost 5.245 GWh/year. The major regions of Iceland that use geothermal energy for electricity generation are Namafjall, Hellisheidi, Husavik, Krafla, Nesjavellir, Reykjanes, and Svartsengi (Ragnarsson 2015).

Philippines
New Philippine legislation includes non-tax and fiscal subsidies to foster the generation of alternative and geothermal energy: 43 geothermal company/operating contracts were issued for extra power of electricity from the projects Maibarara of about 20 MWe and Nasulo of about 30 MWe. The 1870 MWe geothermal installed capacity contributes to fulfill 14% of the total electricity requirement in the Philippines. There are seven major productive poles in the Philippines, namely, Bacon-Manito/Sorsogon/Albay, Mak-Ban/Laguna, Mindanao/Mount Apo, Palinpinon/Negros Oriental, Tiwi/Albay, Maibarara, and Tongonan/Leyte.

Geothermal Energy and Mitigation

Mitigation in the Paris Agreement

The 2015 Paris Agreement concludes the distinction in the form of mitigation measures that defined earlier actions between developed and developing countries. They are substituted by a shared structure that binds all countries to bring forth their best attempts in the years ahead and to improve them. This agreement contains provisions that all parties periodically report on their progress in pollution and execution and undertake international scrutiny.

Key characteristics of the Paris mitigation deal include the following (C2ES 2015):

- Reiterate the objective of reducing global temperature increase below 2°C while encouraging attempts to restrict the increase to 1.5°C
- Create concrete mitigation agreements by all leaders to make "nationally determined contributions" (NDCs) and implement internal mitigation initiatives to achieve those commitments
- Require all nations to report on their emission levels on a routine basis and to "advancement in implementing and accomplishing" their NDCs, and perform global review
- Pledge all nations to send new NDCs within 5 years, specifically expecting them to "document development" beyond their previous goals
- Reiterate developed nations 'legal responsibilities under the UNFCCC to finance developing nations' efforts, while also facilitating voluntary contributions from developing nations for the first time
- Confine the original target of mobilizing $100 billion annually in assistance by 2025, with a new, larger target to be laid for the period after 2025
- Lead to a new framework, similar to the Kyoto Protocol's Clean Development System, allowing for pollution cuts in one nation to be counted against the NDC of another.

GHG Emissions and Mitigation Opportunities from the Geothermal Industry

Carbon emissions per kilowatt-hour of power produced from high-enthalpy thermal springs are substantially lower than those created from fossil fuels. Emissions of greenhouse gases predicted in g kWh^{-1} range between 4 and 740 g, with such a total increase of 122 g kWh^{-1} based on data from 85 geothermal systems in 11 countries (Fridleifsson et al. 2008). Considering that high-enthalpy geothermal energy for electricity generation is mainly confined to seismically active areas, the most successful areas are situated in Central America and the East African Rift Valley with regard to decreased GHG emissions from the growth of high-enthalpy geothermal resources. In 2008 Fridleifsson et al. demonstrated that 39 nations have the capacity to generate nearly 100% of their power needs from geothermal energy. Given the current global prospective and rational investment rate, Fridleifsson et al. (2008) also demonstrated that geothermal resources could meet approximately 8.3% of global electricity consumption by 2050. Considering that this investment will substitute fossil fuel-fired power stations, this investment could reduce carbon dioxide emission by nearly 1 billion tons by 2050. Ogola et al. (2012b) propose a much larger opportunity, or range, of mitigated GHGs, from 1 and 5 billion tons by year 2050.

Direct utilization of geothermal energy from low-enthalpy areas, such as room and water heating in various ways, often offers mitigation advantages and the same relates to heat pumps. Direct-use emissions are low, as they do not surpass standard natural background emissions on average. For instance, the Reykjavík district heating facility emits about 0.5 mg CO_2 kWh^{-1} Fridleifsson et al. (2008), so the mitigation effects can be substantial. Fridleifsson et al. (2008) measured the mitigation capacity of heat pumps to surpass 200 million tons of CO_2/year, as the heat pumps substitute for older fossil-fueled systems. Other direct application potential reductions amounted to more than 90 million tons of CO_2/year.

In addition to mitigation potential from indirect utilization of power generation and heat pumps, the total sustainability potential refers to approximately 12% of total GHG emissions by 2050, if compared with the IEA 2015 reference scenario, without carbon capture and storage.

CCS and CCU

Carbon capture and storage (CCS) is the method of trapping, transferring, and depositing CO_2 from relevant point sources such as power stations to a storage location, usually in an underground landform such as deep aquifers. The objective is to not bring CO_2 into the atmosphere. CCS has been active throughout the growth of the geothermal sector. A new approach called Carbfix has lately been scientifically proven to work by researchers at the University of Iceland, Columbia University, and Reykjavik Energy. The Carbfix program "fixes" carbon by breaking down CO_2 into water and reinjecting it deep into the subsurface into basaltic volcanic rocks. As this happens, the CO_2 becomes a solid mineral (calcite) and thus can be retained for a long time. Recent findings indicate that this approach is successful because 95% of the injected carbon is calcified within 2 years (Matter et al. 2016).

However, the process has its drawbacks because the process requires vast quantities of water and is region specific because it uses basaltic rock. However, it complements other CCS approaches mainly depending on reinjection into sedimentary rock.

Carbon capture and use (CCU) is a method of extracting CO_2 from plants, but instead of retaining it deep underground, the carbon is used in industrial and horticultural applications (Ogola et al. 2012a). Applications for horticulture include use in greenhouses and industrial uses include the transformation of CO_2 using chemical reactions into methanol, which can be used for transportation (Ogola et al. 2012a, b).

The Framework for Adaptation and Mitigation

Geothermal developments contribute to mitigation by eliminating sources dependent on fossil fuel for generating power or heat. These processes also lead to adaptation by cascading usage of climate-sensitive sectors that enhance adaptive capability, although the possibility of enforcing geothermal infrastructure projects in a grid-based converter network, or in an off-grid mini-grid, offers co-benefits for simultaneous mitigation and adaptation. The co-benefits resulting from geothermal adjustment/mitigation programs are shown in Fig. 11.7 (Ogola et al. 2012b).

Figure 11.7 Both geothermal adjustment and mitigation are highlighted, illustrating how these can be derived from a typical geothermal plant using either high- or low-enthalpy energy. Not all geothermal use schemes should be considered direct adaptation programs. Only initiatives aimed at a particular effect on climate change, with clear benefits in terms of adaptation, will qualify as such. An adaptation additionality evaluation as designed for mitigation projects in CDM is needed to reveal such direct impacts (Ogola et al. 2012b)

A conceptual structure, derived from Ogola et al. (2012b), describes the mechanisms associated with establishing and enhancing the efficiencies of adjustment–mitigation in geothermal installations is shown in Fig. 11.8. The potential for geothermal growth to make a significant contribution to climate change-resilient societies is revealed based on this framework. Figure 11.8 has three sections. The first section is scenario setup, the second section shows the process of geothermal energy evaluation, and the third section shows the deployment and thus the period of activities required when connecting adaptation to mitigation.

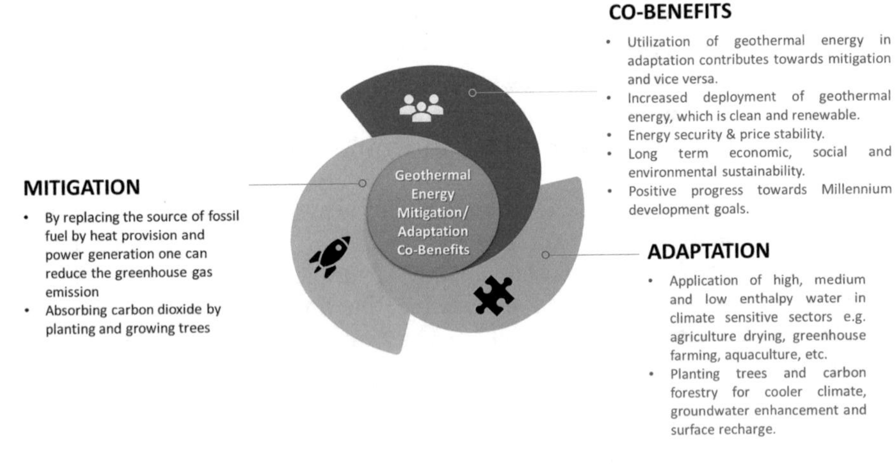

Fig. 11.7 Mitigation and adaptation co-benefits in geothermal projects (Ogola et al. 2012b)

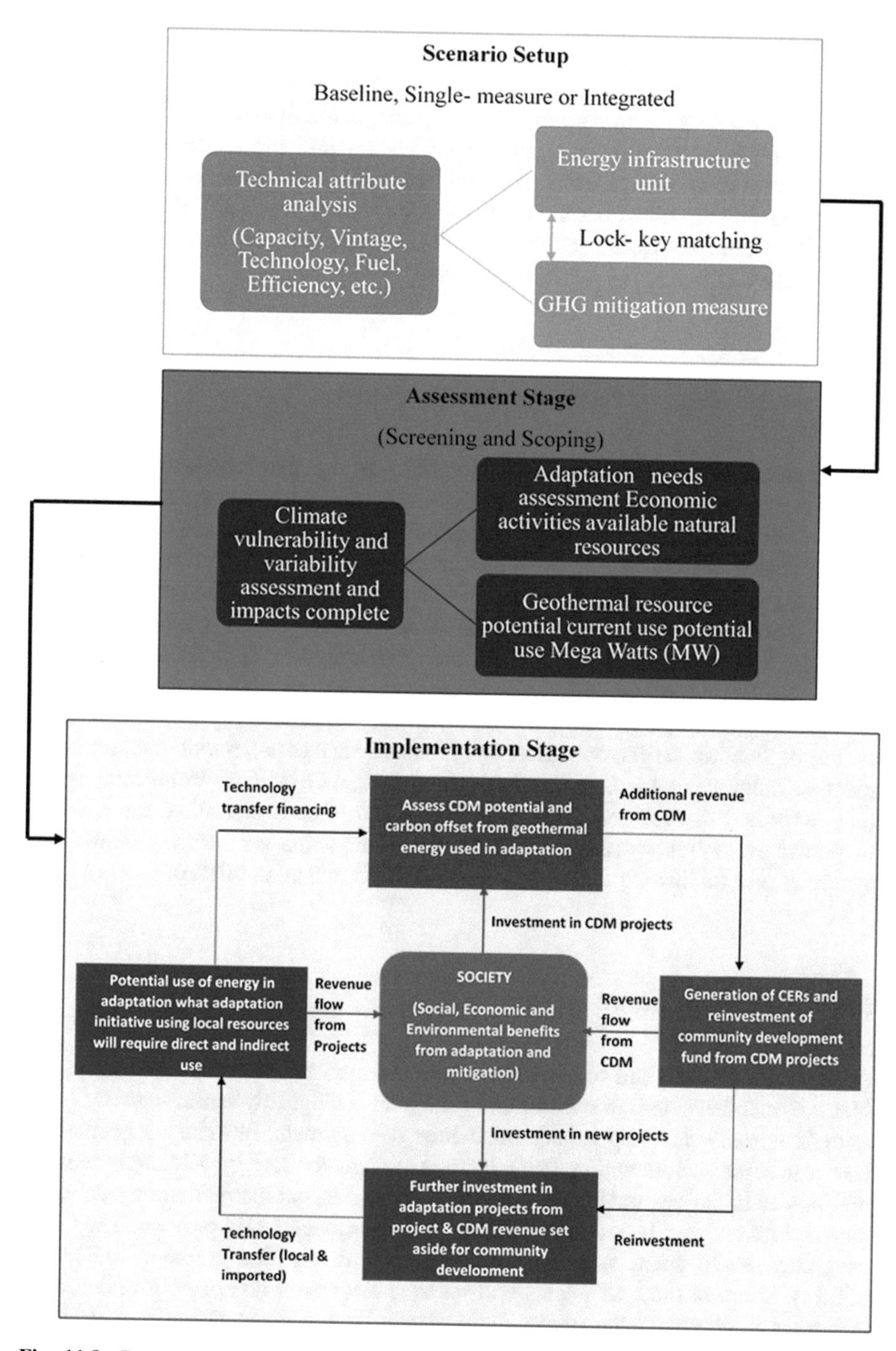

Fig. 11.8 Conceptual framework representing scenario setup, assessment, and implementation benefits for mitigation adaptation of geothermal projects. (Modified after Ogola et al. 2012b)

Scenario Setup

In the first step of scenario setup for mitigation adaptation of geothermal projects, the attributes of infrastructure units should be known and inspected: vintage, fuel, location, efficiency, and capacity. Any unit should be set to have a service lifetime of 30 years. In the case of baseline, all stocks will retire in general. For greenhouse gas emission, retrofitting in geothermal plants is measured at T0, the express time for each measure of individual geothermal system.

Assessment Stage (Screening and Scoping)

At this point, global warming and vulnerability in the community were analysed. Based on the security vulnerabilities analysis, different adaptive capacity studies are required that can be improved by geothermal usage, by finding the supply of raw resources and natural resources within a predefined range, such as agricultural products and fisheries. Evaluation and scoping for vulnerability, adaptation requirements, and availability of geothermal resources should be concluded at this point, and the related sectors/tasks should be clearly identified, with their commitment to adaptation. Upon scoping, different utilization choices can be measured on the basis of technological (e.g., cascaded infrastructure, MW productivity), economic (e.g., cost of production, tariffs, operational and maintenance costs), social (e.g., improvement in cultural well-being, healthy living standards), and environmental issues (e.g., expected decrease of CO_2-eq emissions and other benefits). At the end of a particular evaluation procedure the benefits of adaptation need to be evident. The synergies and trading expected must be explicit (Ogola et al. 2012b).

Implementation Stage

After the monitoring and scoping evaluation, various geothermal application activities are established to improve cultural change in ecologically sensitive sectors such as agriculture, water supply, and aquaculture development. Introducing geothermal power in these sectors will generate income streams for society. Although geothermal power initiatives qualify as mitigation strategies, geothermal energy development initiatives may be recorded as adaptation-portioned CDM projects. The CDM programs would create new revenue sources with certified emission reductions (CERs). Some of the CDM sales will go to society as profits or be invested back in further geothermal adaptation initiatives, resulting in domestic and foreign technology transfer and local development of raw materials (Ogola et al. 2012b). As per the Paris Agreement, a new technique will replace CDM, but its framework is not understood at this time. The new projects were then subsequently assessed for their carbon reduction potential taking in additional financing and technology transfer and maintaining the process while creating resilience for the city. The definition encourages sustainable growth and demonstrates how geothermal energy can boost adaptation as well as mitigation. An extensive list of measures that can be used to track Geo-AdaM geothermal mitigation adaptation activities compared to the baseline includes the following (Ogola et al. 2012b):

- Reduced CO_2-eq per annum
- Sum of megawatts has increased towards the adaptation and mitigation contribution
- Various geothermal resource-based projects and their effect on adaptation and mitigation
- Part of the population benefited from growth, including jobs and increased access to clean water
- Decreased economic losses resulting from extreme climatic conditions
- Increased agricultural production in greenhouses with geothermal heating, crop drying, and preservation of food items
- The number of persons with better health increased and the number of disabled people has decreased
- Significant reduction of human and animal casualties where frequent droughts occur

Life Cycle Assessment

During recent years, many life cycle assessment (LCA) methods have been introduced for renewable energy technologies based on their GHG emissions through power plants. Because several of these studies have assessed the overall adequacy of renewable energy systems, the uncertainty in the results is fairly high. For instance, a recent IPCC study, based mainly on assessment of about 50 LCAs of wind turbines, confirmed a comparatively wide range of carbon emission from 2 to 81 g CO_2-eq/kWh. LCA results from such IPCC studies range from 6 to 79 g CO_2-eq/kWh for geothermal systems (including EGS power stations) (Moomaw et al. 2011).

Moreover, LCA studies recorded throughout the geothermal sector are not numerous, compared to other renewable technologies. In certain cases, the findings reported so far are very common in some cases (Pehnt 2006), or instead, may be specific to a particular geographic area (Hondo 2005). Generally, the results of individual LCAs can only be applicable of a specific technology and can test a number of claims and limitations (e.g., for shipping and the type of accountable materials). More LCA assessment in the geothermal industry currently seeks the expertise of enacting legislation promoting and evaluating promoted plants.

The development of an LCA is very time consuming, particularly by collecting data on the outputs and inputs of resources and energy movements over the evaluated system life cycle. Analysing a panel of engineering concepts in a specified geological province requires shifting for a particular form of strategy to escape continuing to pursue full LCAs of alternative system setups. Furthermore, in light of the growth of the EGS industry, the emergence of a simpler and easier approach than LCA to calculate the GHG emissions may be beneficial for decision makers. A meta-LCA method has tried to globally recognize energy ways forwards (i.e., a collection of processes rather than a single one) in an attempt to settle the efforts of individual-detailed LCAs and has already been frequently implemented in the context of energy in recent years (Warner et al. 2010). The meta-LCA concentrates on an intercomparison of research study: the studies conducted are connected to a particular context (e.g., same life expectancy, same parameters of characterization). One component at a time is assessed for the heterogeneity developed by a single variable, and its relevant information on environmental sustainability is computed. There are two different options: a reduced variety of emission estimates (particularly in relation to a centralized analysis of individual LCAs in research) and, in some situations, meta-models (Lenzen 2008), which enable the research to assess environmental impacts using a linear regression model. However, the results depend primarily on the validity of literature data in both scenarios, and therefore cannot account for the insufficient evidence or relevant case studies published.

The procedure for producing parameterized simulations representing life cycle emissions of GHGs per electricity produced for the EGS system is characterized by the following steps (Padey et al. 2013):

1. Objective description and degree of statistical inference model
2. Proposed technique structure and EGS GHG transfer profile production using Monte Carlo simulations
3. Classification of critical elements via a global sensitivity evaluation, by generalized decay of variability (Sobol indices)
4. Description of generalized model, based on defined key components
5. Comparative analysis of the generalized model results with research and assessment of its reproducibility

To protect a standard EGS system with the comparison model, a large range of possible EGS plants have to be compensated. Any EGS power station will calculate their sustainability impact as follows:

$$\text{GHG performances of EGS} = \frac{\text{GHG emissions } \left[\text{gCO}_{2\text{eq}}\right]}{\text{Electricity Production } [\text{kWh}]} \tag{11.1}$$

The numerator illustrates the cumulative percentage of CO_2 emission demonstrated in the respective CO_2 mass of all life cycle aspects linked to the plant, which can be measured using the IPCC GHG characterization factors from a specific life cycle inventory (LCI) (Bernstein et al. 2008). Such assessment variables are used to quantitatively incorporate every other GHG as per the reference gas, its corresponding CO_2-related climate impact. Further details on this measurement can be seen in Heijungs (1996).

The denominator is the volume of electricity supplied to the grid during its lifetime, measured as

$$\text{Electricity production} = P_{NET} \times LF \times 8760 \times LT \quad (11.2)$$

where the EGS power station ultimately produces power as P_{NET}. PNET is the difference between the P_{ORC} capability constructed by ORC (i.e., the power production of the electric engine minus the electricity consumption by other ORC devices, including the air cooler) and P_{PUMPS} (the energy demand of the geothermal liquid storage and reinjection pumps). The load factor is LF (a percentage of 0 to 1 reflecting the equivalent hours of marginal power operation in 1 year), the amount of hours expended in 1 year is 8760, as well as the LT in the plant's lifetime (represented in years). The statistics used for EGS power stations, LCA, is produced from a proposed methodology of nine parameters: pump power demand, produced flow rate, load factor, drilling depth, fuel usage for drilling, number of wells, life expectancy, intensity increase, and ORC power capacity.

Conclusion

Although commonly recognized as a renewable energy source, geothermal energy development has also led to reduction in pollution levels from greenhouse gas emissions. Numerous studies have also shown that carbon dioxide emissions from geothermal plants occur naturally, often increasing the amount of CO_2 generated from the geothermal system using geothermal energy. Because no combustion methods are employed, geothermal technologies generate little or no carbon dioxide emission. Forecasts of geothermal production for 2050 suggest that greenhouse gas emissions can be mitigated by 100 Mt/year with geothermal electricity production and more than 300 Mt/year with direct applications, much of which could be accomplished by heat pumps. This chapter consists of four major segments: application and status of geothermal energy in the world, geothermal energy and mitigation, a framework for adaptation and mitigation, and life cycle assessment of geothermal power plants. The first segment of application and status of geothermal energy in the world reviews the application of geothermal energy worldwide in forms of direct and indirect applications. This segment also describes the power generation capacity of leading countries such as the USA, New Zealand, Mexico, Iceland, Philippines, Indonesia, and other countries worldwide. The second segment of geothermal energy and mitigation describes the key characteristics of the Paris Agreement with GHG emissions and mitigation opportunities from geothermal industry and carbon capture and storage with carbon capture and use. The third segment is the most important one: the framework for adaptation and mitigation using geothermal energy. In this section we have discussed the modified version of the GeoAdam method for geothermal climate change mitigation. The traditional method of GeoAdam is performed in two stages, as assessment stage and implementation stage. However, through various literature surveys it has been found that it is important to understand the scenario setup before assessment of the geothermal system and implementation of any ideas. The modified version of the GeoAdam method for geothermal climate change mitigation in this chapter describes the three steps: scenario setup, assessment stage, and implementation stage. The last segment of the chapter describes the life cycle assessment of geothermal systems for GHG mitigation.

References

Ahmed I, Rashid A (2010) Study of geothermal energy resources of Pakistan for electric power generation. Energy Sour Part A 32(9):26–38

Bakht MS (2000) An overview of geothermal resources in Pakistan. In: Proceedings of the world geothermal congress, Kyushu-Tohoku, Japan

Bakr MA (1965) Thermal springs of Pakistan. Geological Survey of Pakistan Rec 16:3–4

Basaran A, Ozgener L (2013) Investigation of the effect of different refrigerants on performances of binary geothermal power plants. Energy Convers Manag 76:483–498

Bernstein L, Pachauri RK, Reisinger A, Intergovernmental Panel on Climate Change (2008) Climate change 2007: synthesis report. Intergovernmental Panel on Climate Change, Geneva

Bertani R (2016) Geothermal power generation in the world 2010–2014 update report. Geothermics 60:31–43

Bloomberg NEF (BNEF) and Business Council for Sustainable Energy (2018) Sustainable energy in America Factbook. London/Washington, DC. http://www.bcse.org

Boyd TL, Sifford A, Lund WJ (2015) The United States of America country update 2015. In: Proceedings World Geothermal Congress 2015, Melbourne, Australia, April 19–24, 2015

Bravi M, Basosi R (2014) Environmental impact of electricity from selected geothermal power plants in Italy. J Clean Prod 66:301–308

Bukhari SHS (2000) Country update paper on Pakistan. In: Proceedings of the World Geothermal Congress, Bali, Indonesia

C2ES (2015) Outcomes of the U.N. climate change conference in Paris. Center for Climate and Energy Solutions, Arlington, Virginia

Charlotte Becker (2019) First geothermal site approved by customer, Climeon, 13 March 2019. https://climeon.com/wp-content/uploads/2017/04/Climeon-Tech-Product-Sheet.pdf; Climeon, LinkedIn.com, https://www.linkedin.com/feed/update/urn:li:activity:6468442653817139200, viewed March 2019

Eastland Group (2018) NZ's newest geothermal power plant is full steam ahead, 28 September 2018. http://www.eastland.nz/2018/09/28/nzs-newest-geothermal-power-plant-is-full-steam-ahead/

EIA (2019) Net generating capacity from US Energy Information Administration (EIA). Electric power monthly with data for December 2018. Washington, DC, February 2019, Table 6.2.B. https://www.eia.gov/electricity/monthly/archive/february2019.pdf; Nameplate capacity from EIA, "Preliminary monthly electric generator inventory", December 2018. https://www.eia.gov/electricity/data/eia860M/

Energy Statistics in Iceland (2017) Capacity from Orkustofnun, Reykjavik, 2018. http://os.is/gogn/os-onnur-rit/Orkutolur-2017-enska.pdf; Capacity of 753 MW based on capacity at end-2017 of 708 MW, with addition of 45 MW in 2018; generation share from OECD/IEA, op. cit. note 29

ESDM (2018a) From Indonesian Ministry of Energy and Mineral Resources (ESDM), Pemerintah Tetap Komit, Pemanfaatan EBT Capai 23% Pada Tahun 2025, press release, Jakarta: 9 January 2019. https://www.esdm.go.id/en/mediacenter/news-archives/pemerintah-tetap-komit-pemanfaatan-ebtcapai-23-pada-tahun-2025; Capacity of 1.81 GW at end-2017 from ESDM, "Capaian Sub Sektor Ketenagalistrikan Dan EBTKE Tahun 2017 Dan Outlook 2018", press release (Jakarta: 10 January 2018), https://www.esdm.go.id/en/media-center/arsip-berita/capaian-2017-dan-outlook-2018-subsektor-ketenagalistrikan-dan-ebtke-

ESDM (2018b) Handbook of energy & economic statistics of Indonesia, Jakarta, July 2018. https://www.esdm.go.id/en/publikasi/handbook-of-energy-and-economic

Frick S, Kaltschmitt M, Schröder G (2010) Life cycle assessment of geothermal binary power plants using enhanced low-temperature reservoirs. Energy 35:2281–2294

Fridleifsson IB, Bertani R, Huenges E, Lund JW, Ragnarsson Á, Rybach L (2008) The possible role and contribution of geothermal energy to the mitigation of climate change. In: Hohmeyer O, Trittin T (eds) Proceedings of IPCC Scoping Meeting on Renewable Energy Sources, Luebeck, Germany, pp 59–80

Gudmundsson JS, Lund JW (1985) Direct uses of earth heat. Int J Ener Res 9:345–375

Heijungs R (1996) Identification of key issues for further investigation in improving the reliability of life-cycle assessments. J Clean Prod 4:159–166

Hondo H (2005) Life cycle GHG emission analysis of power generation systems: Japanese case. Energy 30:2042–2056. https://geothermal.org/Annual_Meeting/PDFs/Geothermal_Development_in_Africa-2018.pdf

Hondo H (2017) Capacity data and capacity additions in 2018 from sources in endnote 1, and on sources noted elsewhere in this section. For the purpose of this figure, end-2017 capacity is assumed to be equal to end-2018 capacity less new capacity installed during 2018

Hondo H, Toshiba Corporation and Ormat Technologies Inc (2017) One of the world's largest geothermal power plants commences commercial operation, press release, Reno, NV and Tokyo, 22 March 2017. http://www.toshiba.co.jp/about/press/2017_03/pr2201.htm; Patel S (2017) Sarulla, one of the world's largest geothermal power projects, comes alive with private finance. POWER, 1 December. http://www.powermag.com/sarullaone-of-the-worlds-largest-geothermal-power-projects-comesalive-with-private-finance

IEA (2015) Energy technology perspectives 2015. Organization for Economic Cooperation and Development (OECD)/International Energy Agency (IEA), Paris, France

IEA (2018) Estimates based on the following sources: power capacity data for Italy, Japan, Mexico and New Zealand from International Energy Agency (IEA) Geothermal, 2018 Draft Annual Report. Taupo, New Zealand, February 2019. http://iea-gia.org/publications-2/annualreports/; IEA Geothermal, geothermal power statistics 2017, Taupo, New Zealand: February 2019. http://iea-gia.org/publications-2/working-group-publications/; Power capacity data for Iceland, Indonesia, Philippines, Turkey and the United States from sources noted elsewhere in this section; capacity data for other countries from International Renewable Energy Agency (IRENA), Renewable Capacity Statistics 2019, Abu Dhabi, 2019. https://www.irena.org/publications/2019/Mar/Capacity-Statistics-2019; Additional capacity data by country from Renewable Energy Network for the 21st Century (REN21), Renewables Global Status Report 2018, Paris, 2018. http://www.ren21.net/gsr_2018_full_report_en; Estimated electricity generation in 2018 from Organisation for Economic Co-operation and Development (OECD) and IEA, Renewables 2018, datafiles, Paris, 2018. Heat capacity and output is an extrapolation from five-year growth rates calculated from generation and capacity data for 2009 and 2014, from Lund JW, Boyd TL (2016) Direct utilization of geothermal energy 2015 worldwide review. Geothermics 60(March):66–93. https://doi.org/10.1016/j.geothermics.2015.11.004

Igor Ilic (2019) Croatia seeks to triple renewable energy output. Reuters, 11 April 2019. https://www.reuters.com/article/croatia-energy/croatia-seeks-to-triple-renewable-energyoutput-idUSL8N21T37L; Igor Ilic, Croatia eyes more gas and oil production, use of geothermal sources. Reuters, 26 March 2019. https://af.reuters.com/article/commoditiesNews/idAFL8N21D4AZ

IRENA (2018) Capacity of 1,063.7 MW and 40 plants at end-2017, and 43 plants and 1,199 MW of capacity as of September 2018, from Turkish Electricity Transmission Company (TEİAŞ), http://www.teias.gov.tr, viewed March 2019; capacity of 1,283 MW for 2018 from IRENA, op. cit. note 1

IRENA (2019) Renewable power generation costs in 2018. https://doi.org/10.1007/SpringerReference_7300
Kheli T et al (1979) Geology of Kohistan, Karakoram Himalayas Northern Pakistan. Geological Bulletin 11. University of Peshawar, Pakistan
Kumar A, Garg A, Kriplani S, Sehrawat P (2008) Utilization of geothermal energy resources for power generation in India: a review. In: Proceedings to 7th international conference & exposition on petroleum geophysics
Lacirignola M, Blanc I (2013) Environmental analysis of practical design options for enhanced geothermal systems (EGS) through life-cycle assessment. Renew Energy 50:901–914
Lenie Lectura (2018) Maibarara geothermal puts power plant online. BusinessMirror 12 March 2018. https://businessmirror.com.ph/maibarara-geothermal-puts-power-plant-online; Rivera D, Maibarara Geothermal adds 12 MW power to Luzon. The Philippine Star, 14 March 2018. https://www.doe.gov.ph/energist/maibarara-geothermal-adds-12-mw-power-luzon
Lenzen M (2008) Life cycle energy and greenhouse gas emissions of nuclear energy: a review. Energy Convers Manag 49:2178–2199
Lindal B (1973) Industrial and other applications of geothermal energy. Geothermal Energy. Earth Science UNESCO, Paris, 12:135–148
Lund J, Boyd TL (2015) Direct utilization of geothermal energy 2015 worldwide review. In: Proceedings World Geothermal Congress 2015 Melbourne, Australia, 19–25 April
Masson-Delmotte V, Zhai P, Pörtner H-O, Roberts D, Skea J, Shukla PR, Pirani A, Moufouma-Okia W, Péan C, Pidcock R, Connors S, Matthews JBR, Chen Y, Zhou X, Gomis MI, Lonnoy E, Maycock T, Tignor M, Waterfield T (2018) Global warming of 1.5 °C. An IPCC special report on the impacts of global warming of 1.5°C above pre-industrial levels and related global greenhouse gas emission pathways, in the context of strenghtening the global response to the threat of climate change. IPCC
Matter JM, Stute M, Snaebjörnsdóttir SÓ, Oelkers EH, Gíslason SR, Aradóttir E, Sigfússon B, Gunnarsson I, Sigurdardóttir H, Gunnlaugsson E, Axelsson G, Alfredsson HA, Wolff-Boenisch D, Mesfin K, de la Reguera Taya DF, Hall J, Dideriksen K, Broecker WS (2016) Rapid carbon mineralization for permanent disposal of anthropogenic carbon dioxide emissions. Science 352:1312–1314
Moomaw W, Burgherr P, Heath G, Lenzen M, Nyboer J, Verbruggen A (2011) Annex II: Methodology. In: IPCC special report on renewable energy sources and climate change mitigation. Cambridge University Press, Cambridge/New York
National Power Company of Iceland (Landsvirkjun) (2018) eistareykjastö. komin í fullan rekstur. 18 April 2018. https://www.641.is/theistareykjastod-komin-i-fullan-rekstur/; Landsvirkjun, "Landsvirkjun's 17th power station begins operations at .eistareykir", 21 November 2017. https://www.landsvirkjun.com/company/mediacentre/news/news-read/landsvirkjuns-17thpower-station-begins-operations-at-theistareykir
OECD/IEA (2019) Monthly electricity statistics. Paris, January. https://www.iea.org/media/statistics/surveys/electricity/mes.pdf
Ogola PFA, Davídsdóttir B, Fridleifsson IB (2012a) Potential contribution of geothermal energy in climate change adaptation. A case study of arid and semi-arid eastern Baringo lowlands, Kenya. Renew Sustain Energy Rev 16:4222–4246
Ogola PFA, Davídsdóttir B, Fridleifsson IB (2012b) Opportunities for adaptation–mitigation synergies in geothermal energy utilization: initial conceptual frameworks. Mitigat Adapt Strat Global Change 17:507–536
Ormat (2018a) Sarulla geothermal power plant expands to 330 MW with third and final unit commencing commercial operation. Press release, Reno, NV: 8 May 2018. https://investor.ormat.com/file/Index?KeyFile=393388064
Ormat (2018b) 48 MW McGinnis Hills Phase 3 geothermal power plant in Nevada begins commercial operation, increasing complex capacity to 138 MW. Press release, Reno, NV, 20 December, 2018. https://investor.ormat.com/file/Index?KeyFile=396173220

PABUM News (2018) Penundaan pengeboran sebabkan investasi panasbumi tak penuhi target, 31 October 2018. https://www.panasbuminews.com/berita/penundaan-pengeboran-sebabkaninvestasi-panasbumi-tak-penuhi-target/

Padey P, Girard R, le Boulch D, Banc I (2013) From LCAs to simplified models: a generic methodology applied to wind power electricity. Environ Sci Technol 47:1231–1238

Parisi ML, Ferrara N, Torsello L, Basosi R (2019) Life cycle assessment of atmospheric emission profiles of the Italian geothermal power plants. J Clean Prod 234:881–894

Pehnt M (2006) Dynamic life cycle assessment (LCA) of renewable energy technologies. Renew Energy 31:55–71

Pertamina (2018) PLTP Karaha Unit 1 Terangi 33 Ribu Rumah di Tasik dan Sekitarnya, 7 May 2018. http://www.pge.pertamina.com/News/Detail/1016

Peter Omenda (2018) Update on the status of geothermal development in Africa. Presentation at Geothermal Resources Council Annual Meeting & Expo, Reno, NV, 14–17 October 2018

Power Supply and Demand Highlights (2017) Installed capacity of 1,916 MW as of 2017, from Philippines Department of Energy, Electrical Power Industry Management Bureau, Taguig City, The Philippines, 2018. https://www.doe.gov.ph/electricpower/2017-power-supply-and-demand-highlights-januarydecember-2017

Pratiwi A, Ravier G, Genter A (2018) Life-cycle climate-change impact assessment of enhanced geothermal system plants in the Upper Rhine Valley. Geothermics 75:26–39

Ragnarsson A (2015) Geothermal development in Iceland 2010–2014. In: Proceedings World Geothermal Congress 2015, Melbourne, Australia, April 19–24, 2015

REN21 (2017) Renewables global status report, Paris: 2014–2017 editions. http://www.ren21.net/status-of-renewables/global-status-report/

REN21 (2019) Renewables 2019 global status report. REN21 Secretariat, Paris. ISBN 978-3-9818911-7-1

Richter (2016) India sets ambitious target for geothermal development by 2030. In Think Geoenergy, 9 June. http://www.thinkgeoenergy.com/india-sets-ambitious-target-for-geothermal-development-by-2030/

Richter A (2018) Polaris infrastructure reports record annual geothermal power generation and revenue. Thinkgeoenergy. www.thinkgeoenrgy.com

Rogen B, Ditlefsen C, Vangkilde P, Nielsen LH, Mahler A (2015) Geothermal energy use, 2015 country update for Denmark. In: Proceedings of World Geothermal Congress 2015, Melbourne, Australia, 19–25 April 2015

Rybach L (2010a) Status and prospects of geothermal energy, In: Proceedings of the World Geothermal Congress 2010. Bali 25–29:1–5

Rybach L 2010b) CO_2 emission mitigation by geothermal development – especially with geothermal heat pumps, In: Proceedings of the World Geothermal Congress 2010, Bali, 25–29 Apr 2010, pp 1–4

Sahajpal S, Sircar A, Singh A, Vaidya D, Shah M, Dhale S (2015) Geothermal exploration in Gujarat: case study from Unai. Int J Latest Technol Eng Manag Appl Sci 4:38–47

Schechinger B, Kissling E (2015) WP1: resources. In: Energy from the Earth. Deep geothermal as a resource for the future?

Shah RR, Dutt B (2014) Geothermal energy: an alternative source of energy. Int J Eng Res Appl 4:63–68

Sircar A, Shah M, Sahajpal S, Vaidya D, Dhale S, Chaudhary A (2015) Geothermal exploration in Gujarat: case study from Dholra. Springer Open Access 3:1–25

Sullivan JL, Clark CE, Han J, Wang M (2010) Life-cycle analysis results of geothermal systems in comparison to other power systems. Argonne Nat Lab. https://doi.org/10.2172/993694

Sullivan JL, Clark CE, Yuan L, Han J, Wang M (2011) Life-cycle analysis for geothermal systems in comparison to other power systems: Part II. ANL/ESD/11-12

Tevfik Kaya (2017) An overview on geothermal drilling and projects in Turkey, 2017. Presentation at Geothermal Resources Council Annual Meeting & Expo, Reno, NV, 14–17 October 2018. https://geothermal.org/Annual_Meeting/PDFs/An%20Overview%20on%20Geothermal_Drilling_and_Projects_in_Turkey-2018.pdf

Treyer K, Oshikawa H, Bauer C, Miotti M (2015) WP4: environment. In: Hirschberg S, Wiemer S, Burtgherr P (eds) Energy from the Earth. Deep geothermal as a resource for the future?

Turboden (2018a) Cyrq energy. https://www.turboden.com/casehistories/2127/cyrq-energy, viewed March 2018.

Turboden (2018b) Turboden completed the commissioning of the 17.5 MWe Velika Ciglena Geothermal Plant. Press release, Brescia, Italy: 11 December 2018. https://www.turboden.com/upload/blocchi/X12219allegato1-2X_5269_Turboden_Velika_Ciglena_ENG.pdf; first geothermal plant based on IRENA, Renewable Energy Statistics 2018, Abu Dhabi, March 2018. http://irena.org/publications/2018/Mar/Renewable-Capacity-Statistics-2018

UNFCCC (2015) Adoption of the Paris Agreement. FCCC/CP/2015/L.9/Rev.1. https://doi.org/FCCC/CP/2015/L.9/Rev.1

US National Renewable Energy Laboratory (2019) Geothermal electricity production basics. https://www.nrel.gov/research/re-geo-elec-production.html, viewed April 2019; Geothermal Energy Association, 2016 annual U.S. & global geothermal power production report, Washington, DC, March 2016. http://geo-energy.org/reports/2016/2016%20Annual%20US%20Global%20Geothermal%20Power%20Production.pdf

Varmaorka (2018) Varmaorka: opportunities in low temperature geothermal heat. http://varmaorka.is/en/, viewed March 2018; Climeon, Technical product sheet. https://climeon.com/wp-content/uploads/2017/04/Climeon-Tech-Product-Sheet.pdf, viewed March 2019

Warner E, O'Donoughue P, Heath G (2010) Harmonization of energy generation life cycle assessments (LCA). In: FY2010 LCA milestone report. National Renewable Energy Laboratory, Golden

Wood J, Guth A (2019) East Africa's Great Rift Valley: a complex rift system. https://geology.com/articles/east-africarift.shtml, viewed March 2019

World Bank (2018) Concept project information document-integrated safeguards data sheet – Indonesia Geothermal Resource Risk Mitigation Project (GREM) – P166071, 8 November 2018. http://documents.worldbank.org/curated/en/950171541685249374/Concept-Project Information-Document-Integrated-Safeguards-Data-Sheet-Indonesia-Geothermal Resource-Risk-Mitigation-Project-GREM-P166071

Yadav K, Sircar A (2019) Application of low enthalpy geothermal fluid for space heating and cooling, honey processing and milk pasteurization. Case Stud Therm Eng. 14. (pp. 1–13). Elsevier

Zaigham N, Nayyar ZA, Hisamuddin N (2009) Review of geothermal energy resources in Pakistan. Renew Sustain Energy Rev 13(1):223–232

Zheng K, Dong Y, Chen Z, Tian T Wang G (2015) Speeding up industrialized development of geothermal resources in China–country update report 2010–2014. World Geothermal Congress 2015

Zorlu Energy Group (2018) Zorlu Energy has commissioned the first phase of Kızıldere III geothermal power plant, press release, Istanbul: 5 September 2017. http://www.energyofturkey.com/turkey-energy-news/zorlu-energy-has-commissionedthe-first-phase-of-kizildere-iii-geothermal-power-plant/; Zorlu completes Turkey's largest geothermal plant. Newsbase, 22 March 2018. https://newsbase.com/topstories/zorlu-completes-turkey%E2%80%99s-largest-geothermal-plant

Chapter 12
Trends and Pattern of Rainfall over Semiarid Sahibi Basin in Rajasthan, India

Manpreet Chahal, Pankaj Bhardwaj, and Omvir Singh

Abstract Study of the spatial and temporal trends of rainfall is very important for water resources planning and management. Therefore, this study has been attempted to examine the spatial and temporal trends and pattern of rainfall over the semiarid Sahibi Basin in Rajasthan State of India for a period of 57 years (1961–2017). Rainfall data for nine stations located over the basin have been obtained from the Department of Water Resources, Rajasthan. Statistical methods such as Mann–Kendall test, Sen's slope estimator, and linear regression were employed to detect the rainfall trends. Percent change for rainfall, and years of excess and deficient rainfall, have also been identified. The findings have shown high interannual variability in mean annual rainfall with an average of 633.95 mm (SD = 204.93, CV = 32.33). However, no significant rising or declining trend has been observed in the mean annual rainfall over the basin during the 57-year period, whereas Kotkasim and Tapukara stations have shown significant increasing and decreasing trends, respectively. Further, pre-monsoon season rainfall has shown a significant increasing trend. Spatial analysis has shown an increase in rainfall amount from northeastern and western parts to the southeast. The observed spatial and temporal differences in the amount of rainfall are high. These results may be valuable for efficient and sustainable planning and management of water resources for agricultural purposes in the basin.

Keywords Rainfall · Trend · Pattern · Monsoon · Sahibi Basin · India

M. Chahal · O. Singh (✉)
Department of Geography, Kurukshetra University, Kurukshetra, Haryana, India

P. Bhardwaj
Department of Geography, Government College, Bahu, Jhajjar, India

M. N. Islam, A. van Amstel (eds.), *India: Climate Change Impacts, Mitigation and Adaptation in Developing Countries*, Springer Climate,
https://doi.org/10.1007/978-3-030-67865-4_12

Introduction

Rainfall is the most important climatic parameter affecting the temporal and spatial distribution of available water. Knowledge of the rainfall received over a region assists in estimating water availability for domestic, industrial, and agricultural requirements (Kumar and Jain 2010). In recent decades, the changing climate has significantly disrupted rainfall trends and patterns (IPCC 2014; Loo et al. 2015). The changing rainfall patterns may cause unpredictable hydrological extremes such as floods and droughts (Srivastava et al. 2015). The amount of rainfall controls aquifer recharge and the availability of moisture in the soil for crops (Gupta et al. 2014). Any change in the amount of rainfall may adversely affect agricultural production, food security, and overall growth of the economy of a country such as India (Gadgil and Gadgil 2006; Kumar and Jain 2011; Biswas et al. 2019). The rainfall over a particular region is mainly controlled by both local and large-scale environmental processes and therefore reveals noticeable spatial and temporal variability. The fluctuations in rainfall pattern are not uniform over a region; rather, each has its own distinctive local pattern. Knowledge of the spatiotemporal distribution of rainfall can be beneficial in various hydrological and water resources management applications. Thus, it is vital to examine the trends and pattern of rainfall to evaluate water resources availability in a region.

River basins have been recognized as the principal natural hydrological unit of analysis for planning and development of water resources. Therefore, researchers have extensively studied the rainfall trends and pattern over different river basins the world over (Kliment and Matoušková 2008; Muller et al. 2009; Kamruzzaman et al. 2013; Babatolu et al. 2014; Langat et al. 2017; Ninu Krishnan et al. 2018; Emmanuel et al. 2019; Khaniya et al. 2019; Amoussou et al. 2020). Many reports are also available on the trends and pattern of rainfall over the different river basins of India. For example, Singh et al. (2008) have noticed a rising trend over 8 of 9 river basins in northwest and central India in the past century. Ranade et al. (2008) have not observed any trend in the beginning or termination dates, period, and total amount of rainfall during the wet season over several river basins of India, whereas Kumar and Jain (2011) have witnessed a decline in annual rainfall amounts and number of rainy days in 15 of 22 basins in India. Raj and Azeez (2012) observed a substantial decrease in annual, monsoon, and pre-monsoon season rainfall over the Bharathapuzha River Basin. Deka et al. (2013) noticed a substantial declining trend in monsoon and post-monsoon season rainfall over Barak Basin in the past century. Abeysingha et al. (2014) observed a substantial rising trend in rainfall over upstream areas, whereas a decreasing trend was recorded over downstream areas of Gomti Basin. Taxak et al. (2014) have shown a decline in annual and monsoonal rainfall over Wainganga Basin in the latter half of the twentieth century. Deshpande et al. (2016) noticed declining trends in monsoonal rainfall over Ganga, Narmada, Tapi, Godavari, and other west coast rivers, whereas Krishna and peninsular rivers have shown increasing trends. Gajbhiye et al. (2016) have shown a substantial rise in annual and monsoonal rainfall over Sindh River Basin. Bera (2017) observed diminishing trends in annual, pre-monsoon, and post-monsoon season rainfall over most of the districts in the Kosi, Gandak, and Sone Basins. Sridhar and Raviraj (2017) noticed a rising trend in northeast monsoon rainfall over the Amaravathi River Basin. Jain et al. (2017) found an increasing trend in rainy days over the Baitarani, Brahamani, and Cauvery River Basins but declining trends over the Krishna, Godavari, Mahanadi, Narmada, and Sabarmati Basins. Bisht (2018) found a decreasing trend in seasonal and annual rainfall, although an increasing trend in extreme rainfall events, over 62 of 85 river basins in India after 1970. Biswas et al. (2019) noticed a decline in rainfall amounts over upper Godavari Basin. Singh and Singh (2020) witnessed a decreasing trend in rainfall over the Suketi River Basin. Tirkey et al. (2020) have witnessed both rising and falling trends in monthly and seasonal rainfall over different subwatersheds of the Sutlej River Basin.

Further, in India, Rajasthan State is characterized by the highest likelihood of rainfall deficiencies and drought occurrences (Rathore 2005). The distribution of rainfall in Rajasthan is very uneven and differs considerably from region to region and year to year (Pingale et al. 2014). The state is highly dependent upon rainfall as it is among the important sources of groundwater recharge (Machiwal et al. 2017). Any change in rainfall pattern may affect the water resources and subsequently the agriculture of the state (Yadav et al. 2018). In this arid and semiarid state, the importance of understanding rainfall variability is high when its quantity is less and it is rarer in occurrence, making it challenging to effectively plan and manage the existing water resources (Meena et al. 2019). Therefore, several researchers have investigated rainfall trends and patterns for Rajasthan State (Rao et al. 2011, 2014; Singh et al. 2012; Pingale et al. 2014; Poonia et al. 2014; Mundetia and Sharma 2015; Machiwal and Jha 2017; Srivastava and Pradhan 2018; Meena et al. 2019). These studies have focused on either the entire state or different districts to examine rainfall trends. However, no study has focused on the trends and patterns of rainfall over a particular river basin.

Sahibi River is known for destructive floods in parts of Rajasthan State and the adjoining Haryana State during monsoon months. Such floods often inundate wide areas, causing huge damage to property and agriculture, and bringing misery to the community such as epidemics and loss of life (Parmar 2014). Thus, detailed information related to long-term trends and pattern of rainfall over the Sahibi Basin can be very helpful in the development and management of water resources. Therefore, the present study has been undertaken to examine the long-term trends and patterns of rainfall over the semiarid Sahibi Basin. It is believed that the findings of this study will provide understanding regarding rainfall trends and pattern, which will help policy makers to develop efficient plans to mitigate the risks of droughts and floods by proper management of water resources.

Study Area

The Sahibi River (also known as Sahbi and Sabi) is a tributary of the Yamuna River, located in the northeastern part of Rajasthan State in India. The basin spreads over an area of approximately 4524 km^2 and extends between 27°20′ 50″ to 28°14′ 01″N

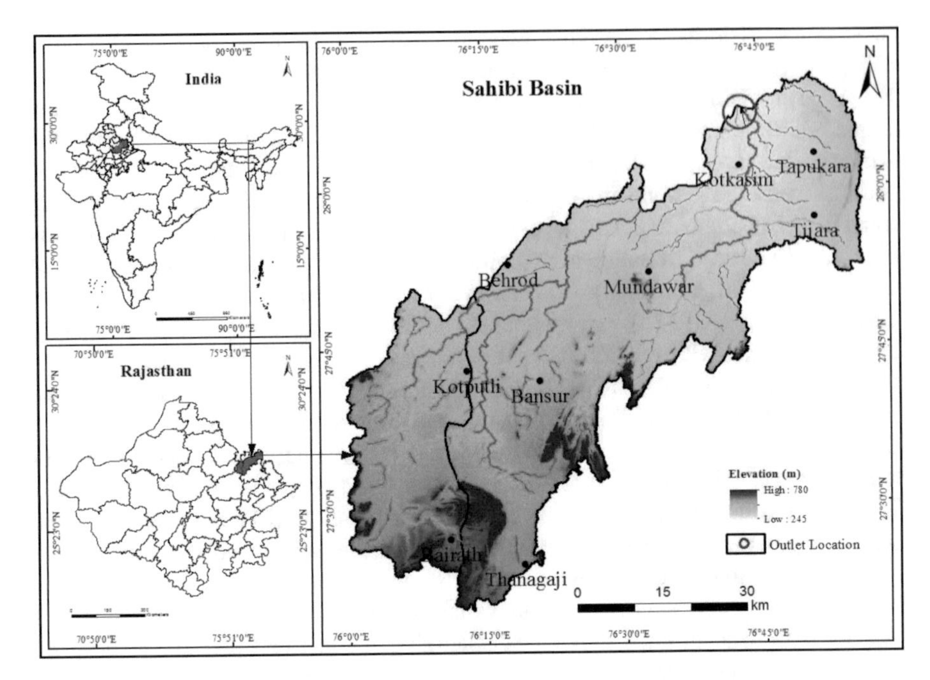

Fig. 12.1 Location of selected nine stations over Sahibi Basin in Rajasthan State during 1961–2017

latitudes and 75° 59′ 07″ to 76° 58′ 50″E longitudes (Fig. 12.1). It is an ephemeral and rain-fed river having an asymmetrical drainage system. The elevation of the basin varies from 189 to 430 m and the general direction of slope is from southwest to northeast. About 80% of the area of the basin is composed of alluvial deposits, windblown sands, and sand deposits. Agriculture and livestock are the leading sources of livelihood in the rural areas. The basin has two main cropping seasons: *kharif* (June–October) and *rabi* (November–March). The *kharif* season crops are highly dependent on rainfall occurrence whereas the *rabi* season crops depend on groundwater. The climate of the basin is semiarid. In summers (April–June), the temperature reaches its peak, ranging from 32° to 46 °C, whereas it ranges between 16.9° and 19.1 °C in winters (November–January). The mean annual rainfall in the basin is approximately 634 mm. The monsoon season (June–September) contributes about 87% to the total annual rainfall.

Materials and Methods

Data Collection and Database Preparation

In the present study, daily rainfall data of nine stations well spread over the Sahibi Basin, for a 57-year period (1961–2017), were used (Fig. 12.1). The data were

Table 12.1 Details of selected nine stations over Sahibi Basin in Rajasthan

Station	District	Latitude	Longitude	Altitude (m)
Bairath	Jaipur	27°27′ N	76°11′ E	430
Kotputli	Jaipur	27°43′ N	76°13′ E	360
Mundawar	Alwar	27°52′ N	76°33′ E	310
Tijara	Alwar	27°57′ N	76°51′ E	293
Kotkasim	Alwar	28°02′ N	76°43′ E	259
Thanagaji	Alwar	27°24′ N	76°19′ E	417
Bansur	Alwar	27°42′ N	76°21′ E	357
Behrod	Alwar	27°53′ N	76°17′ E	323
Tapukara	Alwar	28°03′ N	76°51′ E	275

acquired from the online portal of the Department of Water Resources, Government of Rajasthan, Rajasthan, India (www.waterresources.rajasthan.gov.in). The geographic position (latitude and longitude) and elevation details of the selected nine stations are provided in Table 12.1. The main problem with the rainfall analysis is missing records. The missing rainfall observations were filled in before executing analysis. The normal ratio method has been executed to estimate the missing values by choosing the rainfall amounts of three nearest stations. A distance matrix has been structured for all stations based on their geographic positions to evaluate the closeness of stations.

Further, a data quality check is essential before the investigation, as outliers can greatly affect rainfall trends and patterns (You et al. 2008; Shahid et al. 2016). To ensure homogeneity, the subjective double mass curve method and objective Student's *t* test have been performed for rainfall time-series. The double mass curves of all stations did not show any breakpoint and displayed a more or less straight line, which indicates homogeneity in the data. Similarly, the Student's *t* test did not revealed any breakpoint or statistically significant difference in the rainfall time-series at the 95% confidence level. Apart from this, the daily rainfall data have been summed to obtain monthly, seasonal, and annual values. Seasonal analysis has been completed for four seasons: winter (January–February), pre-monsoon (March–May), monsoon (June–September), and post-monsoon (October–December), as per categorization of the India Meteorological Department.

Methods

After preparation of the dataset, suitable statistical analyses such as sum, frequencies, percentage, mean, standard deviation (SD), coefficient of variation (CV), skewness, and kurtosis have been performed for each station for the period 1961–2017. Then, the excess and deficit rainfall years and heavy rainy days were identified as follows.

Identification of Excess and Deficit Rainfall Years

In this study, a year has been identified as an excess (deficit) rainfall year if rainfall is more (less) than 1 SD from the mean (Pant and Rupa Kumar 1997). Statistically, an excess rainfall year can be expressed as

$$R_i \geq R_m + S_d \tag{12.1}$$

and a deficit year can be expressed as

$$R_i \leq R_m - S_d \tag{12.2}$$

where R_i = rainfall amount in a year i, R_m = mean rainfall, and S_d = SD of rainfall.

Calculation for Percent Change

The percent change has been computed by approximating it with a linear trend, so it is equivalent to the median slope multiplied by the length of the period (57 years). Then, it is divided by the corresponding mean value, given in percentage (Yue and Hashino 2003).

$$\text{Percentage change}(\%) = \frac{\beta \times \text{length of year}}{\text{mean}} \times 100 \tag{12.3}$$

Trend Detection Methods

The nonparametric Mann–Kendall (MK) test (Mann 1945; Kendall 1948) has been employed to detect the trends in rainfall. This MK test has been found to be an excellent tool to examine the possible presence of significant trends in the time-series at various levels of significance (Mayowa et al. 2015; Singh et al. 2020). The standard normal variable Z has been used to detect the trend and its significance level. The positive (negative) values of Z show rising (declining) trends in the time-series. In this study, a trend is considered statistically significant positive or negative at the 95% confidence level. The nonparametric Sen's slope estimator (Sen 1968) has been used to detect the magnitude of trend; it is closely associated with the MK test (Gilbert 1987) and provides a robust estimation of trend (Yue et al. 2002). The parametric simple linear regression has also been used to identify the trend in time-series. These three methods have been used extensively in hydrometeorological studies, and detailed discussion about these methods is available (Deka et al.

2013; Pingale et al. 2014; Jaiswal et al. 2015; Mayowa et al. 2015; Singh et al. 2020). These tests were performed by means of XLSTAT 2017 software.

Spatial Interpolation

To study the spatial pattern of rainfall characteristics, the inverse distance weighting (IDW) interpolation technique has been employed. The IDW technique is a simple and widely used interpolation technique. This spatial interpolation technique considers the role of each input point by a standardized inverse of the distance from the control point to the interpolated point. This technique assumes that each input point has a local impact that lessens with increase in remoteness. It weighs more on points nearer to the processing points than those at a distance, meaning a rainfall or its derived quantity at any desired location is interpolated from the given data using weights that are based on the distance from each rainfall station and the desired location (Burrough and McDonnell 1998). This technique gives a smooth pattern of rainfall along with undesirable troughs and peaks. Interpolation of rainfall characteristics using the IDW technique used ArcGIS 10.2 software.

Results and Discussion

Temporal Variations in Rainfall

Annual Variations

The mean annual rainfall over Sahibi Basin during 1961–2017 is demonstrated in Fig. 12.2a. The mean annual rainfall has shown a very high interannual variability, varying from 217.33 mm in 2002 to 1214.96 mm in 1996, with a mean annual value of 633.95 mm (SD = 204.93, CV = 32.33) in the basin (Table 12.2). The MK test, Sen's slope, and regression analysis have not shown any significant increasing or decreasing trend in the mean annual rainfall over Sahibi Basin during the 57-year period. Likewise, the time-series of mean annual CV has shown an almost constant trend with high variability (Fig. 12.2b). During the 57-year period, a total of 8 rainfall excess years and 7 rainfall deficit years have been observed over the Sahibi Basin (Table 12.3).

The station-wise analysis has shown that the highest mean annual rainfall occurred over Thanagaji station (708 mm/year; SD = 239.48; CV = 33.83), whereas the lowest was over Tapukara station (575 mm/year; SD = 291.94; CV = 50.78) (Table 12.4). The trend lines have shown an increasing trend over Bairath and Kotkasim stations, whereas a decreasing trend occurred over Tijara and Tapukara stations (Fig. 12.3). However, Kotkasim and Tapukara stations have shown significant increasing and decreasing trends, respectively (Table 12.5). Kotkasim station

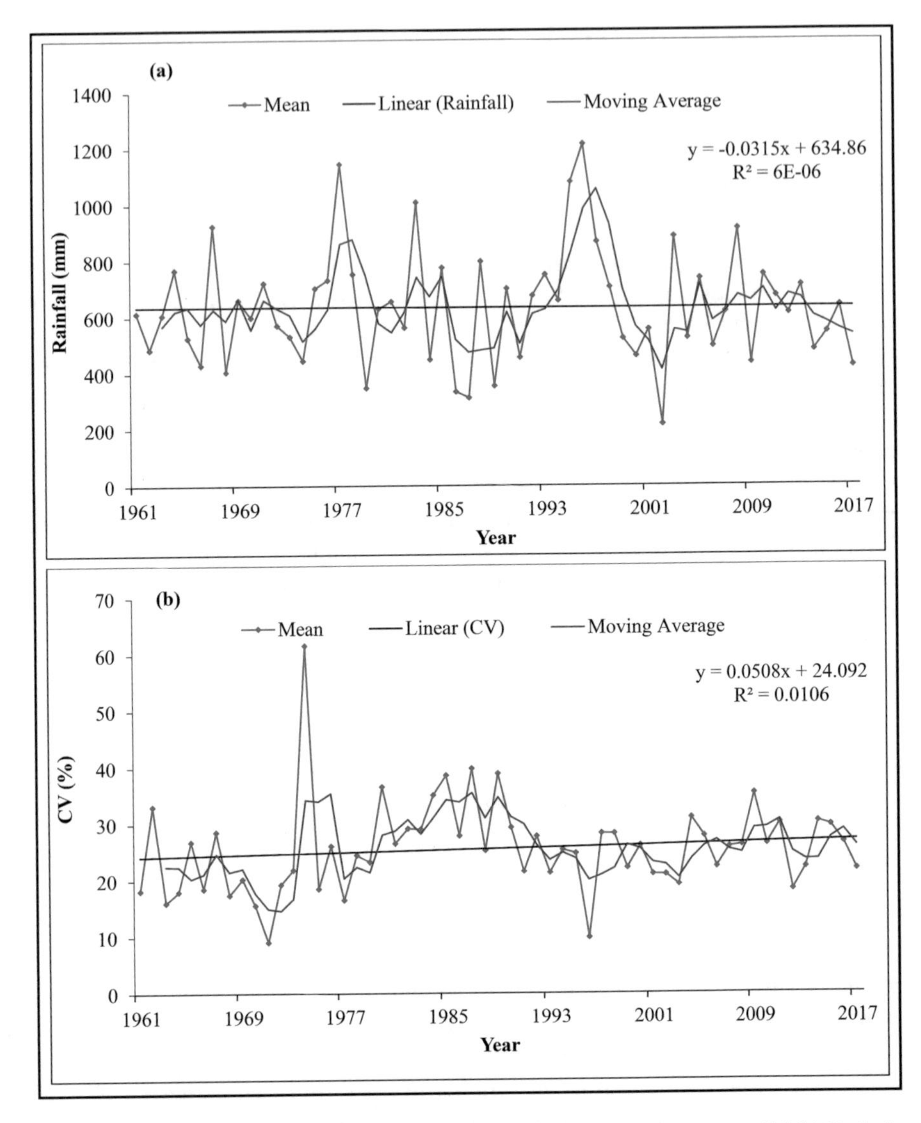

Fig. 12.2 Annual variation in (**a**) rainfall and (**b**) coefficient of variation over Sahibi Basin in Rajasthan during 1961–2017

has shown a positive change of 53.65%, while Tapukara station has shown a negative change of 45.38%. The rest of the stations have shown an almost constant trend during the 57-year study period. Remarkably, SD and CV have shown a high interannual variability over all the stations.

Table 12.2 Mean monthly, seasonal, and annual rainfall statistics over Sahibi Basin in Rajasthan (1961–2017)

Time	Rainfall (mm)	Standard deviation	Coefficient of variation (%)	Kurtosis	Skewness	Contribution to annual rainfall (%)	MK	Sen's Slope	Regression
January	8.99	10.96	121.90	2.68	1.61	1.42	0.98	0.03	0.39
February	13.48	20.08	148.98	7.20	2.52	2.13	0.28	0.00	0.23
March	8.30	16.78	202.23	23.52	4.35	1.31	1.63	0.04	0.03
April	7.29	10.86	148.99	4.17	2.07	1.15	1.21	0.01	0.31
May	22.19	24.31	109.54	6.27	2.28	3.50	1.49	0.19	0.21
June	64.46	66.06	102.47	14.13	3.21	10.17	2.88**	0.82**	0.04
July	198.08	109.12	55.09	1.35	0.92	31.25	−1.09	−0.89	0.21
August	199.96	104.00	52.01	2.82	1.14	31.54	−1.30	−1.06	0.21
September	91.53	65.93	72.04	−0.88	0.35	14.44	0.65	0.35	0.63
October	11.45	22.01	192.32	10.46	3.10	1.81	0.25	0.00	0.99
November	3.71	6.85	184.60	2.68	1.90	0.59	–	–	0.46
December	4.52	9.12	201.75	8.25	2.77	0.71	–	–	0.30
Winter (January–February)	22.47	23.55	104.84	2.87	1.65	3.54	1.27	0.14	0.15
Pre-monsoon (March–May)	37.78	36.02	95.36	4.00	2.00	5.96	2.58**	0.43**	0.03
Monsoon (June–September)	554.03	197.23	35.60	0.94	0.62	87.39	−0.56	−0.66	0.61
Post-monsoon (October–December)	19.68	28.78	146.28	12.16	3.15	3.10	−1.04	−0.09	0.62
Annual	633.95	204.93	32.33	0.66	0.66	100	0.00	0.00	0.98

**Significant at 99% confidence level

Table 12.3 Frequencies of excess and deficient rainfall years over Sahibi Basin in Rajasthan (1961–2017)

Decade	Annual		Winter		Pre-monsoon		Monsoon		Post-monsoon	
	Excess	Deficient	Excess	Deficient	Excess	Deficient	Excess	Deficient	Excess	Deficient
1961–1970	1	1	1	0	0	1	1	1	1	0
1971–1980	1	1	1	0	0	1	1	1	1	0
1981–1990	1	3	2	0	2	2	1	3	1	0
1991–2000	3	0	1	0	0	0	2	0	1	0
2001–2010	2	1	2	0	2	0	2	1	1	0
2011–2017	0	1	1	0	1	0	0	1	0	0

Table 12.4 Station-wise basic characteristics of rainfall over Sahibi Basin in Rajasthan (1961–2017)

Station	Annual rainfall (mm)	Max. annual rainfall (mm)	Min. annual rainfall (mm)	Standard deviation	Coefficient of variation (%)	Kurtosis	Skewness
Bairath	638	1507 (1967)	262 (2002)	236.67	37.11	1.77	1.00
Kotputli	596	1235 (1996)	190 (1989)	232.31	38.96	0.47	0.71
Mundawar	658	1726 (1995)	155 (1987)	300.39	45.63	1.92	1.07
Tijara	603	1139 (1977)	195 (1987)	219.95	36.47	−0.06	0.31
Kotkasim	661	1320 (2008)	197 (1962)	265.90	40.23	−0.20	0.43
Thanagaji	708	1272 (1996)	179 (2002)	239.48	33.83	0.01	0.26
Bansur	673	1490 (1983)	123 (2002)	268.88	39.96	1.40	0.92
Behrod	594	1238 (1996)	216 (1987)	238.36	40.15	0.78	0.90
Tapukara	575	1434 (1983)	188 (2014)	291.94	50.78	1.86	1.43

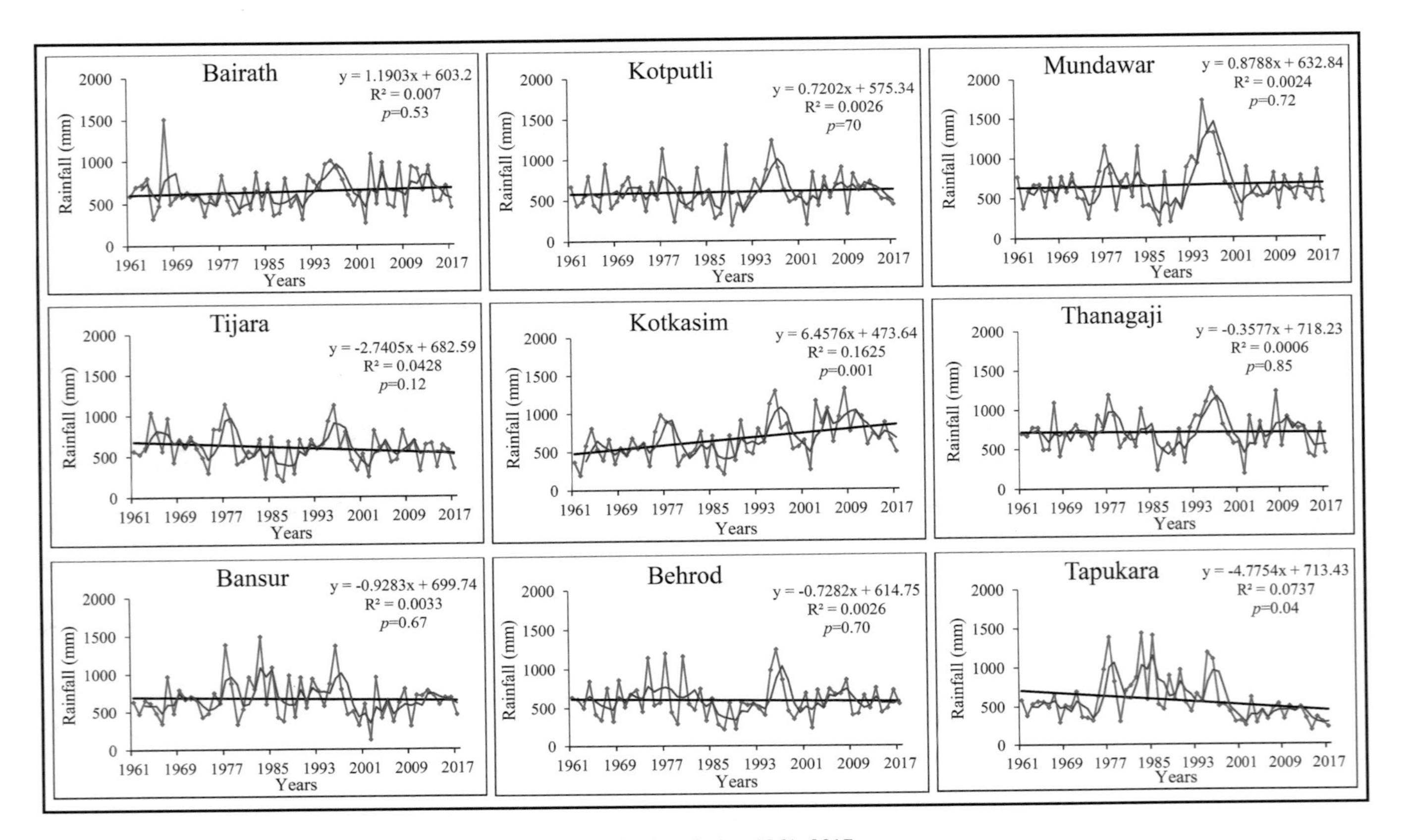

Fig. 12.3 Station-wise mean annual rainfall over Sahibi Basin in Rajasthan during 1961–2017

Table 12.5 Mann–Kendall test statistic (Z), Sen's slope estimator (β) test, and percent change for seasonal precipitation over Sahibi Basin in Rajasthan (1961–2017)

Station	Annual			Winter			Pre-monsoon			Monsoon			Post-monsoon		
	Z	β	% Change	Z	β	% Change	Z	β	% Change	Z	β	% Change	Z	β	% Change
Bairath	0.78	1.86	16.64	0.97	0.02	5.52	2.12*	0.50*	77.39	0.43	0.74	7.55	0.08	0.00	0.00
Kotputli	0.53	1.04	9.95	1.43	0.14	36.56	2.87**	0.65**	102.15	−0.23	−0.29	−3.15	−0.46	0.00	0.00
Mundawar	0.25	0.69	5.99	0.98	0.07	16.08	1.77	0.44	62.63	−0.21	−0.35	−3.44	−0.83	0.00	0.00
Tijara	−1.38	−2.55	−24.05	0.75	0.09	23.33	1.21	0.24	39.38	−1.60	−2.91	−31.44	−1.70	−0.10	−30.73
Kotkasim	2.92**	6.22**	53.65	3.05**	0.44**	96.97	3.21**	0.71**	98.61	2.12*	4.28*	42.48	0.13	0.00	0.00
Thanagaji	−0.07	−0.11	−0.86	0.05	0.00	0.00	0.85	0.17	23.67	−0.42	−0.84	−7.68	−0.87	0.00	0.00
Bansur	−0.33	−0.45	−3.82	1.30	0.06	16.19	2.11*	0.49*	69.05	−0.47	−1.19	−11.54	−0.89	0.00	0.00
Behrod	−0.20	−0.29	−2.82	0.83	0.08	19.41	2.88**	0.63**	89.51	−0.61	−1.19	−13.30	−1.05	0.00	0.00
Tapukara	−2.83**	−4.58**	−45.38	0.31	0.00	0.00	0.56	0.00	0.00	−2.84**	−4.49**	−50.62	−1.50	0.00	0.00

*Significant at 95% confidence level
**Significant at 99% confidence level

Seasonal Variations

The analysis has shown large seasonal variations in the occurrence of rainfall over Sahibi Basin during 1961–2017. The mean rainfall during monsoon, pre-monsoon, winter, and post-monsoon seasons is about 554 mm (SD = 197.23, CV = 35.60), 38 mm (SD = 36.02, CV = 95.36), 22 mm (SD = 23.55, CV = 104.84), and 20 mm (SD = 28.78, CV = 146.28), respectively (Table 12.2). This analysis indicates that the monsoon season alone contributes about 87% to the mean annual rainfall, whereas the contribution of pre-monsoon, winter, and post-monsoon season rainfall is about 6%, 4%, and 3%, respectively.

The season-wise mean annual variations are shown in Fig. 12.4a–d. The mean seasonal rainfall has shown a slightly rising trend during winter and pre-monsoon seasons while the monsoon and post-monsoon seasons have shown slightly decreasing trends. Among these trends, the pre-monsoon season trend was found to be statistically significant (Table 12.2). These findings are consistent with those of Kumar and Jain (2011) and Bisht et al. (2018). Further, the trend line of CV values has shown a trend of increasing rainfall during the winter season, whereas opposite trends have been witnessed in CV values for mean rainfall in other seasons.

The station-wise seasonal analysis has shown declining trends over all stations during the monsoon season except Bairath and Kotkasim stations, which have experienced increasing trends (Table 12.5). However, the increasing trend is statistically significant only for Kotkasim station. Similarly, the declining trend is statistically significant only for Tapukara station. During the pre-monsoon season, all the stations have shown a statistically significant positive trend in rainfall, except Mundawar, Tijara, Thanagaji, and Tapukara stations. During the winter season, only Kotkasim station has shown a statistically significant positive trend. During the post-monsoon season, no station has shown a statistically significant declining trend in the Sahibi Basin (Table 12.5).

Further, seven excess and seven deficient monsoon rainfall seasons have been observed during the 57 years (Table 12.3). The annual excess and deficient rainfall years are associated with the monsoon season excess and deficient rainfall years, respectively. During pre-monsoon seasons, 5 rainfall excess and 4 rainfall-deficient years have been observed. During winter and post-monsoon seasons, 8 and 5 years have been observed as rainfall excess years, respectively. Remarkably, no rainfall-deficient years have been witnessed in these two seasons (Table 12.3).

Monthly Variations

Monthly analysis has shown that the maximum amount of rainfall occurred in August, followed by July, September, and June (Table 12.2). The least rainfall occurred in the month of November, followed by the months of December and April. The months of August and July accounted for approximately 63% of total annual rainfall. However, these months have shown nonsignificant declining trends in amount of rainfall (Fig. 12.5). The month of June has shown a significant rising

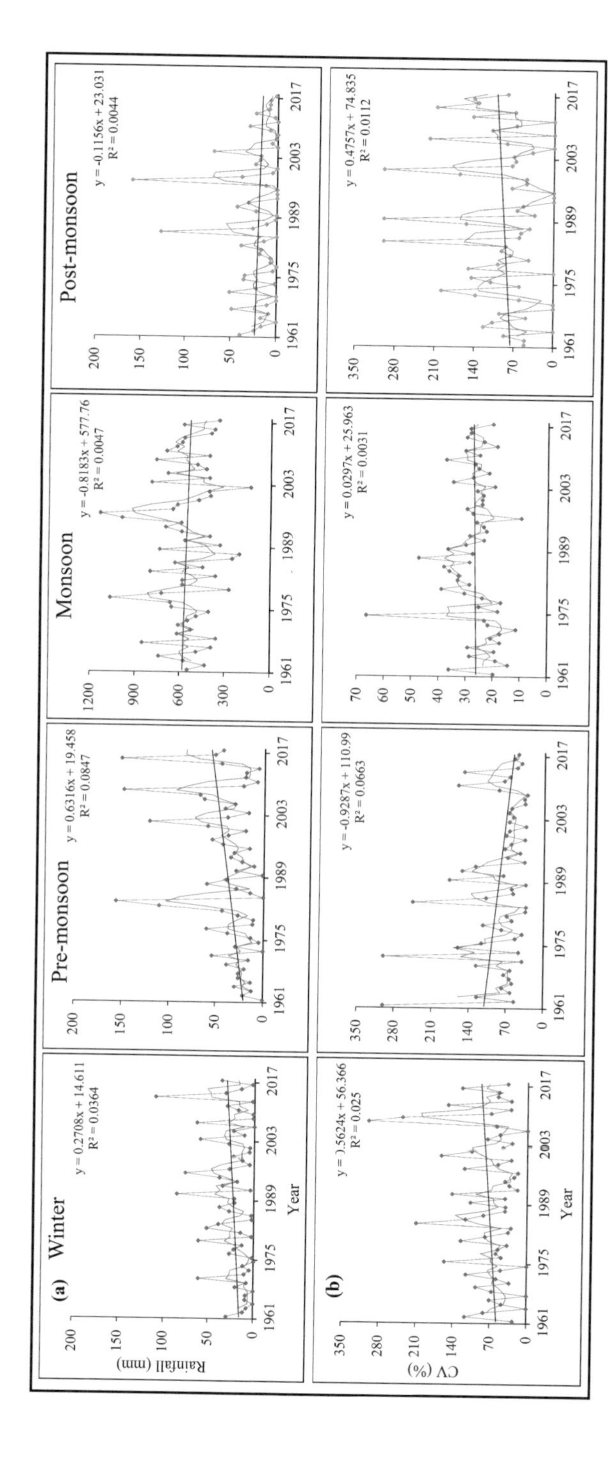

Fig. 12.4 Season-wise annual variation in (**a**) rainfall and (**b**) coefficient of variation over Sahibi Basin in Rajasthan during 1961–2017

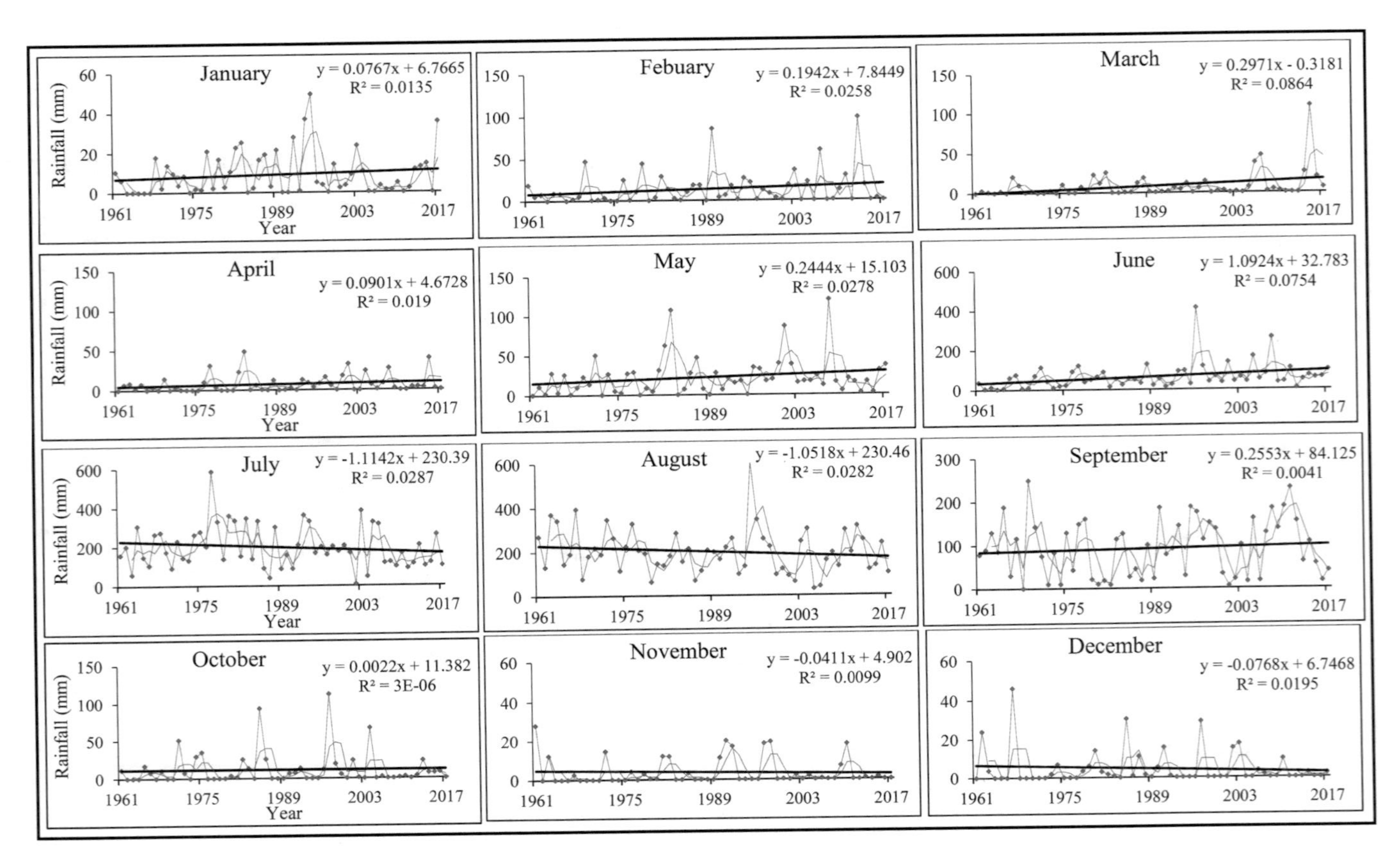

Fig. 12.5 Monthly variation in rainfall over Sahibi Basin in Rajasthan during 1961–2017

trend in the amount of rainfall over Sahibi Basin (Table 12.2). Similarly, most of the stations have shown significant rising trends in the amount of rainfall, except Thanagaji and Tapukara stations in June. In July, all the stations have shown decreasing trends except Kotkasim, whereas Thanagaji and Tapukara stations have shown significant declining trends. In August, Tapukara, Tijara, and Kotputli stations have shown significant decreasing trends. Similar results have been observed by Meena et al. (2019) for western Rajasthan. In May, Behrod, Kotkasim, and Kotputli stations have shown significant increasing trends. However, no significant rising or declining trend has been observed over different stations during other months.

Spatial Variations in Rainfall

Annual Variations

The spatial pattern of mean annual rainfall over Sahibi Basin is demonstrated in Fig. 12.6a. An increase in rainfall from northeastern and western parts to the southeast has been observed. This increase in rainfall southeastward may be attributed to the existence of the Aravalli Mountains. The northeastern and western parts have received less rainfall. The maximum mean amount of rainfall has occurred over Thanagaji station (about 708 mm) in the south, whereas the least occurred over Tapukara station (about 575 mm) in the north (Table 12.4). This difference in the magnitude of rainfall is high for such a small basin. Further, CV values have shown an opposite pattern to the mean annual rainfall (Fig. 12.6b). The rainfall variability is relatively less over the stations receiving higher rainfall and vice versa. High rainfall variability can be seen over the northern parts, whereas less variability has been witnessed over the southern parts. The minimum annual CV value has been observed over Thanagaji station (about 34%) in the south, whereas the maximum occurred over Tapukara station (about 51%) in the north (Table 12.4).

The pattern of percent change in amount of rainfall has not shown any unique pattern over Sahibi Basin (Fig. 12.6c). In the northeast, Tapukara and Tijara have shown a negative change in mean annual rainfall. Exceptionally, the Kotkasim station located near Tapukara and Tijara stations has shown a positive change following an increasing trend in rainfall (Fig. 12.6d). Bairath and Kotputli stations have shown positive change as these two stations have shown nonsignificant increasing trends. The remaining stations have not shown any significant change in the amount of rainfall over Sahibi Basin.

Seasonal Variations

The spatial pattern of mean seasonal rainfall is shown in Fig. 12.7. The mean amount of rainfall during the winter season is much less (between 20 and 26 mm) over all

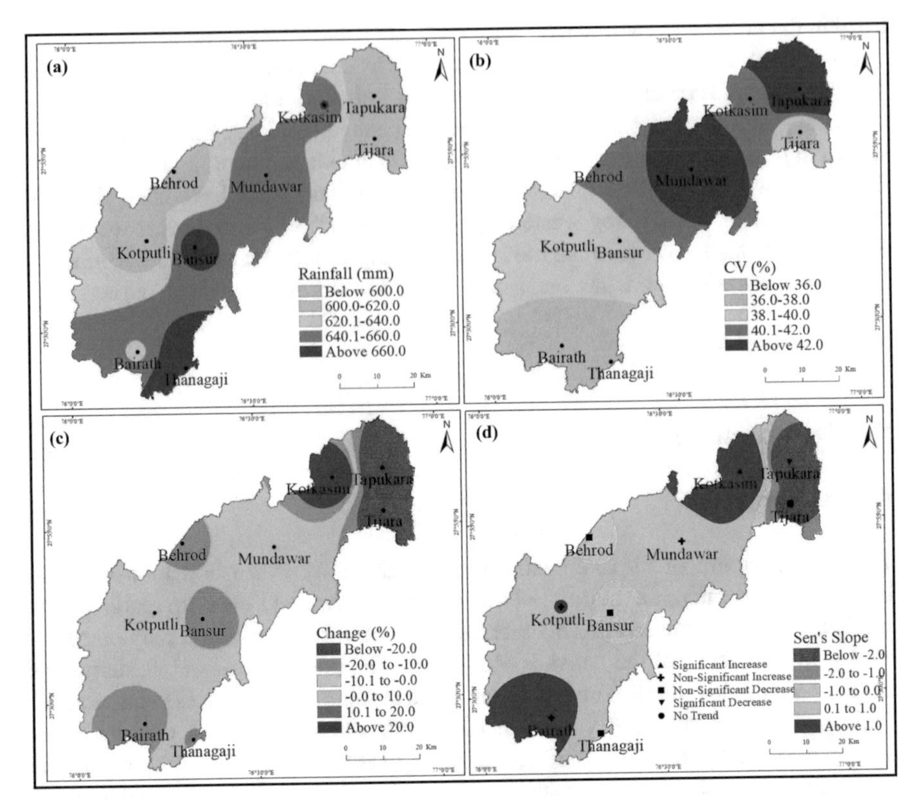

Fig. 12.6 Annual pattern of (**a**) rainfall, (**b**) coefficient of variation, (**c**) percent change, and (**d**) Mann–Kendall test and Sen's slope estimator over Sahibi Basin in Rajasthan during 1961–2017

parts of the Sahibi Basin (Fig. 12.7a; Table 12.6). Only the northern and northwestern parts observed slightly higher amounts of rainfall than other parts, which may be attributed to rainfall associated with western disturbances. During the pre-monsoon season, rainfall increased slightly (between 29 and 41 mm) over the entire basin, especially over the southeastern parts. After that, a sharp increase in rainfall (between 506 and 624 mm) was witnessed over the entire basin during monsoon season. During the post-monsoon season, the entire basin received the least rainfall (between 18 and 23 mm). The northeastern part received the least rainfall in all the seasons. Apart from this, the CV has shown an opposite pattern to the mean rainfall in each season (Fig. 12.7b). The lowest CV values have been witnessed in the monsoon season, followed by post-monsoon, winter, and pre-monsoon, over the entire Sahibi Basin (Table 12.6).

Further, the entire basin has witnessed a positive percent change in mean rainfall, as all the stations observed a rising trend during winter and pre-monsoon seasons over the Sahibi Basin (Fig. 12.7c–d). However, a negative change has been observed during the monsoon season over large parts of the basin. During the post-monsoon season, no significant positive or negative percent change in rainfall has been observed.

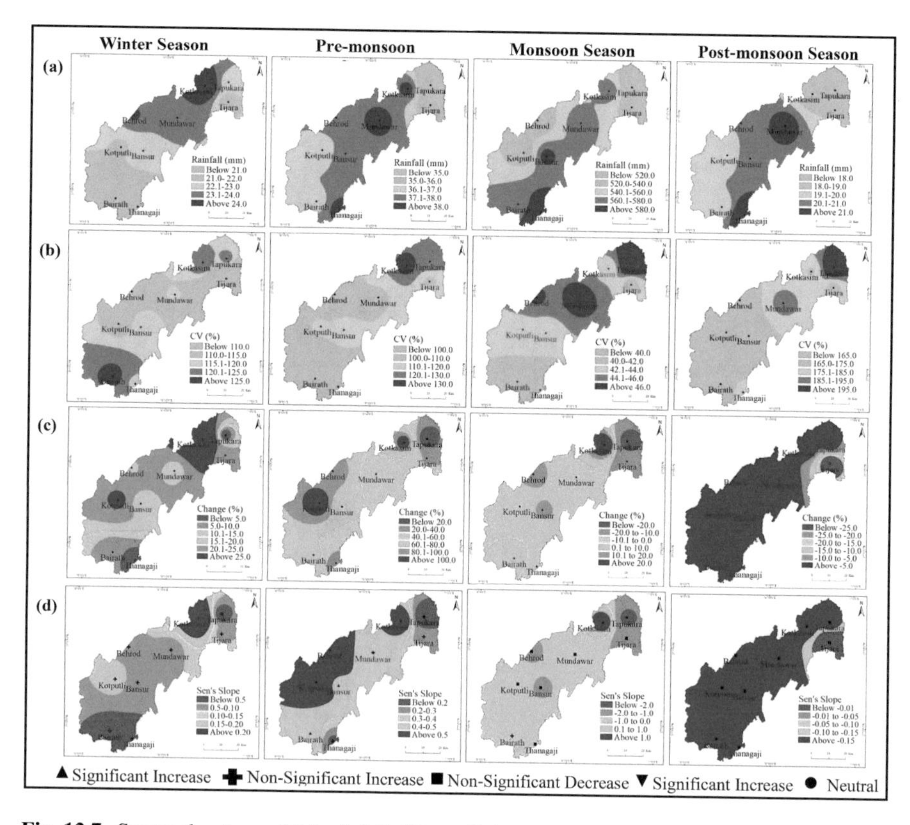

Fig. 12.7 Seasonal pattern of (**a**) rainfall, (**b**) coefficient of variation, (**c**) percent change, and (**d**) Mann–Kendall and Sen's slope estimator over Sahibi Basin in Rajasthan during 1961–2017

Monthly Variations

The mean monthly rainfall over the Sahibi Basin during 1957–2017 is shown in Fig. 12.8. This figure shows wide spatial variations in the amount of rainfall during different months over the Sahibi Basin. During January, February, and March, the amount of rainfall is much less over the entire basin. Only the northwestern parts received slightly higher rainfall, whereas the northeastern and southern parts received the lowest rainfall in these 3 months. In April, the amount of rainfall is relatively less, as in the earlier 3 months; however, the pattern of rainfall changed slightly and the southeastern parts have also witnessed an increase in rainfall. In the months of May and June, a sharp increase in rainfall was observed over the entire basin, especially over the southern areas. Then, maximum rainfall amounts have been witnessed over the entire basin during July and August because of the moisture-laden monsoonal winds. After the month of September, rainfall decreased rapidly, and the least amount of rainfall was received in November over the entire basin. On the other hand, the CV values have shown an opposite pattern to the mean monthly rainfall (Fig. 12.9). The northeastern parts of the basin experienced high variability

Table 12.6 Station-wise seasonal statistics of rainfall of selected nine stations over Sahibi Basin in Rajasthan (1961–2017)

Station	Mean				Standard deviation				Coefficient of variation (%)			
	Winter	Pre-monsoon	Monsoon	Post-monsoon	Winter	Pre-monsoon	Monsoon	Post-monsoon	Winter	Pre-monsoon	Monsoon	Post-monsoon
Bairath	19.87	36.82	562.09	18.94	25.33	34.61	227.95	30.27	127.50	94.00	40.56	159.85
Kotputli	21.79	36.33	519.46	18.64	25.47	40.28	226.20	28.97	116.86	110.89	43.55	155.40
Mundawar	23.64	40.86	571.99	21.83	27.92	44.18	273.31	41.53	118.10	108.12	47.78	190.22
Tijara	21.07	34.55	528.23	19.26	20.22	37.15	214.36	32.84	95.97	107.50	40.58	170.50
Kotkasim	26.39	41.16	575.66	17.70	32.98	55.90	244.26	30.78	124.94	135.83	42.43	173.85
Thanagaji	19.97	40.82	624.17	22.90	23.79	43.86	232.10	41.41	119.14	107.44	37.18	180.86
Bansur	22.70	40.45	589.14	20.52	26.86	45.92	251.47	34.32	118.33	113.52	42.68	167.19
Behrod	24.14	40.25	509.52	19.72	28.41	38.63	236.43	32.40	117.66	95.98	46.40	164.28
Tapukara	22.61	28.73	506.02	17.58	27.53	36.98	262.91	38.30	121.76	128.69	51.96	217.86

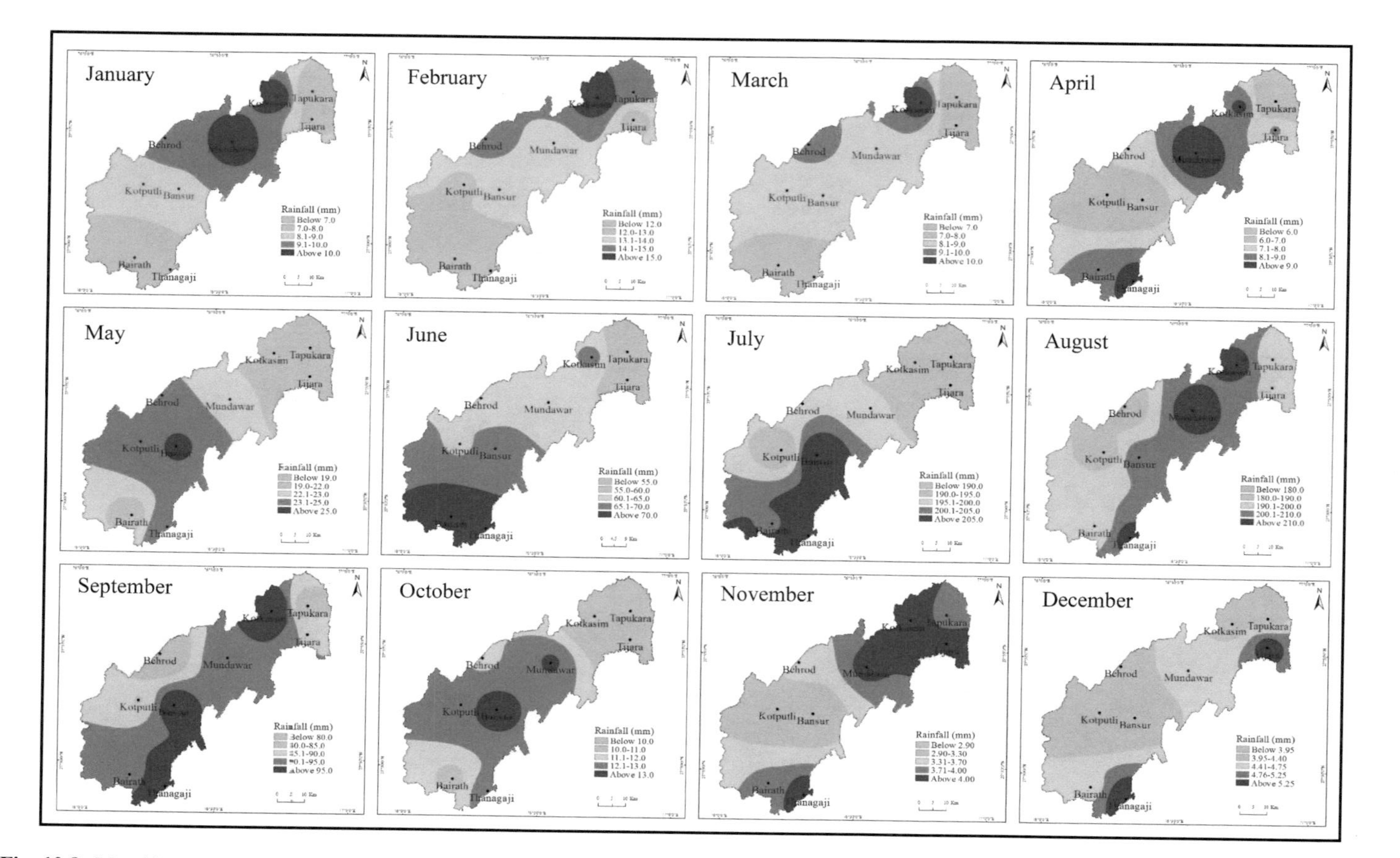

Fig. 12.8 Monthly pattern of rainfall over Sahibi Basin in Rajasthan during 1961–2017

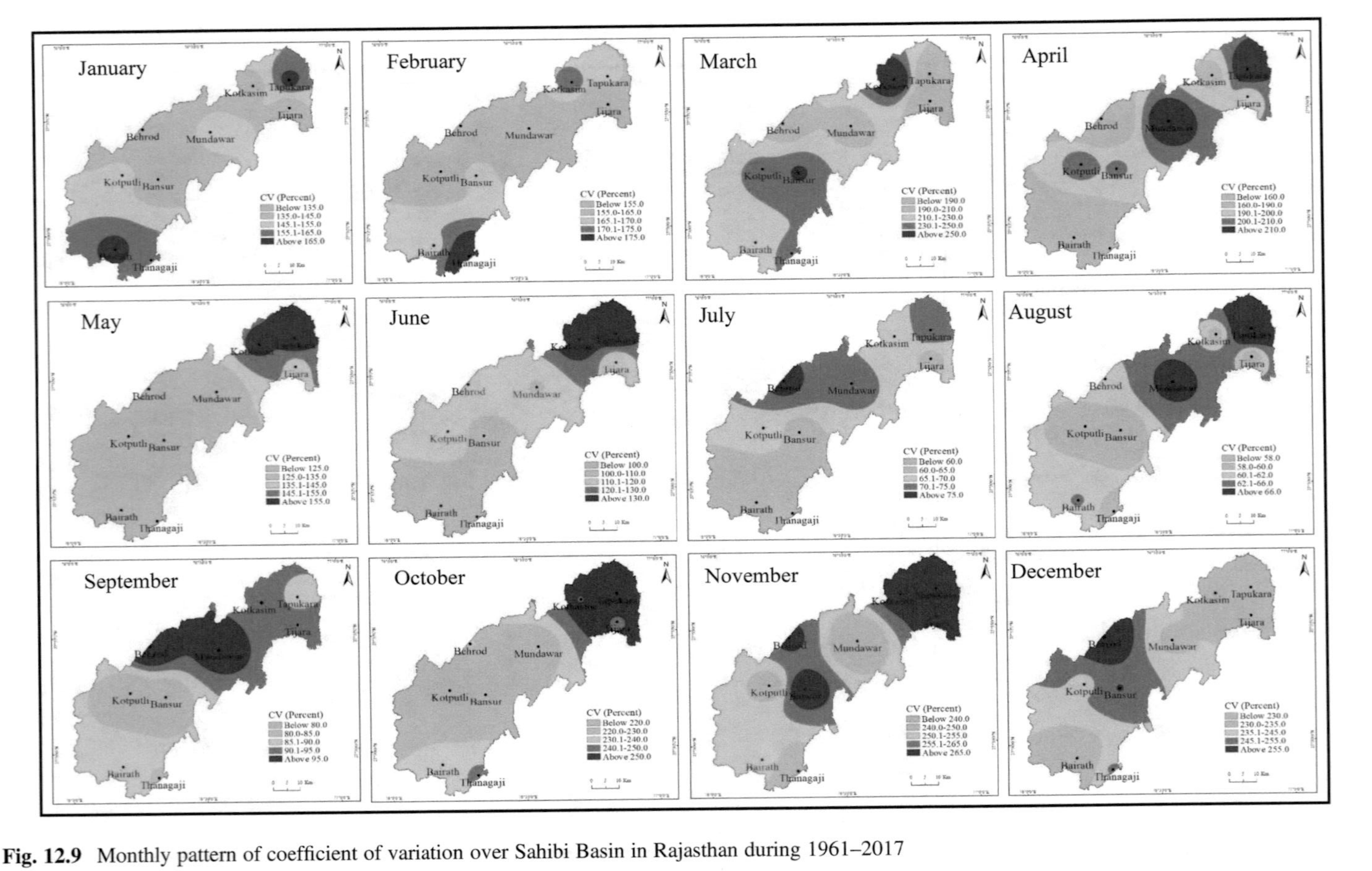

Fig. 12.9 Monthly pattern of coefficient of variation over Sahibi Basin in Rajasthan during 1961–2017

whereas the least variability occurred over the central to southern parts in most months.

Conclusions

In this study, an attempt has been made to examine the spatial and temporal trends in rainfall over Sahibi Basin in Rajasthan State of India for a period of 57 years (1961–2017). The results have shown a very high interannual variability in mean annual rainfall, ranging from 217.33 mm in 2002 to 1214.96 mm in 1996, with a mean annual value of 633.95 mm (SD = 204.93, CV = 32.33). However, no significant increasing or declining trend was observed in mean annual rainfall over Sahibi Basin during this 57-year period. Kotkasim and Tapukara stations have shown substantial rising and declining trends, respectively. The monsoon season contributes about 87% of mean annual rainfall. Pre-monsoon season rainfall has shown a significant increasing trend; the winter season rainfall has shown an insignificant increasing trend. Conversely, monsoon and post-monsoon season rainfall has shown insignificant decreasing trends. The months of July and August account for approximately 63% of the total annual rainfall. However, these months have shown nonsignificant declining trends. The month of June has shown a significant increasing trend in the amount of rainfall over Sahibi Basin. Spatial analysis has shown an increase in rainfall from the northeastern and western to southeastward areas. This increase in rainfall toward the southeast may be attributed to the existence of the Aravalli Mountains. The northeastern and western parts of the basin have received less rainfall. The maximum mean amount of rainfall (about 708 mm) occurred over Thanagaji station in the south, whereas the least (about 575 mm) was seen over Tapukara station in the north. The observed spatial and temporal variability in the amount of rainfall is high for such a small basin. From these findings, it can be inferred that rainfall differs significantly during seasons and months with great spatial and temporal variability. Finally, the results of this study will be useful for management of water resources for future planning of irrigation and agricultural practices over the semiarid Sahibi Basin.

References

Abeysingha NS, Singh M, Sehgal VK, Khanna M, Pathak M (2014) Analysis of rainfall and temperature trends in Gomti River basin. J Agric Phys 14:56–66

Amoussou E, Awoye H, Totin Vodounon HS, Obahoundje S, Camberlin P, Diedhiou A, Kouadio K, Mahé G, Houndénou C, Boko M (2020) Climate and extreme rainfall events in the Mono River Basin (West Africa): investigating future changes with regional climate models. Water 12:833

Babatolu JS, Akinnubi RT, Folagimi AT, Bukola OO (2014) Variability and trends of daily heavy rainfall events over Niger River Basin Development Authority area in Nigeria. Am J Clim Change 3:1–7

Bera S (2017) Trend analysis of rainfall in Ganga Basin, India during 1901–2000. Am J Clim Change 6:116–131

Bisht DS, Chatterjee C, Raghuwanshi NS, Sridhar V (2018) Spatio-temporal trends of rainfall across Indian river basins. Theor Appl Climatol 132:419–436

Biswas B, Jadhav RS, Tikone N (2019) Rainfall distribution and trend analysis for upper Godavari Basin, India, from 100 years record (1911–2010). J Indian Soc Remote Sens 47:1781–1792

Burrough PA, McDonnell RA (1998) Principles of geographical information systems: spatial information systems and geostatistics. Oxford University Press, Oxford

Deka RL, Mahanta C, Pathak H, Nath KK, Das S (2013) Trends and fluctuations of rainfall regime in the Brahmaputra and Barak basins of Assam, India. Theor Appl Climatol 114:61–71

Deshpande NR, Kothawale DR, Kulkarni A (2016) Changes in climate extremes over major river basins of India. Int J Climatol 36:4548–4559

Emmanuel LA, Hounguè NR, Biaou CA, Badou DF (2019) Statistical analysis of recent and future rainfall and temperature variability in the Mono River watershed (Benin, Togo). Climate 7(1):8

Gadgil S, Gadgil S (2006) The Indian monsoon, GDP and agriculture. Econ Polit Wkly 41:4887–4895

Gajbhiye S, Meshram C, Singh SK, Srivastava PK, Islam T (2016) Precipitation trend analysis of Sindh River basin, India, from 102-year record (1901–2002). Atmos Sci Lett 17:71–77

Gilbert RO (1987) Statistical methods for environmental pollution monitoring. Van Nostrand Reinhold, New York

Gupta M, Srivastava PK, Islam T, Bin Ishak AM (2014) Evaluation of TRMM rainfall for soil moisture prediction in a subtropical climate. Environ Earth Sci 71(10):4421–4431

IPCC (2014) Summary for policymakers. In: Climate change 2014: impacts, adaptation, and vulnerability. Contribution of Working Group II to the Fifth assessment report of the Intergovernmental Panel on Climate Change. Cambridge University Press, Cambridge

Jain SK, Nayak PC, Singh Y, Chandniha SK (2017) Trends in rainfall and peak flows for some river basins in India. Curr Sci 112:1712–1726

Jaiswal RK, Lohani AK, Tiwari HL (2015) Statistical analysis for change detection and trend assessment in climatological parameters. Environ Process 2:729–749

Kamruzzaman M, Beecham S, Metcalfe AV (2013) Climatic influences on rainfall and runoff variability in the southeast region of the Murray–Darling basin. Int J Climatol 33:291–311

Kendall MG (1948) Rank correlation methods. Griffin, London

Khaniya B, Jayanayaka I, Jayasanka P, Rathnayake U (2019) Rainfall trend analysis in Uma Oya basin, Sri Lanka, and future water scarcity problems in perspective of climate variability. Adv Meteorol 2019:3636158

Kliment Z, Matoušková M (2008) Long-term trends of rainfall and runoff regime in upper Otava River basin. Soil Water Res 3:155–167

Kumar V, Jain SK (2010) Trends in rainfall amount and number of rainy days in river basins of India (1951–2004). Hydrol Res 42(4):290–306

Langat PK, Kumar L, Koech R (2017) Temporal variability and trends of rainfall and streamflow in Tana River Basin, Kenya. Sustainability 9(11):1963

Loo YY, Billa L, Singh A (2015) Effect of climate change on seasonal monsoon in Asia and its impact on the variability of monsoon rainfall in Southeast Asia. Geosci Front 6:817–823

Machiwal D, Jha MK (2017) Evaluating persistence, and identifying trends and abrupt changes in monthly and annual rainfalls of a semi-arid region in western India. Theor Appl Climatol 128:689–708

Machiwal D, Dayal D, Kumar S (2017) Long-term rainfall trends and change points in hot and cold arid regions of India. Hydrol Sci J 62:1050–1066

Mann HB (1945) Nonparametric tests against trend. Econometrica 13:245–259

Mayowa OO, Pour SH, Shahid S, Mohsenipour M, Harun SB, Heryansyah A, Ismail T (2015) Trends in rainfall and rainfall-related extremes in the east coast of peninsular Malaysia. J Earth Syst Sci 124:1609–1622

Meena HM, Machiwal D, Santra P, Moharana PC, Singh DV (2019) Trends and homogeneity of monthly, seasonal, and annual rainfall over arid region of Rajasthan, India. Theor Appl Climatol 136:795–811

Muller M, Kaspar M, Matschullat J (2009) Heavy rains and extreme rainfall-runoff events in Central Europe from 1951 to 2002. Nat Hazards Earth Syst Sci 9:441–450

Mundetia N, Sharma D (2015) Analysis of rainfall and drought in Rajasthan state, India. Global NEST J 17:12–21

Ninu Krishnan MV, Prasanna MV, Vijith H (2018) Fluctuations in monthly and annual rainfall trend in the Limbang River Basin, Malaysia: a statistical assessment to detect the influence of climate change. J Clim Change 4:15–29

Pant GB, Rupa Kumar K (1997) Climates of South Asia. John Wiley and Sons, Chichester, UK

Parmar RS (2014) Watershed management and mitigation strategies: a case study of Sahibi River (Haryana). Acad Discourse 3(1):47–51

Pingale SM, Khare D, Jat MK, Adamowski J (2014) Spatial and temporal trends of mean and extreme rainfall and temperature for the 33 urban centres of the arid and semi-arid state of Rajasthan. India Atmos Res 138:73–90

Poonia S, Rao AS, Singh DV, Choudhary S (2014) Rainfall characteristics and meteorological drought in Hanumangarh district of arid Rajasthan. Ann Plant Soil Res 16:279–284

Raj PPN, Azeez PA (2012) Trend analysis of rainfall in Bharathapuzha River basin, Kerala, India. Int J Climatol 32:533–539

Ranade A, Singh N, Singh HN, Sontakke NA (2008) On variability of hydrological wet season, seasonal rainfall and rainwater potential of the river basins of India (1813–2006). J Hydrol Res Dev 23:79–108

Rao VB, Arai E, Franchito SH, Shimabukuro YE, Ramakrishna SSVS, Naidu CV (2011) The Thar, Rajputana desert unprecedented rainfall in 2006 and 2010: effect of climate change? Geofis Int 50:363–370

Rao AS, Poonia S, Purohit RS, Choudhary S (2014) Rainfall characteristics and meteorological drought condition in Jhunjhunu district of western Rajasthan. Ann Plant Soil Res 15:110–113

Rathore MS (2005) State level analysis of drought policies and impacts in Rajasthan, India, international Water Management Institute, working paper 93, drought series paper no. 6. Colombo, Sri Lanka

Sen PK (1968) Estimates of the regression coefficient based on Kendall's tau. J Am Stat Assoc 63:1379–1389

Shahid S, Wang XJ, Harun SB, Shamsudin SB, Ismail T, Minhans A (2016) Climate variability and changes in the major cities of Bangladesh: observations, possible impacts and adaptation. Reg Environ Change 16:459–471

Singh J, Singh O (2020) Assessing rainfall erosivity and erosivity density over a western Himalayan catchment, India. J Earth Syst Sci 129:97

Singh P, Kumar V, Thomas T, Arora M (2008) Basin-wide assessment of temperature trends in the northwest and central India. Hydrol Sci J 53:421–433

Singh B, Rajpurohit D, Vasishth A, Singh J (2012) Probability analysis for estimation of annual one day maximum rainfall of Jhalarapatan area of Rajasthan, India. Plant Archives 12:1093–1100

Singh O, Kasana A, Singh KP, Sarangi A (2020) Analysis of drivers of trends in groundwater levels under rice-wheat ecosystem in Haryana, India. Nat Resour Res 29:1101–1126

Sridhar SI, Raviraj A (2017) Statistical trend analysis of rainfall in Amaravathi river basin using Mann–Kendall test. Curr World Environ 12:89–96

Srivastava K, Pradhan D (2018) Real-time extremely heavy rainfall forecast and warning over Rajasthan during the monsoon season (2016). Pure Appl Geophys 175:421–448

Srivastava PK, Islam T, Gupta M, Petropoulos G, Dai Q (2015) WRF dynamical downscaling and bias correction schemes for NCEP estimated hydro-meteorological variables. Water Resour Manag 29(7):2267–2284

Taxak AK, Murumkar AR, Arya DS (2014) Long term spatial and temporal rainfall trends and homogeneity analysis in Wainganga basin, Central India. Weather Clim Extremes 4:50–61

Tirkey N, Parhi PK, Lohani AK, Chandniha SK (2020) Analysis of precipitation variability over Satluj Basin, Himachal Pradesh, India: 1901–2013. J Water Clim Change. https://doi.org/10.2166/wcc.2020.136

Yadav SK, Gautam S, Rawat S (2018) Rainfall variability estimation for western Rajasthan, India. Int J Curr Microbiol App Sci 7:4344–4348

You Q, Kang S, Aguilar E, Yan Y (2008) Changes in daily climate extremes in the eastern and central Tibetan 31 plateau during 1961–2005. J Geophys Res 113:D0710

Yue S, Hashino M (2003) Temperature trends in Japan: 1900–1996. Theor Appl Climatol 75:15–27

Yue S, Pilon P, Cavadias G (2002) Power of the Mann–Kendall and Spearman's rho tests for detecting monotonic trends in hydrologic series. J Hydrol 259:254–271

Printed in the United States
by Baker & Taylor Publisher Services